Geomorphology in Deserts

Geomorphology in Deserts

Ronald U. Cooke
and
Andrew Warren
Lecturers in Geography,
University College London

B. T. Batsford Ltd
London

First published 1973

Printed by the Anchor Press, Tiptree and bound by William
Brendon and Son Ltd, Tiptree for the publishers
B. T. Batsford Ltd 4 Fitzhardinge Street, London W.1

ISBN 0 7134 2104 5

Contents

Preface

Our aim in this book is to examine the nature of landforms, soils and geomorphological processes in deserts. Since deserts comprise over twenty per cent of the earth's land surface, the task is important and the prodigious literature shows that its significance has long been realized. But the extent and variety of deserts and their geomorphological literature present a formidable challenge to those who wish, as we do, to make generalizations in a survey of the field. Perhaps this is why there are few volumes covering the subject of desert geomorphology. Walther's classic *Das Gesetz der Wüstenbildung* (1924) has been followed only by Cotton's shorter survey in his *Climatic Accidents in Landscape Making* (1942) and by Tricart and Cailleux's review, *Le Modelé des Régions Sèches* (1961). Since we began to write this volume, the study of desert geomorphology has been advanced by the publication of the University of Arizona's bibliographic compendium, *Deserts of the World* (1968), and by the appearance of K. W. Glennie's monograph, *Desert Sedimentary Environments* (1970). Both of these books are complementary to our own—the first as a source of further reading, the second as a guide to ancient and modern desert sediments.

We cannot claim that our study is comprehensive or impartial. We have drawn on our own experience of deserts in North America, Chile and Peru, North Africa and Pakistan, and from time to time we have included some of our own previously unpublished research. Although we have reported material written in French, German, Russian and Spanish, we have used English-language material most extensively. We have also been selective in the themes we pursue, since our aim has been to consider landforms and soils and the processes currently at work in deserts in the context of geomorphological systems at different scales. In adopting this approach, we believe we are reflecting much contemporary thinking in geomorphology; but a consequence of our emphasis on present conditions is that we have found little space for much of the work on regional chronologies of landform development in deserts.

Many desert geomorphologists still feel that they are working in a field as bizarre as medieval cosmology. Much of the literature is less concerned with field evidence than with the cumulated speculations of generations of workers who were intent on validating vaguely conceived hypotheses. Perhaps the best-known of these march hares is the idea of parallel slope retreat, but others include notions of pediment development, the concept of pervasive wind erosion and the view that temperature change is important in rock weathering. We hope this book may provide acceptable generalizations concerning some of these ideas and that it may illuminate others; but at least we hope that we may have asked some useful questions and established a baseline from which subsequent research may proceed.

Various parts of the volume have been reviewed for us by Dr C. Vita-Finzi, Dr I. G. Wilson, Dr D. Yaalon and Professor T. J. Chandler. We have adopted many of their suggestions, and we greatly appreciate their help. We are most grateful for the financial support of many organizations which has allowed us to work in deserts: the American Council of Learned Societies, the U.K. Ministry of Overseas Development, The Royal Society, University College London, and the Central Research Fund of the University of London. Our thanks are also due to the staff of the Drawing Office, Department of Geography, University College London, for their skilful execution of the diagrams. Dr C. Vita-Finzi, Mr D. N. Hall, Dr I. G. Wilson, Dr E. C. F. Bird, Hunting Aerosurveys Ltd, and the Institut Géographique National kindly allowed us to reproduce photographic material. Finally, there are colleagues and friends too numerous to mention individually who have helped in so many ways to make our field work in deserts a success: we extend our sincere thanks to them all.

Ronald U. Cooke
Andrew Warren

Acknowledgments

We would like to thank the following for permission to reproduce figures and photographs from copyright works. We regret that it has proved impossible to contact a few authors and publishers whose diagrams we have used. Figure numbers refer to this book.

Professor S. A. Schumm and Princeton University Press for Fig. 1.7 (from H. E. Wright and D. G. Frey (eds.), *The Quaternary of the United States*); Professor C. V. Haynes and the University of Utah Press for Fig. 1.8; Professor J. A. Mabbutt and the Australian National University Press for Fig. 2.1; Dr D. Dragovich and Gebrüder Borntraeger for Fig. 2.3; Professor W. Meckelein and Georg Westermann for Fig. 2.4; Professor S. A. Schumm, Mr R. J. Chorley and Gebrüder Borntraeger for Fig. 2.5; Macmillan (Journals) Ltd. (*Nature*) for Fig. 2.7; Professor H. Jenny, Dr R. M. Scott and the Clarendon Press for Fig. 2.10; Professor R. LeB. Hooke and the University of Chicago Press (*Journal of Geology*) for Fig. 2.11; Professor M. Glasovskaya and Angus and Robertson (U.K.) Ltd for Fig. 2.12; Keter Publishing House Ltd. for Fig. 2.13; Professor D. Yaalon and Butterworth and Co. (Publishers) Ltd. for Fig. 2.14; Dr L. H. Gile and the Soil Science Society of America for Fig. 2.16; Professor R. J. Arkley and Williams and Wilkins Co. (*Soil Science*); Professor H. Jenny for Fig. 2.18a; Professor J. A. Mabbutt, Professor M. E. Springer and the Soil Science Society of America for Fig. 2.21; Librairie Armand Colin for Fig. 2.22 and Fig. 2.23; Dr C. D. Ollier and Macmillan (Journals) Ltd. (*Nature*) for Fig. 2.24; Captain J. T. Neal and the *U.S.A.F.* Office of Aerospace Research for Figs. 2.25, 2.26 and 2.27; Dr B. E. Lofgren for Fig. 2.28; Professor C. B. Hunt for Fig. 2.30; Dr W. W. Emmett for Fig. 3.1; Dr A. Schick for Fig. 3.2; Dr K. G. Renard and the American Society of Civil Engineers for Fig. 3.3; Professor R. LeB. Hooke and the *American Journal of Science* for Fig. 3.6; Dr C. S. Denny and the *American Journal of Science* for Fig, 3.7; Dr L. K. Lustig for Fig. 3.8; Professor S. A. Schumm and the Geological Society of America (*Bulletin*) for Fig. 3.11; Dr R. Twidale and Thomas Nelson (Australia) Ltd. for Fig. 3.12; Dr B. P. Ruxton and the University of Chicago Press (*Journal of Geology*) and C.D.U.E. Editions SEDES for Fig. 3.13; Professor Yi-Fu Tuan and the Association of American Geographers for Fig. 3.14; Professor W. S. Motts and the University of Chicago Press (*Journal of*

Geology) for Fig. 3.15; Dr G. I. Smith and the University of Utah Press for Fig. 3.16; the late Dr I. G. Wilson for Figs. 4.2, 4.3, 4.26, 4.35, 4.39, 4.43 and 4.44; Brigadier R. A. Bagnold and Methuen and Co. Ltd. for Figs. 4.4, 4.5, 4.12, 4.14, 4.16 and 4.18; Professor D. Yaalon for Fig. 4.6; Librairie Armand Colin for Fig. 4.7; Dr A. T. Grove, Professor J. H. Wellington, the Cambridge University Press and the Royal Geographical Society (*Geographical Journal*) for Fig. 4.8; Dr W. A. Price and the Society of Economic Palaeontologists and Mineralogists (*Journal of Sedimentary Petrology*) for Fig. 4.9; Professor K. Horikawa and the U.S. Department of the Army, Coastal Engineering Research Center; Dr M. Williams and the Elsevier Publishing Company (*Sedimentology*) for Fig. 4.13; Professor R. P. Sharp and the University of Chicago Press (*Journal of Geology*) for Fig. 4.15; Dr S. L. Hastenrath and Gebrüder Borntraeger for Figs. 4.19, 4.20 and 4.30; Dr H. and Dr K. Lettau and Gebrüder Borntraeger for Figs. 4.21 and 4.22; Dr F. S. Simons and the University of Chicago Press (*Journal of Geology*) for Fig. 4.23; Professor J. A. Mabbutt and the Institute of Australian Geographers for Fig. 4.24; Professor T. Monod and I.F.A.N. for Fig. 4.25; Dr A. Clos-Arceduc, Gauthier-Villars and the American Association of Petroleum Geologists for Fig. 4.31; Dr A. Clos-Arceduc and Gauthier-Villars for Fig. 4.32; Professor R. L. Folk, Professor E. K. Walton and the Elsevier Publishing Company for Fig. 4.33; Professor W. S. Cooper and the Geological Society of America (*Memoirs*) for Fig. 4.37; Professor T. Monod and the Institut Géographique National, Paris, for plates.

PART 1

The Desert Context

1.1 Nature of Desert Research

Man's progressive penetration into generally unwelcoming environments has been the result of two allied motives: exploration and exploitation. Most scientific research in deserts has been accomplished within one or both of these contexts.

The urge to explore is a complex and deep-rooted need created by the romance of exotic and unknown places, the desire for geographical knowledge and new resources, the appeal of escape and the challenge of harsh environments. As the area of unexplored desert has been reduced and the problems of travel and survival have been solved, the exploration urge has had fewer valid outlets. But desert expeditions retain their attraction, and scientific research has increased in importance among their activities. The scientific reports from early desert explorations were sometimes formidable memorials to courageous exploits. Witness the observations of Blake (1858), Powell (1875), Gilbert (1875), McGee (1896) and more recently Bryan (1925*a*) in the south-western United States, or the perceptive geological record of Darwin (1891) and the geographical vision of Bowman (1924) in the Atacama Desert. Huntington (1907) invested experiences in the Gobi Desert with his consistent philosophy of environmental determinism; and Hedin's (e.g. 1904) scientific record of his Asian forays remains a valuable source of information on an area still little-known in the West. In the Sahara, successive French, British and Egyptian expeditions brought a wealth of knowledge and speculation to the metropolitan scientific societies. Such were the expeditions of Gautier (e.g. 1909), Flammand (e.g. 1899), Tilho (e.g. 1911), Bagnold (e.g. 1933), Hassanein Bey (1925) and Newbold (e.g. 1924). An almost apostolic succession of gifted explorers infiltrated Australian deserts from the middle of the nineteenth century into the early twentieth century (Cumpston, 1964), and several expeditions yielded important scientific information on the arid lands in addition to much basic geographical knowledge (e.g. Madigan, 1936*a*; Spencer, 1896; Wells, 1902). Usually, exploration has been accomplished by group expeditions; but there have been great lone desert travellers who have played a significant role in desert exploration, such as Thesiger (e.g. 1949) and Monod (e.g. 1958).

A surprising number of the early desert travellers were motivated by romantic notions. The Egyptian Desert west of the Nile was repeatedly searched for lost oases such as Zerzura (e.g. Bagnold, 1931; de Lancey-Forth, 1930; Hume, 1925). Passarge (1930) investigated the geomorphology of southern Tunisia as an aside in his attempt to ascertain the location of Atlantis. And in India, the search for the Vedic Sarasvati, and for reasons behind the decay of the ancient Indus civilization gave the initial impetus to scientific enquiry into the geomorphology of the area (Raverty, 1898).

The great exploratory desert expeditions are almost a thing of the past; but their tradition persists, mainly in the form of seasonal forays, often organized from university centres. Many of these contemporary expeditions focus their attention on scientific objectives. A few examples from many are the Atacama Desert Expedition (Hollingworth, 1964), the British Expedition to Niger in 1970 (Hall *et al.*, 1971), and the German Saharan expeditions of 1954/55 and 1969 (Meckelein, 1959; Mensching *et al.*, 1969). Increasingly, however, the exploration is of specific scientific problems rather than of areas in general; and some former deserts of exploration are now becoming accessible tourist playgrounds (e.g. Stevens, 1969).

The exploitation of deserts has been generated by a variety of pressures: the growth of population and the demand for *lebensraum*; imperialism, colonialism and military adventurism; the scarcity of mineral resources; and expansion within established national boundaries. Much scientific intelligence has come both from expeditions and surveys deliberately established to plan exploitation and development, and from the environmental analyses by frontier settlers. The exploitation incentive has certainly been a factor behind the desert experience for millennia (e.g. Lattimore, 1951), but as the pressures on land have increased, so resource evaluation of deserts by scientists has assumed a more important role.

Until recently, the influence of national groups has largely been confined to particular deserts. Three aspects of this political segregation are important. Firstly, scientific research in deserts has reflected the prejudices and intellectual attitudes of the national group concerned; secondly, communication amongst scientists, partially blocked by barriers of language and competition, has been restricted; and thirdly, much of the research was accomplished, until recently, by aliens working in unfamiliar environments. Despite the efforts of UNESCO and other organizations to remove communication barriers, they still exist and are most clearly expressed in the scientific literature.

German influence in South West Africa is reflected in the writings of Passarge (1904), Jaeger (1921) and many others; and it is also seen in geomorphological studies of deserts in South America (e.g. Brüggen, 1950; Mortensen, 1927; Penck, 1953) where political links were close though not colonial. In north Africa, long-standing French interests have been accompanied by an enormous literature; many contributions illustrate a preoccupation with evidence of climatic change and with detailed ideographic and encyclopaedic regional description (e.g. Capot-Rey, 1953 *a* and *b*; Coque, 1962; Daveau, 1966; Dresch, 1953; Gautier, 1935; Monod, 1958; Rognon, 1967; Urvoy, 1936). In recent years, French interest in deserts has extended far beyond the Sahara (e.g. Dresch, 1961 and 1970). The brief Italian occupation of north Africa also yielded a scientific literature on parts of the Sahara (e.g. Desio, 1968 and 1969). The British were concerned with deserts in the eastern Sahara and Arabia, in southern and eastern Africa, in the Indian sub-continent and in Australia, but the geomorphological work accomplished during their periods of occupation is not impressive in quantity and it is strongly rooted in the traditions of British geology. (e.g. Blandford, 1877; Hume, 1925; Jutson, 1934). The outstanding British geomorphological contribution, interestingly separate from the geological mould, is undoubtedly Bagnold's *Physics of Blown Sand and Desert Dunes* (1941). Similar in their isolation to the work of national groups are investigations by private companies, in connection with the exploitation of resources such as oil (e.g. Holm, 1960; Kerr and Nigra, 1952).

Frontier movements into arid areas within established nation states have also been important in extending scientific knowledge of deserts.

The new settlers have had to accommodate themselves to alien environments and scientific knowledge has been acquired in the process. The settlement of the western United States is a good example of this, and more recently there have been settlement movements in Israel, Australia, South Africa and China. The latter half of the nineteenth century in the United States saw an enormous number of resource evaluations in desert areas. Some of the earliest were the systematic searches by Mormons for settlement sites in their expanding *Deseret* (Meinig, 1965), and the detailed surveys for railroad routes (e.g. Blake, 1858; Parke, 1857). Powell's (1878) more general but classic *Report on the Lands of the Arid Region of the United States* was the precursor of a library of resource evaluations. Amongst more recent evaluations are the regional surveys of the C.S.I.R.O. in Australia (e.g. Perry *et al.*, 1962), and environmental appraisals for military purposes (e.g. Howe *et al.*, 1968; Mitchell and Perrin, 1967). Many resource evaluations incorporate valuable geomorphological information, and some soil surveys too have laid important foundations for later geomorphological research (e.g. Hunting Technical Services, 1961; Jessup, 1961; Mitchell, 1959; Mitchell and Naylor, 1960; Robertson and Lebon, 1961; Ruhe, 1967–70).

Today there is a growing body of desert research workers who are motivated more by a desire for scientific enquiry and for knowledge about optimum human adjustment to aridity than by the simpler motives of exploration or exploitation. Many operate from permanent research organizations within desert areas—such as *the University of Arizona, the Negev Institute for Arid Zone Research, L'Institut Fondemental de l'Afrique Noire,* the *Universidad del Norte* (Antofagasta, Chile), the *Turkmanistan Academy of Sciences*, and the *Academy of Sciences of the People's Republic of China* (UNESCO, 1953). Publications in the field of desert research are widely disseminated through numerous systematic and regional journals, but in recent years the Arid Zone Research series of UNESCO has provided an important focus of desert publication, and bibliographic work at the University of Arizona (McGinnies *et al.*, 1968) has helped to provide cohesion to the burgeoning literature.

1.2 Geomorphological Studies in Deserts

The large and scattered literature on desert geomorphology has several distinctive characteristics. We have mentioned the pursuit of desert studies within different national contexts. A consequence of this is a multilingual and confused terminology (Stone, 1967). The literature is liberally peppered with *bornhardts* and *monadnocks*, with *playas*, *sebkhas* and *chotts*, and with *hamada*, *gibber plains* and *pavements*.

More significant are features that arise from the conditions under which the research was done, and from the aims of the investigation.

Geomorphological descriptions of deserts have tended to be superficial for several reasons. Firstly, geomorphology has rarely been the exclusive concern of exploratory expeditions. On many desert journeys, there is competition between the concern for survival in an insecure and potentially hazardous environment, and the collection of scientific information, usually to the detriment of the latter. Secondly, where an expedition does not include a committed geomorphologist, the landforms may be described in an amateurish fashion. Thirdly, because exploratory expeditions are usually concerned with areas about which very little scientific information of any sort is available, all manner of environmental observations may be recorded so that the geomorphological information may be incidental, and the results may be published as an unpalatable farrago of descriptive data. Fourthly, although many resource surveys include sections on geomorphology, other considerations are commonly more prominent. A fifth reason is that many expeditions and surveys cover large areas quickly, and observations tend to be only of a reconnaissance nature.

Nevertheless, landform description occupies a considerable proportion of the desert literature. There are two main reasons for this: landforms attract attention in the general absence of vegetation and cultural features; and they may appear strange to an explorer from more temperate areas. Indeed, desert landforms have often been described in terms of particular cultural preconceptions translated from the homeland. The first illustration of the Grand Canyon, for instance, was drawn in a 'Gothic' manner by the German topographer F. W. von Eglottstein (Chorley, Dunn and Beckinsale, 1964). In other cases, explorers confronted with alien arid landforms have responded with original and imaginative analyses appropriate to the new environment. Perhaps the outstanding example is Gilbert's *Report on the Geology of the Henry Mountains* (1877), the product of a fertile mind and two months' field work. A common response to the novelty of landforms in deserts has been for the observer to focus his attention on individual, and often spectacular or strange features. Thus desert geomorphology has come to be characterized by concern for specific landform types (such as zeugen, yardangs, dreikanter or barchans) which may be neither very common nor of great significance.

An emphasis on description rather than on analysis of desert landforms arises mainly from the primary need for basic information. The preoccupation with description carries with it the corollary that the processes of landform sculpture have rarely been observed and recorded in deserts. This results partly from fashion and partly from the peculiar nature of desert enquiries. In particular, the seasonal reconnaissance journey is inadequate for monitoring desert processes which, many believe, either operate very slowly over long periods or are rare and catastrophic. In the case of the former, expeditions are unlikely to make realistic recordings; and in the case of the latter, the event may not be observed, or the observer may not live to tell the tale. To this day, for instance, there are few detailed descriptions of sheetfloods (McGee, 1898; Rahn, 1967; Gavrilović, 1970). Without direct observation of processes, the desert geomorphologist has tended to deduce their details from the evidence of forms and deposits.

In many ways, deserts are still the resort of the reconnaissance geomorphologist. The annual increment of 'preliminary remarks' and *tournées rapides* is prodigious. Such contributions are increasingly based on the use of aerial photographs, and the advent of space photography is likely to be particularly important to this kind of work (e.g. Pesce, 1968), especially in the identification of large-scale landforms (Morrison and Chown, 1965).

But it would be wrong to suggest that all desert geomorphology consists of superficial descriptions of landforms and deposits based on reconnaissance surveys. There is a growing number of works based on the precise and extended investigation of landforms and processes. This is especially true in the United States, where there is a long history of important contributions to geomorphological thought from the arid lands. In this heritage, the work of Blackwelder, Bryan, Bull, Gilbert, Hunt, Leopold, Schumm and Sharp is outstanding. And today, studies of arid and semi-arid environments remain in the vanguard of geomorphological progress in the United States.

Finally, the search for generalizations has not been absent in desert geomorphology. Perhaps the cardinal concern has been with landscape evolution. There have been numerous efforts to define 'the cycle of arid erosion' (Cotton, 1942) or 'erosion cycles in arid and semiarid climates' (Birot, 1968). We shall consider these efforts below (section 1.4.1). We need only say here that the primary concern has been to establish relations between landforms and desert climates, and fundamental to this are the recurring and allied themes of climatic change and landform inheritance.

1.3 The Distinctiveness and Diversity of Desert Conditions

In our view, the formulation of simple and comprehensive generalizations about the nature of desert geomorphology, especially if they are based on the recognition of relations between desert landforms and desert climates, will be difficult, and in the present state of knowledge is impossible. The major obstacles to such attempts lie in the enormous climatic, edaphic, biological and hydrological diversity of deserts, the impress of past climatic changes, and the roles of endogenetic processes, rock types and geological structures. In addition, generalization is restricted at present by lack of basic data, and by ignorance of the links between landforms and climate. In this section we touch briefly on some of these themes and attempt to identify singular features of desert environments which may contribute towards the formation of distinctive landforms.

1.3.1 DESERTS AND CLIMATE

(a) Definition and distribution of deserts. Definitions of deserts are legion. A desert is regarded by some simply as a barren area capable of supporting few lifeforms; by others, it is defined with greater precision by climatic criteria based directly or indirectly on the nature of vegetation or the availability of water. Most recent climatic classifications employ aridity or moisture indices (Wallén, 1967), and we have adopted Meigs' widely accepted scheme based upon Thornthwaite's moisture index, which involves a consideration of potential evaporation and water balance (Meigs, 1953; McGinnies *et al.*, 1968). In this classification, the outer limit of dry lands is taken to be at the −20 value of the moisture index. Areas bounded by the −20 and −40 values are designated *semi-arid*, and those areas less than −40 are called *arid*. Within the arid boundary, Meigs recognized *extremely arid* areas, which are defined as those where at least 12 consecutive months without rainfall have been recorded and where there is no regular seasonal rhythm of rainfall. The semi-arid, arid and extremely arid zones are further classified according to the period of the year when precipitation occurs and the mean temperature of the warmest and coldest months. The pattern produced by this scheme is shown on Fig. 1.1. Deserts are usually considered to comprise the arid and extremely arid areas, and most of our discussion will concern these areas. But semi-arid lands are often geomorphologically similar to the more arid lands, and they are called deserts locally, so we shall consider semi-arid areas and draw appropriate examples from them. The classification also includes some high-latitude cold deserts but, like Meigs, we have excluded such deserts from our survey.

The arid areas shown on Fig. 1.1 occupy a third of the earth's land surface. Some four per cent of the land surface is extremely arid, 15 per cent is arid, and 14·6 per cent is semi-arid. These estimates are similar to

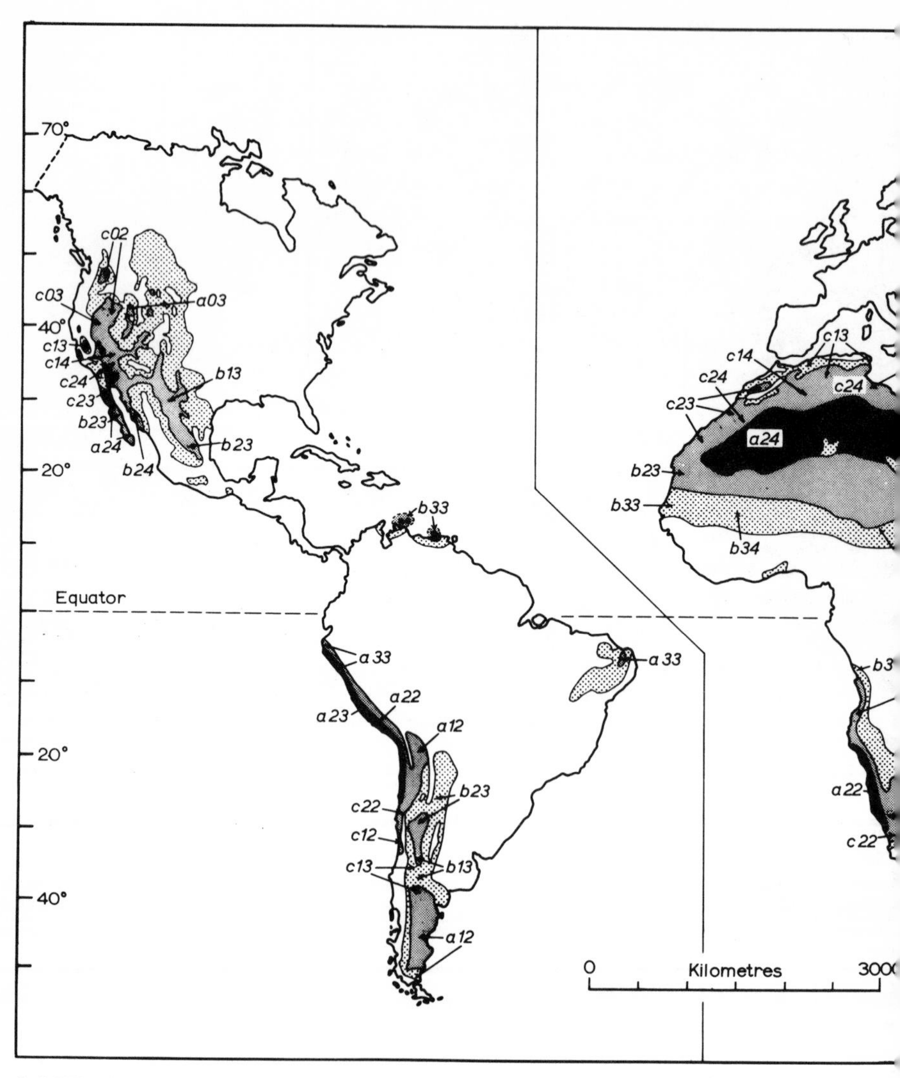

1.1 World distribution of arid lands (after Meigs, 1953).

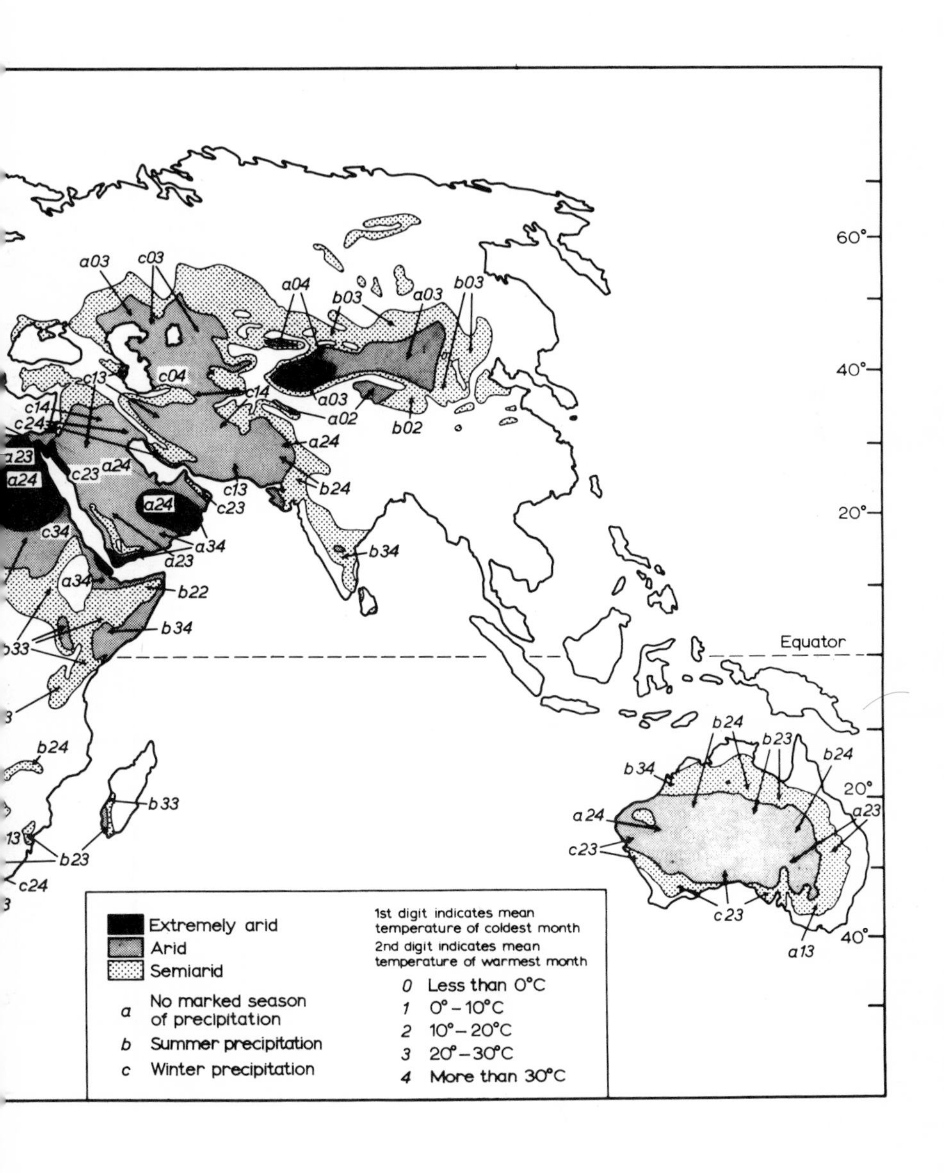

Extremely arid
Arid
Semiarid
a No marked season of precipitation
b Summer precipitation
c Winter precipitation
1st digit indicates mean temperature of coldest month
2nd digit indicates mean temperature of warmest month
0 Less than 0°C
1 0° – 10°C
2 10° – 20°C
3 20° – 30°C
4 More than 30°C
Equator
60°
40°
20°
20°
40°

those based on vegetation occurrence (Shantz, 1956). The pattern of deserts is dominated by five great continental areas of aridity—North Africa–Eurasia, South Africa, North America, South America and Australia—each surrounded by semi-arid zones.

Desert areas, generally low in precipitation and in relative humidity and high in temperature, are associated mainly with divergent air flows at low altitudes, with atmospheric subsidence and stability, with the occurrence of high pressure cells near the 30th parallels, and with only occasional penetration of rain-bearing atmospheric wave disturbances common to the circum-polar westerlies and the convergence zones of the tropics (Hare, 1961). Departures from the basic distribution of aridity are numerous and are associated with a number of factors such as distance from the sea and the distribution of mountain ranges. One consistent departure occurs along the western littorals of the major continental areas where high humidities, low daily temperature ranges and low precipitation are the results mainly of the reinforcement of atmospheric stability by cold offshore currents and the upwelling associated with them (Lyddolph, 1957). The continuity of the earth's arid zone is broken by the penetration of the monsoonal pattern in subtropical Asia and by the formation of oceanic high pressure cells in the lower troposphere which deflect moist air streams over what might otherwise have been desert areas (Hare, 1961).

(b) Climatic extremes and diversity. Deserts lie at that extreme of the climatic continuum characterized in general by high temperatures, excess of potential evaporation over precipitation, and the limited availability of water for plant growth. (We are excluding circumpolar and high-altitude deserts where physiological aridity is associated with low temperatures). Within the climatic zones with these characteristics, variations of climatic conditions are immense, and the detailed response of landforms to such variations is inevitably complex. To classify the variations, with their effects on landforms in mind, is a difficult task. As good an attempt as any is that of Tricart and Cailleux (1964) which distinguishes between extremely arid, arid and semi-arid regions and subdivides them according to whether or not freezing is rare. Regions where freezing is rare are essentially inter-tropical deserts, and these are further classified according to whether precipitation is seasonal or sporadic, or the atmosphere humid. The deserts where freezing occurs are chiefly high-altitude deserts within the tropics or extratropical deserts, and two categories are distinguished, based on whether the freezing is seasonal or sporadic. The criteria used in this classification are very general, but they may be related in a gross way to geomor-

phological processes, such as mechanical weathering and debris movement on slopes. Examples of deserts which fall within the categories defined by Tricart and Cailleux are shown in Table 1.1.

TABLE 1.1 *Examples of Different Desert Climates*

(a) Dry, hot regions where freezing is rare or absent	*Semi-arid*	*Arid*	*Extremely Arid*
Seasonal precipitation	Australian Bush Desert	Australian Shrub Desert	—
Sporadic precipitation	Northeast Brazil	Southwest Arizona	Central Sahara
Humid atmosphere	Chaparral coast of California	Southwest Africa	Coastal Desert of Chile and Peru

(b) Dry regions where freezing is seasonal or sporadic	*Semi-arid*	*Arid*
Seasonal freezing	Canadian Prairie	Chinese Turkestan
Sporadic freezing	Turkish Steppes	Syria

SOURCE: Tricart and Cailleux, 1964.

Within the context of this classification we may attempt to identify some of the climatic singularities of deserts that may be of geomorphological importance. At the outset, however, it is necessary to emphasize that, although we are concerned with one extreme of the climatic spectrum (Table 1.2), it is not necessarily the extreme events in any particular area which are of the greatest importance in moulding landforms. It may be the smaller events of more frequent recurrence which accomplish most (Wolman and Miller, 1960), but as many of these events are less frequent in deserts than in humid climates they are likely to lead to more sporadic change (section 1.3.5). Desert areas experience longer periods without precipitation than more humid regions, periods when the surface of the ground is relatively dry and vulnerable to the activity of aeolian processes. (Active dunes, for instance, are found only in areas with less than 150 mm mean annual rainfall.) It also seems likely that the proportion of runoff-producing rains is smaller in deserts than elsewhere, so that a greater proportion of precipitation is only involved in geomorphological or pedological

TABLE 1.2 *Examples of Extreme Climatic Conditions in Deserts*

(*a*) *Temperature* (Air, °C)	
Absolute Maximum temperatures	Tindouf (Algeria) 57° William Creek (Australia) 48·3°
Absolute Minimum temperatures	Tibesti (Libya) −7° Yuma (Arizona) −7·2° San Simon (Arizona) −20°
Amplitude of temperature	*Annual* (mean) Central Sahara 25° Central Australia 34° *Daily* (absolute) Aïr Mountains (Niger) 34°
(*b*) *Precipitation* (mm)	
Lowest recorded mean annual precipitation	Sahara: Dakla (Egypt) 0·4
Range (variability) of annual precipitation	Swakopmund (S.W. Africa) 1–148 Turpillo (Chile) 0–390 Araouan (Mali) 11·7–260 Tamanrasset (Algeria) 6·4–159
Maximum number of years without precipitation	Swakopmund (S.W. Africa) 15
Intensity of precipitation	Tamanrasset (Algeria) 44 mm in 3 hrs Port Etienne (Mauritania) 300 mm in 24 hrs Larger Saharan storms: commonly 1 mm in 1 minute
(*c*) *Relative humidity*	
Low relative humidity	20% of Australia has average relative humidity of less than 40%
Low absolute humidity (average)	Béchar (Algeria) 6·3 mm (Jan), 14·2 mm (July)
Seasonal variation of relative humidity	Béchar (Algeria) 23% (July), −56% (Dec)
Daily variation of relative humidity	Tanezrouft (Algeria) 12% (July), −29% (Nov–Dec)
(*d*) *Evaporation*	
Ratio of potential evaporation to precipitation (approx. figures)	Australian Desert: 2,100–2,500 mm/pa: 240–280 mm/pa Sahara: 2,500–6,000 mm/pa: 5–170 mm/pa

SOURCE: Capot-Rey, Rougerie, Tricart and Cailleux, and others.

activity at the place where it falls. Insolation is higher (except in deserts with humid atmospheres) because of latitudinal position and the feeble filtration by dry, clear air. Terrestrial re-radiation is also high so that daily and seasonal temperature variations are great, both in the air and in surface material. This fact is certainly of consequence to evaporation and humidity at the desert surface, and is reflected in the nature of soil development and superficial weathering processes.

Absolute humidities are often low (except in humid deserts along western littorals), but relative humidities may be extremely variable, both temporally and spatially. Dew is common in some deserts and may be a relatively significant source of water for rock weathering in environments where other sources are rare. Another consequence is the great variability in vegetative production in deserts.

1.3.2 VEGETATION AND BARE GROUND

Because of climatic conditions, deserts have a sparse plant cover. At this vegetational extreme, however, there is enormous diversity. There is a variety of species and of species richness; of adaptations to drought; of relative proportions of trees, shrubs, succulents, herbs and grasses; of ground coverage; and of changes through time and in space, especially in sympathy with changes in the availability of moisture. This variety is illustrated by Hastings and Turner (1965) in their account of *The Changing Mile* in the Sonoran Desert of Arizona and Mexico. They showed that vegetation responds to latitudinal, longitudinal and altitudinal climatic gradients, as well as to local edaphic and microclimatic conditions. The three major vegetation zones that they described are the desert, the desert grassland and the oak woodland; at higher elevations there occur zones of coniferous forest. The species diversity of vegetation in these North American deserts is large. In the desert grassland zone, for example, there are at least 48 grass species, and the discontinuous grass cover is studded with several low-growing woody and semi-woody plants. Both species diversity and the diversity of lifeforms increase from the woodland towards the desert (Whitaker and Niering, 1965).

The great range in biomass, productivity, species composition and chemistry of desert vegetation is described for Soviet deserts and the Syrian Desert by Rodin and Bazilevich (1965). Biomass varies enormously from the sub-boreal deserts in the north to the richer deserts of Khazakstan. Within the latter the biomass may vary several-fold, from open wormwood and saltwort desert to sparse woodlands in wetter areas. Species richness also varies widely from desert to desert. In the Chinese deserts the eastern areas are dominated by poor Mongolian flora, whereas the western areas are dominated by the richer flora of Khazakstan (Petrov, 1962). The Sahara has a richer Mediterranean flora on its northern edge and a poorer Sudanic flora in the south (Cloudsley-Thompson and Chadwick, 1965). Monod (1954) drew a distinction between the 'contracted' vegetation pattern of the south Sahara and the 'diffuse' pattern of the north. The tropical deserts are evidently richer in tree species than cool deserts, such as those in America and central Asia, which are often characterized by low, woody shrubs (McCleary, 1968). Australian deserts are said to be more vegetated than old-world deserts of similar aridity because of their long protection from domestic grazing, and they appear to be richer in grasses than other deserts (McCleary, 1968).

Several other characteristics of desert vegetation deserve emphasis in a geomorphological context. Firstly, desert plants tend to have extensive, near-surface rooting systems which may serve to bind surface material and limit the full impact of surface erosion, even though the percentage cover of above-ground vegetation may be slight (Chew and Chew, 1965; Cloudsley-Thompson and Chadwick, 1965).

Secondly, in an environment where water is the most important limiting factor, the pattern of desert vegetation responds very sharply to variations in soil moisture. In Israel, for instance, there are dense stands of vegetation in wadis; scrub is found on sandy soils, which retain moisture because of deep percolation, and on stony soils whose stones provide a surface mulch which hinders evaporation by insulating the underlying soil; and much more sparse vegetation covers 'loessal plains' where low infiltration capacity and high *p*F makes moisture scarce and difficult to extract (Hillel and Tadmor, 1962). An enormous variation in biomass between wetter and drier areas close together has been measured by Gillet (quoted by Bourlière and Hadley, 1970) and noted by Bocquier (1968) in semi-arid Tchad. In Egypt too the great differences in biomass and cover have been noted and attributed chiefly to soil depth and water-holding capacity (Kassas and Imam, 1954).

A third point of significance is that much desert vegetation is annual or ephemeral. This has an important bearing on litter-fall which may be considerable in deserts, although it may not serve greatly to protect the surface, since the rate of litter destruction may also be high. In addition, seasonal variations of percentage ground cover may be great.

Probably the most significant general feature of plant cover in deserts is its sparsity: the proportion of bare ground is greater than in other environments. Precise data on ground cover appear to be rather scarce. In Arizona, Whitaker *et al.* (1968) found that deserts of the lower mountain slopes had only a 30–50 per cent cover, whereas woodland cover on nearby mountain slopes was 60–80 per cent; in addition bare rock increased as a percentage of the ground area from 8–12 per cent in woodland to 30–60 per cent in the deserts. In south-west Arizona on shallow soils over caliche, a *Larrea tridentata* community covers only about 20·7 per cent of the ground (Chew and Chew, 1965). Several important geomorphological consequences may follow from the generally high proportion of bare ground. In the first place, the interception and surface-protection roles of vegetation are reduced so that there should be a simplified and more direct relationship between climatic phenomena and the ground surface. In particular, raindrop impact and rainsplash erosion may be effective over much of the surface (although high root-density near the surface may restrict their effect), thermal changes at the surface tend to be large (Sinclair, 1922), and winds may act with little hindrance on surface material. Secondly, organic matter contents within the surface mantle are low, since the rate of destruction is relatively high, so that the rate of weathering is reduced as a result of smaller amounts of organic acids. A general conclusion from these observations might be that, because weathering rates may be reduced and erosion rates may be increased, weathering mantles on bedrock surfaces tend to be thin or non-existent more frequently than in moister areas. If this is so, two further geomorphological conclusions are apparent: rock lithology and structure is likely to play a more important role in determining the detail of surface topography; and erosion-controlled slopes are likely to be more common.

1.3.3 WINDS

Since we shall discuss winds at greater length in section 4.1.2, the following short section merely introduces the general and distinctive character of desert winds. Wind patterns in deserts may be examined at the level of the continental circulations and at the level of much smaller-scale air movements.

In the most general models of global circulation, deserts are found near the tropics on the larger continents. The equatorial sides of these zones are associated with the trade winds, which blow in a clockwise direction around high-pressure cells in the northern hemisphere, and in an anti-clockwise direction in the southern hemisphere. The cells dip eastward, so that the wind pattern is more evident at the surface on the eastern sides of the continents whereas the cells are open on the western sides. These winds are reflected by the patterns of dunes and wind erosion features in the Sahara, Arabia and Australia (e.g. Dubief, 1953; Madigan, 1945).

Such winds are, of course, only one part of a yearly cycle of change,

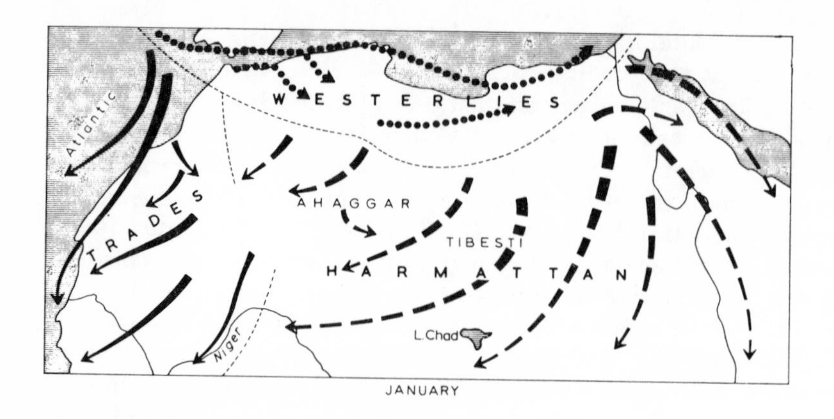

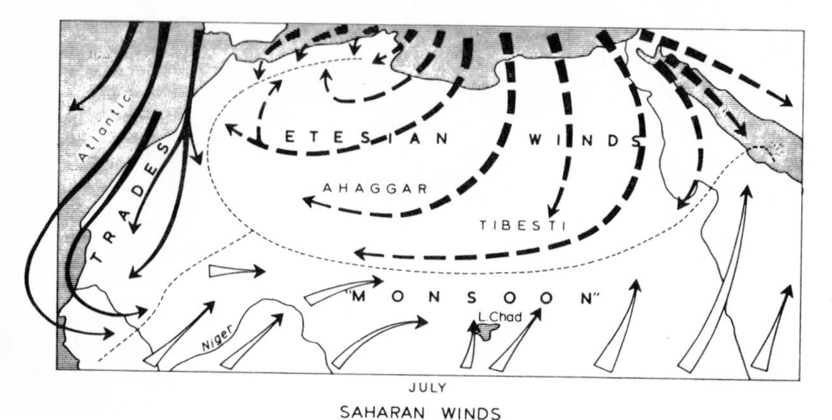

SAHARAN WINDS

1.2 The seasonal pattern of Saharan winds (after Dubief, 1953).

as Dubief (1951) has explained for the Sahara (Fig. 1.2). In the winter months the *harmattan* blows from the southern desert into West Africa whereas the northern desert is invaded by fronts of the zone of westerlies. The winds associated with these fronts blow dust north-eastwards to the Libyan coast as the *ghibli*, and the passage of the fronts in the Algerian Sahara is often marked by intense *vents de sable* (sandstorms) (Plate 1.1; Dubief, 1952 and 1953). In summer the harmattan moderates as the southern desert is invaded by the intertropical convergence zone, whose winds are fickle and gentle (e.g. Powell and Pedgley, 1969). In the north, the north-easterly winds migrate into Algeria, and in the north-east they are augmented by the Etesian winds that are warmed and dried as they enter the desert to blow south-westward as a summer continuation of the harmattan. In Australia, the southward migration of northern tropical air in summer appears to be associated with the strongest winds of the year (Brookfield, 1970), but in the Sahara it is undoubtedly the winter months, both in the south and in the north-west, that have the most frequent and intense sandstorms. (Dubief, 1953; Warren, 1971*b*). In Australia the wind circulation around the high-pressure cell is well developed except on the western side, in that winds blow round the southern, eastern and northern desert margins; the winds in the central area seem to be light and variable (Brookfield, 1970; Mabbutt, 1968). The Sahara-Arabian cell has no such central belt since the northern limb of the cell is broken into by the zone of westerlies (e.g. Wilson, 1971*b*). The western Sahara is invaded by the Atlantic trade winds (Dubief, 1951).

Most of the mid-latitude deserts are found in the zone of westerlies and many studies of aeolian sediment transport have noted a general westerly movement in such areas. In the south-western United States the prevailing winds are westerly (e.g. Clements *et al.*, 1963; Hack, 1941; McKee, 1966; Melton, 1940; Sharp, 1964 and 1966), but there are locally important spatial and temporal variations in the pattern. The 'Nevada High' and the 'Mojave Low' produce distinct winter-wind circulations. In the central Asian deserts, it seems that only on the northern and southern margins is there a dominant westward drift of aeolian sediments, for example in Siberia and in Persia (Gabriel, 1958; Suslov, 1961). In the Kara-Kum the prevailing winds are from the north-east, blowing sand across the course of the Amu Darya (e.g. Doubiansky, 1928; Heller, 1932; Suslov, 1961). The Chinese deserts are dominated in winter by light outward flow from an intense high-pressure zone, and are penetrated in summer by the south-east monsoon: in consequence, the wind patterns are complex (Petrov, 1962). Hedin (1904) maintained that the dune-building winds in the Tarim Basin came from the east-north-east (Fig. 4.34).

There are other deserts with wind patterns that are anomalous to the general model of the global climate. The Thar Desert in India and Pakistan and the deserts of the Horn of Africa experience their most effective sand-moving winds during the summer monsoon, when winds blow north-eastward in the Thar (Verstappen, 1968) and northward in the Horn. In the southern Peruvian Desert a generally south-easterly wind blows for most of the year (Finkel, 1959).

At a smaller scale the intense heating and cooling of deserts, and the contrasts between the different thermal properties of parts of the desert surface, generate a variety of very local winds.

Heating and cooling effects are amongst the better recorded of these phenomena. For example, it has been noted that Death Valley, California, has a 'respiratory' wind pattern. There are inward-moving winds during the day when there is heating, and outward-moving winds at night when there is cooling (Clements *et al.*, 1963). The juxtaposition of basins with mountains is very often associated with local intense winds. In the San Luis Valley, Colorado, the prevailing south-westerly winds are met at some seasons by a strong north-easterly wind blowing off the Sangre de Cristo Mountains (McKee, 1966).

Coastal areas are also likely to have distinct thermal contrasts. Many desert coasts experience constant, relatively strong, onshore winds. The western coast of Baja California and the coast of Chile are examples of areas where there is an important inland transport of sediment by coastal winds (Inman *et al.*, 1966; Segerstrom, 1962). On the north coast of Peru, some of the onshore winds are so strong that small pebbles can be built into ripple-like features by the wind (Newell and Boyd, 1955).

At a yet smaller scale the local 'dust-devil' is a well-known feature of arid areas (Plate 1.2). It seems to be related to very intense local heating and instabilities. Winds of up to 55 kph can be generated within these disturbances, but the winds are of short duration and unreliable direction, and are therefore unlikely to be of much importance in determining the overall directions in sediment transport (Clements *et al.*, 1963).

One of the wind effects most easily observed in deserts, where winds very often have a distinct topographic expression, is the intensification and deflection of winds round topographic obstacles. The most spectacular examples of this can be seen as wind-produced landforms on the satellite photographs of the Tibesti Mountains (e.g. Pesce, 1968; Verstappen and Zincam, 1970). We return to a more detailed discussion of the structure of the wind in section 4.1.2.

1.3.4 HYDROLOGICAL CONDITIONS

Precipitation in deserts is variable in frequency, intensity, duration and 'spottiness', but it is possible that, although the return period between events may be longer and the total precipitation may be smaller, the degree of variability may not always be greatly different from that in more humid areas. In the south-western deserts of the United States, for example, Russell (1936) demonstrated that precipitation may be less spotty in its distribution than it is in Wisconsin and other more humid regions, and that its apparently greater spottiness is a false impression given by small totals and wide vistas.

The variability of precipitation is reflected in patterns of runoff. Unfortunately, data on desert runoff are even more scarce than data on climatic conditions; and it is difficult to estimate runoff from climatic

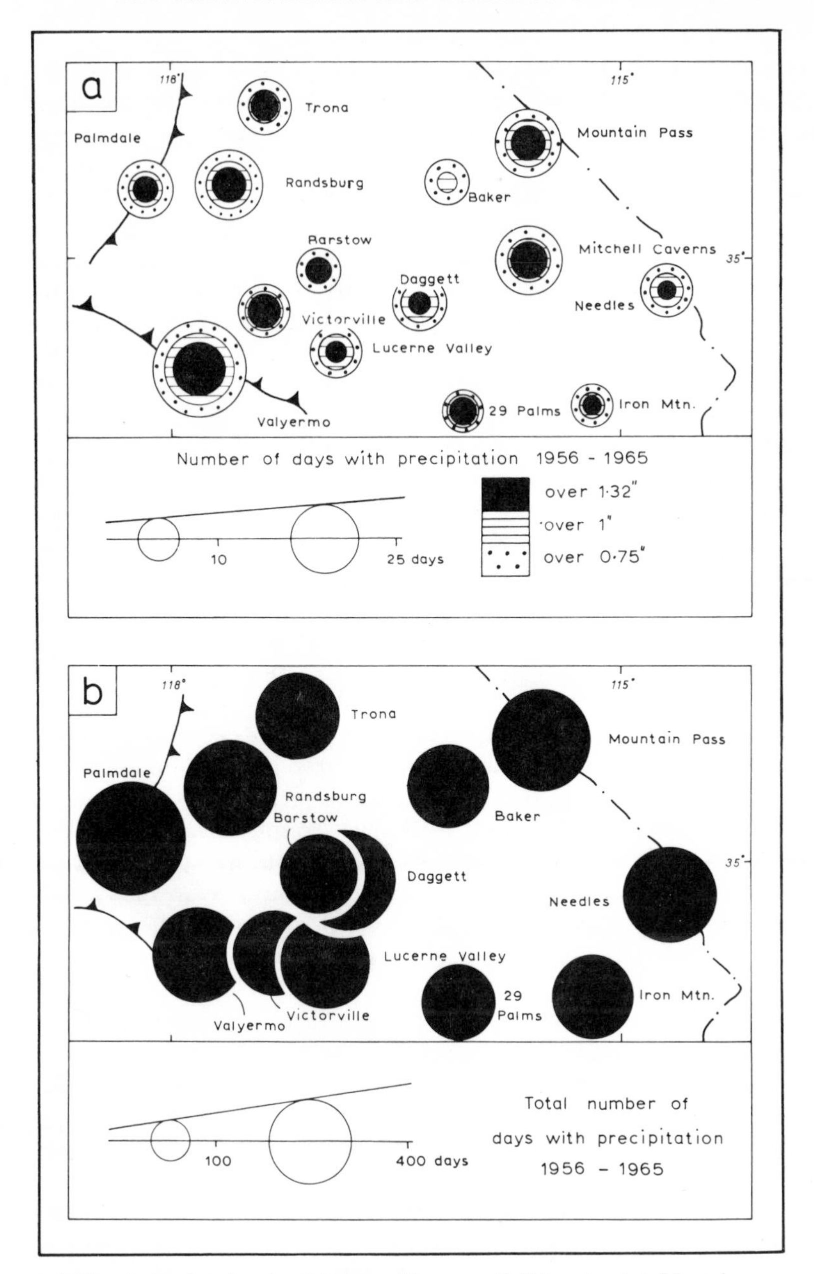

1.3 Precipitation in the Mojave Desert, California. (*a*) Number of days in a ten-year period with precipitation totals above 19 mm, 25·4 mm and 34 mm. (*b*) Total number of days with precipitation.

records because it is a precipitation residual, affected by rates of precipitation, infiltration capacities, water deficiencies in surface material, and local topography, all of which are usually unknown variables. A very crude, general impression of runoff frequencies may be gained by examining the frequency of daily precipitation totals above certain limits. We have shown for the Mojave Desert, California (Fig. 1.3a) the number of days in a 10-year period with precipitation totals above 19 mm, 25·4 mm and 34 mm (the last threshold was selected because it was associated with runoff in 1965). If runoff occurs when daily precipitation exceeds 19 mm, its frequency will be once a year in many places and almost everywhere once in two years; if the 25·4 mm level is taken, the frequency would be approximately once in every three years; and if there is runoff only when daily precipitation exceeds 34 mm, then it will be experienced only once or twice in a decade. Analysis of the precipitation record in the Mojave Desert yields other points of geomorphological interest. Runoff is more likely in winter because most heavy rainfalls occur at this time, when temperatures and evaporation are lower. In this area precipitation, and presumably runoff associated with summer thunderstorms, is relatively unusual, for an average of only 15 per cent of rain days in which precipitation exceeds 19 mm occur from July to September. Another feature is that winter precipitation usually affects a large area, a fact which may be attributed to its origin from Pacific air masses; heavy summer precipitation, in contrast, not only occurs infrequently, but it is also more localized. The seasonal contrast in daily precipitation distribution is exemplified in Fig. 1.4. Total precipitation declines, and the proportion of summer precipitation increases towards Arizona in the east. Thus it would seem likely that across the Mojave Desert there is an eastward decline in the amount and frequency of runoff, and an increase in its localization.

Detailed studies of the relations between rainfall and runoff have been carried out in the 'summer convective storm' desert of southern Arizona at the Walnut Gulch experimental watershed near Tombstone. Schreiber and Kincaid (1967) showed from a multiple regression study of small runoff-plot data that average runoff was proportional to precipitation quantity; that the taking into account of crown spread of vegetation improved runoff prediction, but only slightly, and that antecedent soil moisture was relatively unimportant. The three independent variables accounted for 72, 3 and 0·5 per cent, respectively, of the prediction variance. Their general equation indicates that 6·6 mm of rain must fall before runoff begins. In a later study, Osborn and Lane (1969) extended the investigation to small watersheds (0·2–4·4 h) in the same area. Their multiple regression analysis showed that runoff

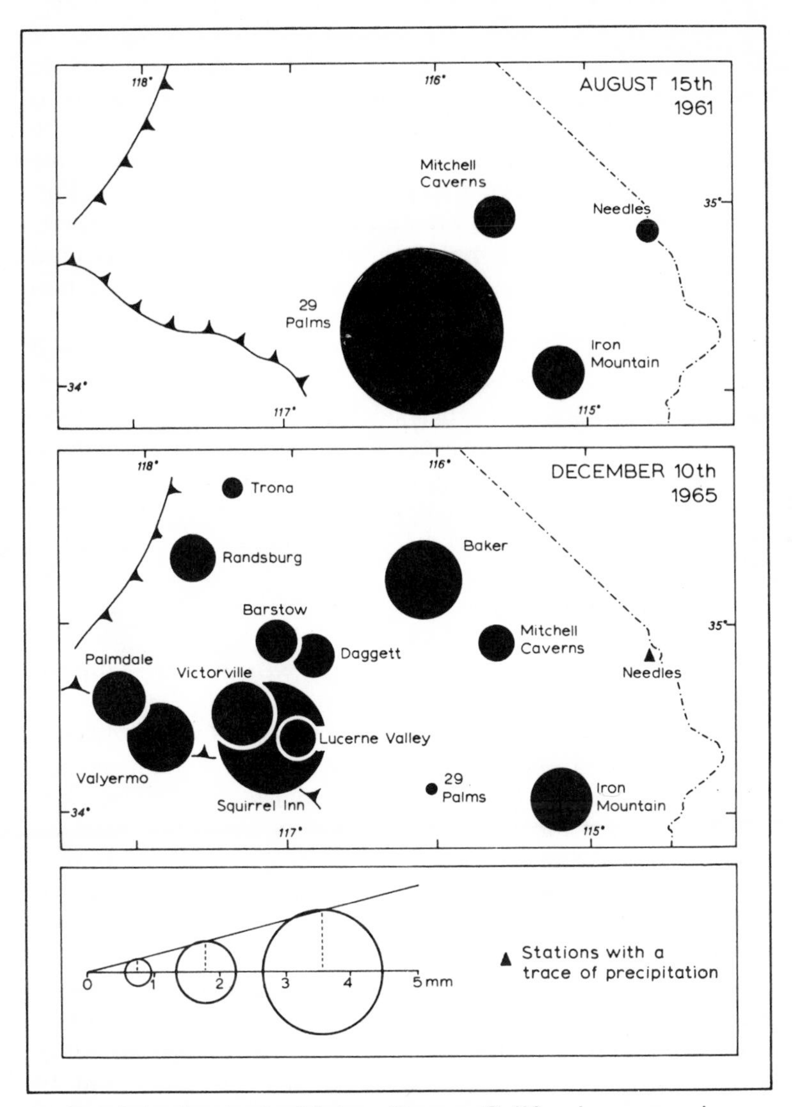

1.4 Precipitation in the Mojave Desert, California: comparison of precipitation from a summer and a winter rain period.

volume was again most strongly correlated with total precipitation; that peak rate of runoff related most closely to the maximum 15-minute depth of precipitation; that flow duration was best correlated with watershed length; and that lag time was most strongly correlated with watershed area. The independent variables accounted for approximately 70, 70, 50 and 30 per cent, respectively, of the prediction variance.

Some Saharan floods have been documented almost by chance. In 1960 Medinger (1961) was able to observe a flood and relate it to rainfall on the Tadmaït Plateau where wadis are only lightly incised into the very flat pavement surface. Mean annual rainfall cannot be more than 18 mm, but on 7 December 1960 a flood of 120 m^3/sec passed down a fourth-order stream near Fort Mirabel, probably reaching a maximum discharge of 1,600 m^3/sec. The rainfall which generated this flood does not appear to have been more than 16 mm as measured on an upland rain gauge. Medinger calculated that perhaps 28 per cent of the rain falling on the catchment contributed to the flood—a very high figure. The measured solids in this flood reached 1,712 gm/litre. Other floods in the Sahara are documented in several volumes of the *Travaux de l'Institut de Recherches Sahariennes*, and by Dubief (e.g. 1953).

On a world scale, there are useful distinctions to be drawn between endogenic and allogenic drainage, and between endoreic and exoreic drainage (de Martonne and Aufrère, 1928). Endogenic drainage is that derived from precipitation within the desert area; allogenic drainage originates outside the desert and flows into it. An example of an allogenic stream is the Mojave River, in California, which rises in the San Bernardino Mountains and flows north into the Mojave Desert (Plate 1.3); another is the Oued Saoura, which flows south to the Sahara from the Atlas Mountains in Algeria. Geomorphologically, allogenic streams may be associated with landforms and deposits in deserts which are alien to the desert environment and reflect conditions more common in humid areas (Fig. 1.5). Much desert drainage, whether it be endogenic or allogenic in origin, is endoreic; that is to say, it does not reach the sea, mainly because of high water losses due to evaporation and surface absorption, and because much drainage is within enclosed basins. In areic areas surface drainage is practically nil. Some deserts do have areas of exoreic drainage, where flowing water reaches the sea: in the southern Atacama Desert, for example, three perennial allogenic streams flow from the Andes across the desert to the Pacific. The world pattern of endoreic, exoreic and areic areas is shown in Fig. 1.6. Geomorphologically, these distinctions are important, for characteristic landforms may be associated with the different types. Changes of sea-level may be reflected in the landforms developed by exoreic drainage, and much sediment may be removed to the sea from such basins (Fig. 1.5; Plate 1.4); enclosed basins, on the other hand, may experience progressive accumulation of sediment, and may remain unaffected by changing sea-level (e.g. Mulcahy, 1967).

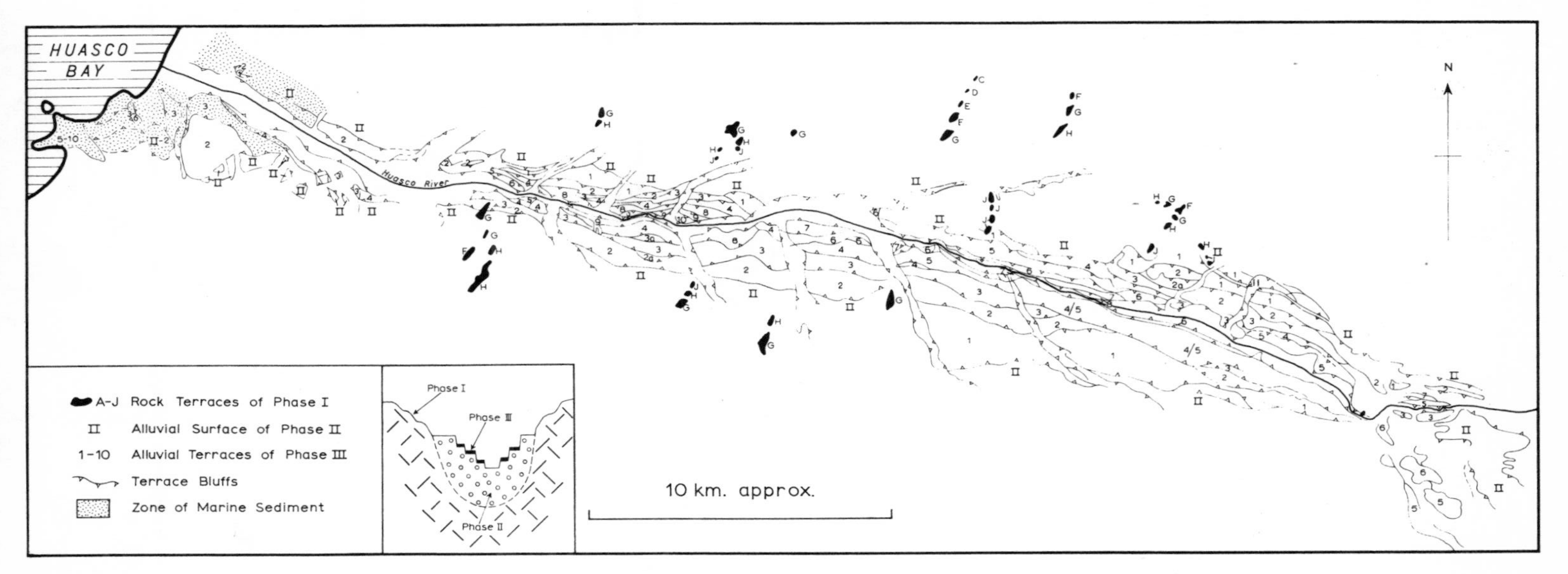

1.5 A complex system of river terraces along the Huasco River, southern Atacama Desert, Chile (Cooke, 1964a). The river is sustained by snowmelt in the High Andes, and flows to the Pacific across the arid zone.

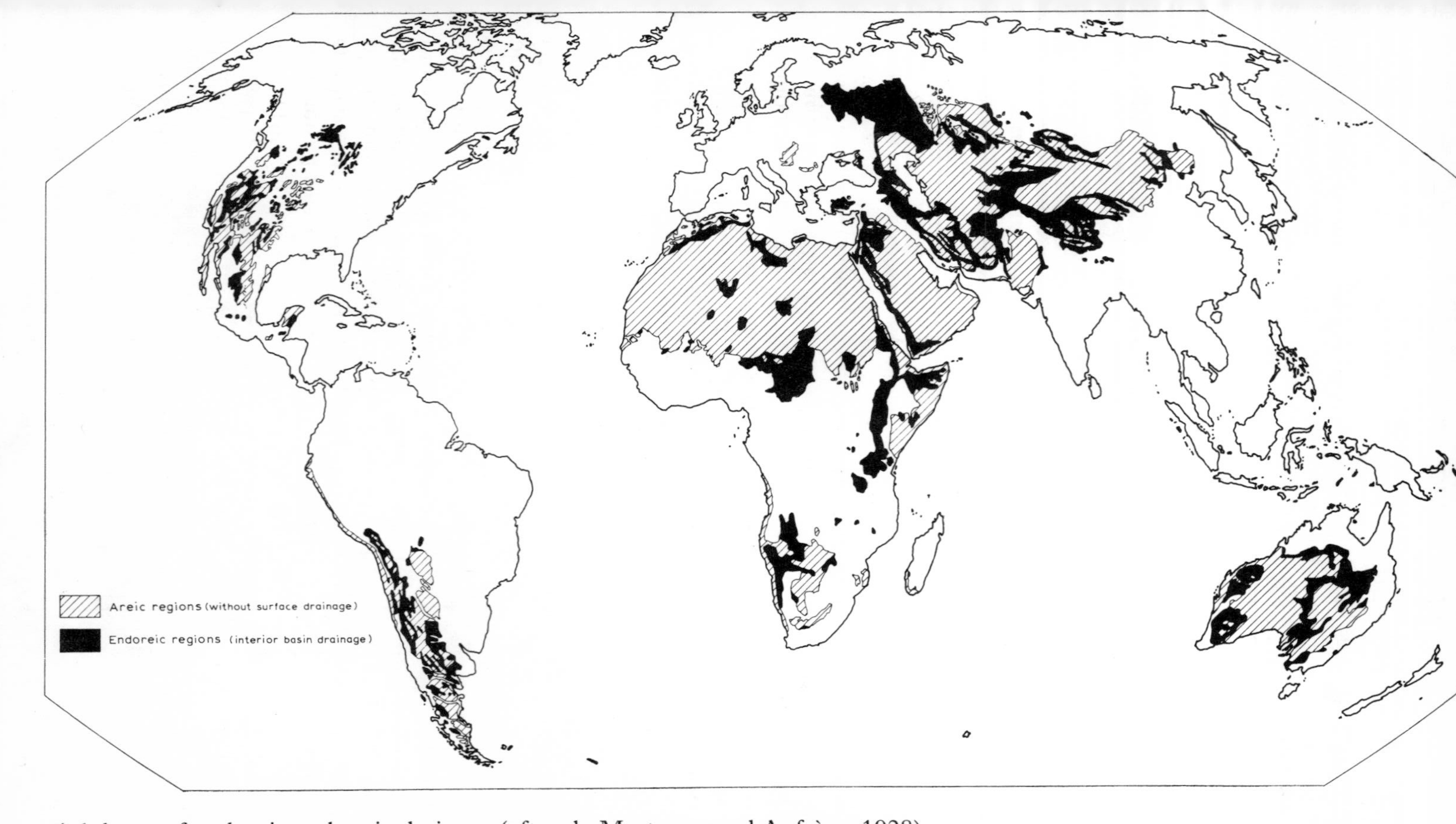

1.6 Areas of endoreic and areic drainage (after de Martonne and Aufrère, 1928).

There are several distinctive features of drainage within desert basins. Although relatively little is known of water movement in the zone between the watershed and channel, it is certain that slopewash in thin sheets or rills occurs. Observations of surface-particle movements over several years by Cooke and Reeves in the Mojave Desert, by Leopold *et al.* (1966) in New Mexico and by Sharon (1962) in Israel, all confirm this view. Tricart and Cailleux (1964, vol. 1) also place great qualitative emphasis on this process; they cited the limited absorption of water by plants and desert soils, and the intensity of desert storms, as being significant factors affecting the phenomenon; and they suggest that rainwash occurs above a relatively low threshold of 5 mm total rainfall and an intensity of at least 0·5 mm per minute. It is not altogether clear that slopewash is relatively more important as a geomorphological process in deserts than elsewhere, but this could be the case. Kirkby (1969), for instance, argued that, of the two types of hillside waterflow, overland flow may be most common in semi-arid climates (where soil and vegetation covers are thin, and rainfall intensities may be high), whereas throughflow of water may be predominant in humid and humid-temperate regions. However, in section 2.3.2 we show that there is some evidence for important throughflow components in some deserts.

Flow of water in natural channels may be perennial, intermittent, or ephemeral. Perennial flow is rare in deserts, except when the drainage is allogenic. Intermittent flow—in which there is an alternation of dry and flowing reaches along a drainage channel—is more common. Most channels in deserts, however, only carry water during storms, and flow in them is therefore ephemeral. As a result, flow in most river systems is only rarely, if ever, integrated. Furthermore, the location and size of the sector affected by runoff within a basin may vary from storm to storm. In addition, streams may be 'underfit'—they may be too small for the valleys in which they flow. Underfit streams are common in many environments (Dury, 1964); in deserts, streams may be underfit partly because of the contemporary variability of runoff events, and in part for reasons of climatic change (e.g. Peel, 1966).

Equally important, the *actual* area of a drainage basin may be considerably greater than the *effective* area of drainage within it. For example, in a desert basin which has suffered a degree of desiccation since the last pluvial period, the effective drainage area may only occupy the mountains and higher piedmont plains near the watershed of the basin. A particularly striking example of such a situation is the Dalol Mauri drainage system which extends from the Aïr Mountains in Niger south to the River Niger (Fig. 1.9); flow in this system rarely extends far beyond the mountains. Another feature of ephemeral desert streams is that they are less dependent on ground-water sustenance, and receive a greater proportion of their water directly from overland flow by comparison with their temperate-land counterparts. It might be supposed that these hydrological phenomena result in distinctive features of desert geomorphology, and to some extent this is true. For example, there may be discordance between tributary valleys and main channels due to filling of the lower parts of tributary valleys with sediment (Schumm and Hadley, 1957) and localized erosion in the main channels. Very rapid runoff across unvegetated surface debris may lead to high sediment concentration, and to the creation of distinctively viscous flows, such as debris flows (Sharp and Nobles, 1953); and sediment load may increase downstream faster than discharge because of channel absorption. Drainage-channel density too, tends to be higher where the proportion of bare ground is greater (e.g. Melton, 1957).

1.3.5 PROCESSES AND CHANGE

(a) Sediment yield. One useful measure of the extent of geomorphological activity in deserts is sediment yield, expressed in $m^3/km^2/yr$. Broadly, this index is designed to be a measure of the amount of material removed by all processes from a given area over a period of time; in reality it is extremely difficult to measure the total yield because of problems in estimating the extent of wind erosion, bedload movement, density of weathered mantles and deposits and other components. Thus, sediment yield is usually based on measurement of suspended sediment from drainage basins or on monitoring the extent of sediment accumulation in reservoirs. Measurements are required over a period of years in order to give realistic average values, for sediment yields may be subject to large annual variations. The index does not reveal the nature and relative importance of different processes at work within drainage basins; and to ignore wind erosion and dissolved load, both of which may be relatively important in arid and semi-arid lands, may significantly affect the reliability of the estimates from desert areas.

Over the world as a whole, the total erosion of material from deserts may be small compared with that from areas in other climates. Several attempts have been made to estimate regional and world rates of erosion (Stoddart, 1969). Corbel (1964), for instance, estimated that in tropical arid areas (under 200 mm precipitation p.a.), a rate of 1·0 $m^3/km^2/yr$ characterizes mountains, and in the plains a rate of 0·5 $m^3/km^2/yr$ is a reasonable figure. These rates are lower than those in any other region he considered. At the other extreme, Corbel estimated a rate of 2,000 $m^3/km^2/yr$ in humid, glaciated polar regions. Corbel also estimated that total erosion from warm arid lands amounted to only about seven per cent of erosion from all arid lands, and erosion from *all* arid lands (warm, temperate and cold) accounted for only some nine per cent of the erosion in unglaciated lands and about four per cent of total world erosion. Other estimates, such as that of Fournier (1960), diverge from those of Corbel, but there appears to be general agreement on the assertion that total erosion from deserts is relatively small.

An important statement on sediment yield is that by Langbein and Schumm (1958) relating annual sediment yield, based on sediment-station data from approximately 3,900km^2 drainage basins and on reservoir data from approximately 78 km^2 drainage basins in the United States, to mean annual effective precipitation (the amount of precipitation required to produce a known amount of runoff under specified temperature conditions). Their original curves of this relationship were based on a reference average annual temperature of 50°F (10°C); in Fig. 1.7 we show a family of curves based on several temperatures (Schumm, 1965). The effect of increasing average annual temperature is to move peak sediment yields to higher totals of mean annual effective

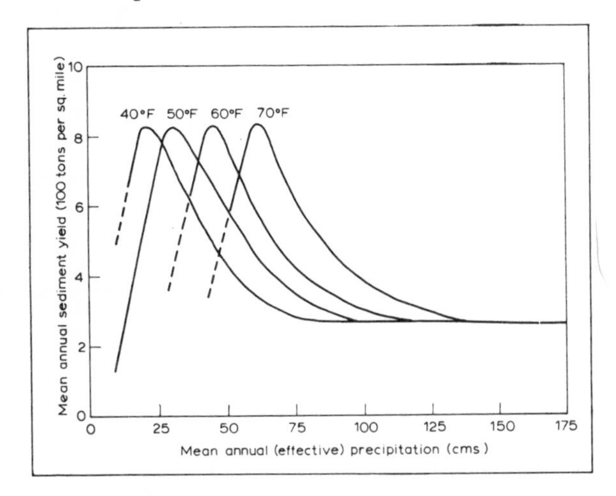

1.7 The effect of temperature on the relation between mean annual sediment yield and mean annual precipitation (after Schumm, 1965).

precipitation. (It should be emphasized, as these curves have been widely accepted, that they are extrapolated from the 50°F (10°C) curve which is itself based on only six group averages of sediment-station data.) The most significant feature of these curves is that sediment yield reaches a maximum between 254 and 625 mm (depending on annual temperature) and that yield falls off sharply towards areas of more or less effective precipitation. 'This variation in sediment yield with climate can be explained by the operation of two factors, each related to precipitation. The erosive influence of precipitation increases with its amount, through its direct impact in eroding soil and in generating runoff with further capacity for erosion and for transportation. Opposing this influence is the effect of vegetation, which increases in bulk with effective annual precipitation' (Langbein and Schumm, 1958, pp. 1079–80)*. Most deserts, as we define them, would occur on the rising limbs of the curves. In these areas, as precipitation rises from zero, sediment yield increases at a rapid rate because more runoff becomes available to move sediment and there is still bare ground susceptible to rainwash erosion. As the peak is approached, the desert scrub progressively gives way to grasses which reduce the vulnerability of the surface to erosion. If these curves are typical of sediment yield-climate relationships on a world scale it would seem reasonable to draw a conclusion which accords well with our previous observations on the diversity of deserts: the world's dry lands include a wide range of sediment yields, from virtually nil to over 300 $m^3/km^2/yr$.

Langbein and Schumm's work emphasizes the relations between sediment yield and climate. Their data need to be qualified by reference to other factors affecting the yields. Firstly, the data are derived from relatively small drainage basins of about 3,900 km^2 or less; yield from larger basins may be lower. Secondly, erosion rates may often increase with the ratio of drainage basin relief to drainage basin length (Schumm and Hadley, 1961). Thirdly, the correlation between sediment yield and precipitation is improved if the seasonality of precipitation is considered. Erosion rates tend to be higher in those areas of seasonal rainfall such

*These opposing factors can be represented by the equation:

$$S = aP^m \frac{1}{1+bP^n}$$

where S = annual sediment yield (in tons per sq. mile), P is effective annual precipitation, m and n are exponents and a and b are coefficients. aP^m describes the erosive factor and $(1/1+bP^n)$ represents the vegetation-protection factor.

This equation can be solved graphically, and converting to metric units, the approximate result is:

$$S = \frac{1{\cdot}631(0{\cdot}03937P)^{2{\cdot}3}}{1+0{\cdot}0007(0{\cdot}03937P)^{3{\cdot}3}}$$

where S = sediment yield in m^3/km^2 and P = effective precipitation in mm (Douglas, 1967).

as north-east Queensland (Douglas, 1967) where the erosive impact of intense storms is increased by the fact that vegetation is relatively sparse because of seasonal drought. In some semi-arid areas of seasonal precipitation, therefore, very high sediment yield may be expected. Fourthly, interference by man in natural ecosystems often leads to increased sediment yields, so that the data used in contemporary estimates of erosion may be unnaturally high. In particular, as Douglas stated (1967), the peak of Langbein and Schumm's curve at an effective precipitation of about 304 mm (at 10°C)—in the areas of transition between desert scrub and grassland—may be exaggerated, because in these areas removal of vegetation may cause great erosion by the few storms which produce significant runoff. Finally, sediment yield data will clearly be affected by the erodibility of materials within the areas under consideration.

(b) Periodicity of events. Periodic variations in the energy inputs and in the geomorphological processes that depend on them are not well understood. The principal gaps in our knowledge are probably about the storage components in the landscape, and about the relaxation times that are involved. Another serious problem that must be disentangled before we can understand periodicity is the problem of the scales of change. When, for example, does a run of drought years become a climatic change?

As we shall explain in a later section, we are concerned in this book only with desert landforms up to the scale of a basin-and-range province. This means that for most purposes we are concerned only with the periodicity of climatic events, and only very marginally with tectonic periodicity. There is ample evidence that there is a form of periodic change in the climatic inputs (e.g. Hare, 1961; Lamb, 1966). Whether these are the result of regular changes in the solar energy input into the atmosphere or whether they arise from random variations in the input, and the free play of a large storage element, does not concern us here (see Curry, 1962). We are concerned with the nature of the response of the landscape to these changes.

It is useful to adopt Curry's (1962) framework in looking at this response. Every element of the landscape can be said to have a memory of a certain length. Curry used a vegetation example, where a grass cover has a shorter memory than a tree cover, the trees will respond to changes of a different magnitude to those that lead to changes in the grass cover. This scheme can be applied to geomorphological features. A small proportion of a slope will respond markedly to one storm, whereas a valley may need a succession of wet years to induce a noticeable response, and a desert basin would need a far longer period—one, in fact, which might be termed a period of climatic change. In the aeolian system a similar progression can be charted. A ripple will have a form that will change almost instantaneously with the wind, a dune changes its form with the seasons, and a massive *draa* ridge will change only in response to changes of the order of its reconstruction time, namely a few thousand years (Wilson, 1970).

The problem of periodicity is not peculiar to desert landscapes. What we wish to know is whether the nature of the memories and responses is different in deserts. Scanty though the evidence is, we would claim that there is no distinctiveness about deserts in this respect. In examining the problem in general, Wolman and Miller (1960) came to the conclusion that it was events of the order of bankfull discharge that were responsible for the main outlines of the forms of channels. They remarked that since these events were far further apart in deserts than in humid climates, desert stream channels were the result of rarer events. However, they also noted that 'for small drainage basins, differences between humid and arid regions appear to be slight' in respect to the percentages of load carried by flows of the same frequency (Wolman and Miller, 1960, p. 60).

Most processes in the landscape only take place in response to an energy input that is larger than a certain threshold value. As this threshold value falls, more work can be done by events of a lesser magnitude. Since the thresholds for erosion in deserts are probably lower than in temperate climates, it could be concluded that events of the same frequency as in temperate areas, but of lower magnitudes, could accomplish the same work. This may account for Wolman and Miller's findings. Our discussion of the patterns of rainfall and runoff in deserts have shown that, although rainfall is indeed spotty and infrequent, it may be no more so than in humid climates. Clearly this immensely important subject needs much further investigation in deserts.

We should emphasize the importance of the threshold to the discussion of periodicity. In the Ténéré Desert in the Republic of Niger, most of the desert surface is covered with coarse sands of low undulating dunes that have a bimodal grain-size distribution with an important coarser mode at about 0·6 mm. In the wind records of the area it is found that winds which are capable of moving these sands do not occur more than about ten times in the year, and that since the sandflow is a power function of windspeed, the winds which move most of the sand are much rarer than this. Nearby sief dunes are composed of unimodal sands with a grain-size of about 0·25 mm. These sands can be distributed by much more frequent winds and the magnitude-frequency relationship means that winds of a lower magnitude move the greater proportion of the finer sands (Warren, 1971*a*).

(c) Endogenetic and exogenetic processes. Processes responsible for creating landforms comprise two principal groups: the endogenetic processes, related to tectonic and volcanic activity, and the exogenetic processes, associated with ice, wind and water. Sediment yield data represents in a general way the rate of landform denudation by exogenetic processes, and we have suggested that in deserts there may be a great diversity of rates. Similarly, there is reason to believe that there is great variability in the nature and rates of endogenetic processes in deserts: on the one hand there are tectonically stable deserts, of which the Australian and Kalahari deserts are examples; on the other, there are tectonically unstable deserts such as those in California and Chile (Plates 1.5–1.8). Aspects of the variability of endogenetic processes include their general rate (Schumm, 1963*a*), frequency, intensity, duration, location and direction.

In any one area, the relationship between the two groups of processes may, in large measure, determine the nature of the topography. In extremely arid deserts experiencing intense tectonic activity, the topography may be dominated by the consequences of earth movements, as exemplified by the numerous fault scarps, disrupted, tilted and folded surfaces, and blocked and diverted drainage of the southern part of the Great Basin Desert in California. Contrasted with this area are the plains and complex suite of aeolian features in the extremely arid, but tectonically stable, north-eastern Sahara. The increase of precipitation and erosional activity southward through the tectonically unstable Atacama Desert may help to explain why landforms produced by endogenetic processes are so much more clearly displayed in the extremely arid northern part of the area. It is at once apparent, therefore, that a diversity of desert landforms may result from variations in the relationship between the two groups of processes, making even more difficult the task of realistic generalization.

(d) Climatic change and fossil forms. For desert areas located on the rising limbs of the curves shown in Fig. 1.6, the effect of changing climate will clearly depend on the initial climatic conditions, and on the extent and direction of the change itself. To take a simple example, if an area with an average annual temperature of 10°C which has an average annual effective precipitation of less than approximately 300 mm experiences an increase of effective precipitation to produce a total of no more than 300 mm or so, it is likely to increase its sediment yield, and

sediment concentration would decline. A decrease in average annual temperature would probably have a similar effect. To increase temperature or decrease precipitation would have the reverse consequences: a decline of sediment yield, an increase of sediment concentration and aggradation, and appropriate adjustments in channel geometry (Langbein and Schumm, 1958). Changes of both temperature and precipitation occurred in many deserts during the Quaternary era, and the variety of permutations of climatic variables is infinite, involving as it may have done, seasonal, annual and long-term movements of varying magnitude and direction which, in turn, may have varied in frequency and duration. At one extreme, there are some desert areas which appear to have been more or less permanently arid throughout the Quaternary—Mortensen's (1927) *Kernwüste* of the Atacama Desert, the eastern coast of Baja California (Walker, 1967), and Butzer's (1961) arid core of the Sahara are examples. The persistence of a tropical and subtropical arid zone during the Quaternary, albeit contracted in size during pluvial periods, is a probability indicated by Büdel (1959), Butzer (1961 and 1965) and Warren (1970). At the other extreme, some deserts have experienced a succession of climatic changes. Smith (1968), for instance, described detailed evidence of numerous climatically controlled lake-level fluctuations in the Searles Basin of the southern Great Basin Desert, California, which correlated with the late Quaternary glacial history of the Sierra Nevada. The southern margins of the Sahara, too, have seen several 'cycles' of aridity and humidity (Grove and Warren, 1968; Warren, 1970); and Jessup (1961) among others charted a complex history of change in Australia.

It is established that many deserts are more arid today than they have been in the past, particularly in the 'pluvial' conditions of the late Pleistocene (Plate 1.10). In such deserts there has, therefore, commonly been reduction of erosion and sediment yield, and increased local aggradation. The single most important and pervasive consequence of this fact is that landforms produced under previous, different conditions stand a good chance of persisting or being preserved in the new, less active, environment. Thus it is that the problem of 'fossil' or 'relict' landforms has become central to studies of desert geomorphology (e.g. Peel, 1966).

The response of desert landforms to the onset of aridity, or more pronounced aridity, may take several forms, and in the following paragraphs we attempt to identify and exemplify them.

There are at least three main types of landform preservation: *(i)* burial, *(ii)* abandonment and *(iii)* preservation by protective veneers.

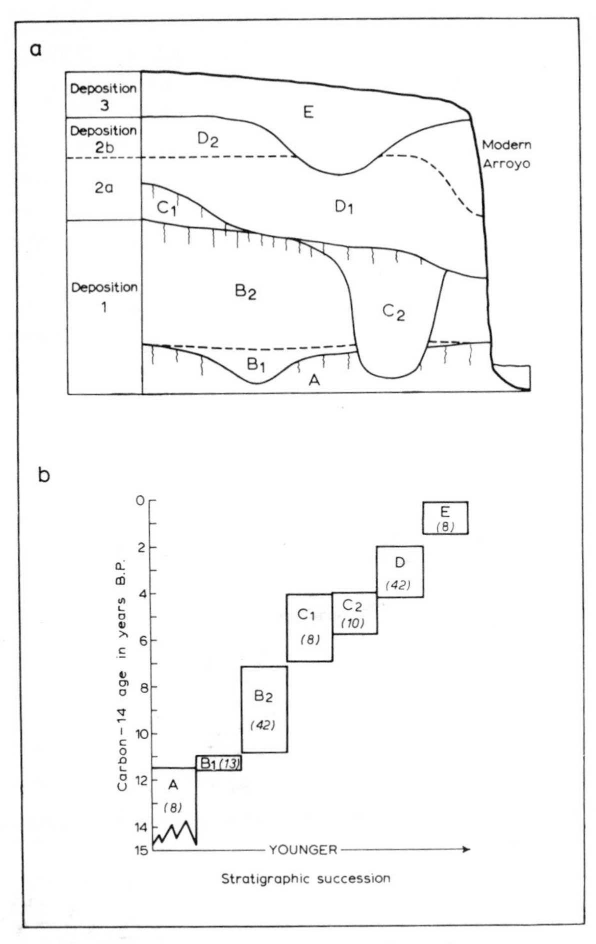

1.8 (*a*) Diagrammatic cross-section of an arroyo bank in the south-western United States showing idealized relationship of depositional units. Vertical wiggly lines indicate weathering. Not to scale. (*b*) Depositional units arranged according to C14 dates. Number of C14 dates in parentheses (after Haynes, 1968).

(i) An alternating sequence of climatic changes in deserts may be accompanied by an alternating sequence of erosional and depositional episodes, and it is clearly possible that landforms created under one set of circumstances may be buried under sediment laid down under the

other. It is likely that the changes will be most evident along drainage lines; indeed, former drainage channels are probably the most common feature to be preserved by burial. If, as we have argued, increased aggradation in drainage basins accompanies progressive desiccation of climate, then the channels would be inherited from more pluvial conditions. But this is a controversial and oversimplified assertion, which has been considered at length in the semi-arid south-western United States. Here Haynes (1968) and many others have described an alluvial stratigraphy consisting of aggradational phases separated by episodes of arroyo-cutting. Haynes's comprehensive interpretation is shown in Fig. 1.8. The same problem has been discussed for the semi-arid margins of the Mediterranean by Vita-Finzi (1969).

(ii) A more common form of preservation occurs where a landform or landscape produced under a given climatic regime is merely abandoned by the processes which formed it with the advent of aridity. The degree of abandonment may vary greatly: in a river system, for instance, drainage may change from perennial to ephemeral or intermittent, or ephemeral drainage may become even more sporadic. In this situation, the channel system will probably appear to retain its form, although effective drainage area and drainage density may be reduced, and there may be appropriate changes of channel cross-section and gradient.

There is extensive evidence of fluvial systems and landforms on a scale too great to have been created or maintained under present climatic conditions. The Chad Basin and the north-eastern part of the Niger Basin are a remarkable example of this in West Africa (Fig. 1.9): formerly Lake 'Mega-Chad' and its predecessors in the Chad and Bodelé depressions were sustained by drainage from the surrounding massifs. Today the lakes are represented only by fluvio-lacustrine deposits and strandline features; the great drainage ways are generally dry; and aeolian landforms have been created throughout much of the basin (Grove and Warren, 1968). Similar sequences have been described in central Australia (Wopfner and Twidale, 1967). In the southern Atacama Desert of Chile, dating of ignimbrites on extensive pediplain landscapes high in the Andes reveals that since the Upper Miocene the only significant modification to the topography has been its dissection with canyons which are only about 100 m deep (Clark *et al.*, 1967). Fig. 1.10 shows a reconstruction of the drainage system of the south-western Great Basin Desert during the Wisconsin period. At this time, or for a short period within it, rivers flowed from the Sierra Nevada and ranges to the south through a series of lakes into Death Valley, and there were many small, isolated lakes. Today the system is clearly preserved with numerous examples of shoreline features, overflow channels

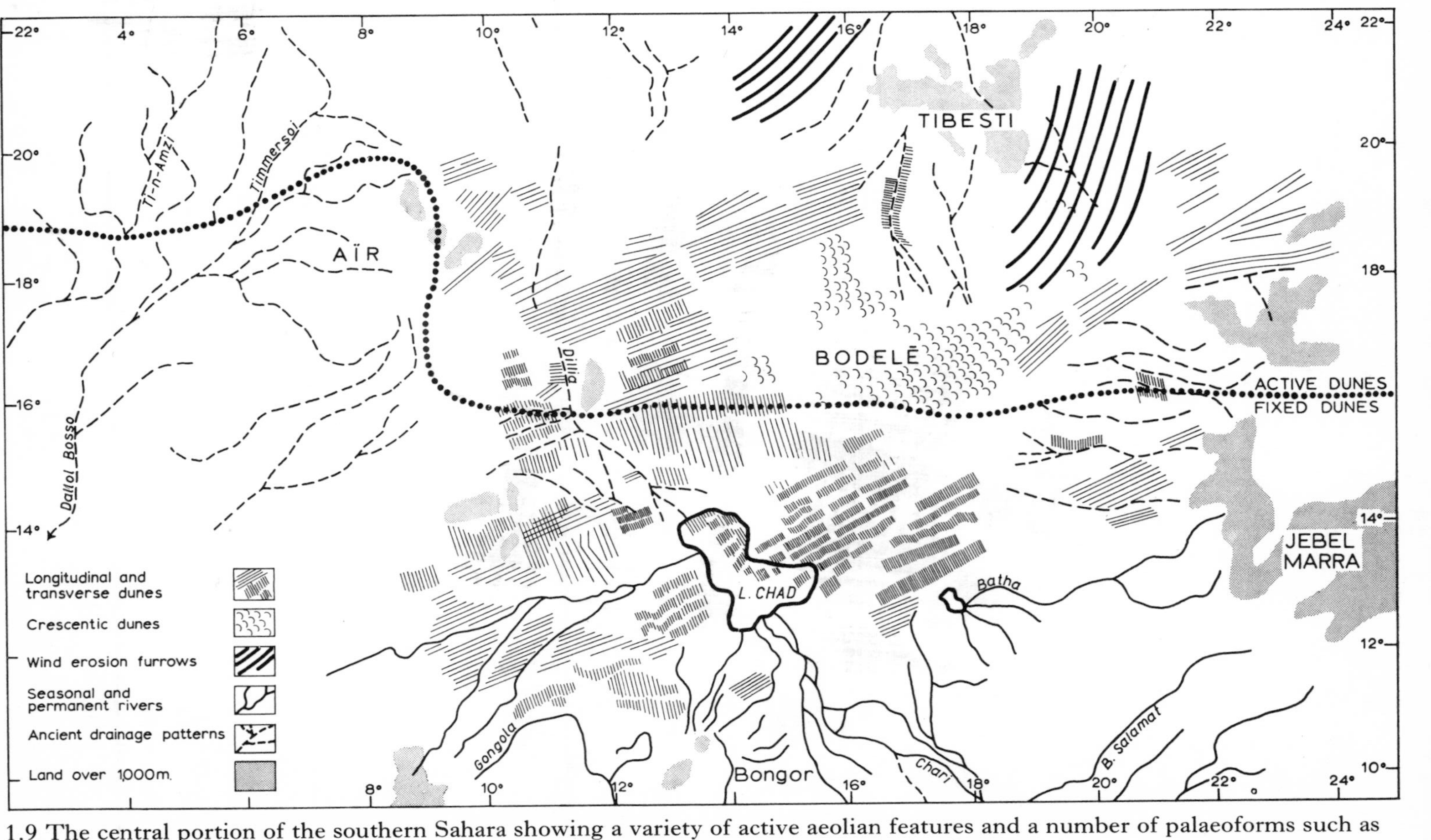

1.9 The central portion of the southern Sahara showing a variety of active aeolian features and a number of palaeoforms such as ancient drainage systems and fixed dunes (after Grove and Warren, 1967, and others).

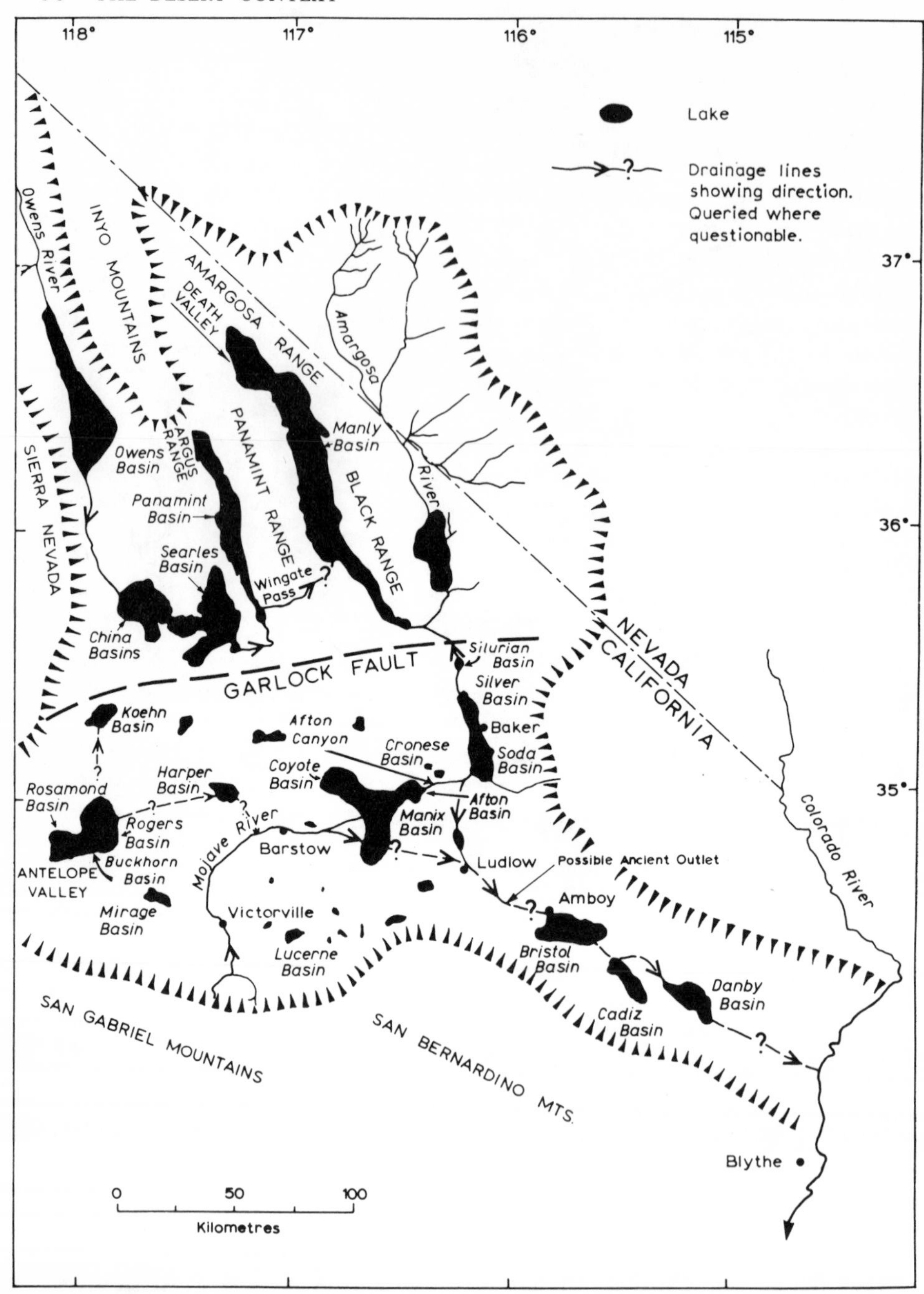

2.10 Drainage network in the south-western Great Basin during the Wisconsin period (after Blackwelder, 1954, and others).

(Plate 1.9) and terraces, but drainage rarely extends into the desert beyond the Owens and Silver basins. Similar systems of fossil river valleys have been described in the central Kalahari Desert (Boocock and van Straten, 1962), in Western Australia (Mulcahy, 1967), and in central Sudan (Warren, 1966).

(iii) The preservation of landforms by protective veneers is both common and distinctive. On a large scale, the formation of duricrusts—which may be composed of silica, gypsum, calcium carbonate, or the sesquioxides (section 2.3.2)—may serve to protect associated surfaces from further erosion. In these circumstances, no climatic change is necessary to produce the protective effect, but such a change is often involved.

In addition to landform preservation, there are perhaps three additional types of response by landforms in deserts to a climatic change towards aridity: *(i)* continued development without apparent change of form, *(ii)* changes leading to replacement or superimposition of landforms and *(iii)* metamorphosis of one feature into another.

(i) Particular landforms or erosion systems may respond to climatic change merely in the rate of their development. For example, an erosion system composed of mountains, pediments and alluvial plains may develop more slowly under conditions of increased aridity and may experience no distinguishable change of form. It is true that such systems do often show evidence of adjustments to climatic change but such adjustments are by no means essential. In the case of the pediplain landscape of the high Andes in the southern Atacama Desert, for example, it is possible that mountain-front retreat has continued in the undissected areas throughout much of Pliocene and Quaternary time. In such circumstances the landforms may be descended from features formed under different conditions and they may display no characteristics diagnostic of those conditions or, for that matter, of contemporary conditions.

(ii) Where one erosion system is replaced by another, landforms characteristic of the new system may replace those of the old, or be superimposed upon them. Wright (1958) cited an example from the Syrian (Arabian) Desert. He described a wadi system leading from the Arabian Plateau down to the Mesopotamian Lowland, the floors of which are crossed by a pattern of stripes transverse to their trend. Active features which produce the micro-relief are ridges or 'boils' about 0·7 m wide and 2·5–5 cm high; they are probably raised by swelling of the clay-rich soil on the wadi floors with subsurface water. In places, the wadi system is masked by wind-blown sand. This, and the development of the stripes in a stable soil formation suggests that the wadis were

formed under different, probably 'pluvial' conditions in the past. The onset of more arid conditions has led to the replacement of the fluvial system by a relatively stable pedogenic system and an aeolian system, and by a superimposition of forms associated with the latter on to those characteristic of the former. Some good examples of this kind of change can be taken from soil profile development. For example the initial growth, probably under more humid conditions than those of the present, of clayey *B* horizons is superceded, in New Mexican profiles, by the upward growth of *K* (carbonate) horizons. The *K* horizons eventually mask the lower *B* (Gile, 1966*a*).

(*iii*) Metamorphosis of one feature into another following a change of climate is not an easy concept to exemplify because, by definition, the nature of the original feature is unknown. An example might be the solodization of saline soils after a change in rainfall conditions.

To conclude, a very important qualification is necessary. Implicit in much of this discussion of landform responses to changing climate has been the idea that landforms may acquire a state of equilibrium with one set of environmental circumstances, and that a change in those circumstances causes a short period of readjustment followed by the establishment of a new equilibrium. This view is especially apposite to channel changes, for example. It is quite possible, however, that some landforms may develop an equilibrium condition which is not only relevant to the initial circumstances, but is sufficiently secure to be maintained when the situation changes. For example, stone pavements might develop to a stable condition under a semi-arid climate but retain their form virtually unchanged when the climate becomes arid.

1.4 Frameworks for Geomorphological Generalization

Studies of desert landforms in the last hundred years have been cast in terms of one of a few explanatory models. In this section we review briefly some of the more important of these models, and outline the framework within which we shall present our material.

1.4.1 EXPLANATORY MODELS

There has always been a need in geomorphology for comprehensive explanatory models which satisfactorily explain the landscape of an area and the individual features within it. Such schemes facilitate field observation by directing attention to particular phenomena and by allowing apparently irrelevant material to be discarded or ignored. In deserts, most schemes are what Chorley described as 'historical analogue models' which 'group together geomorphic phenomena with regard to their positions in time-controlled sequences, on the assumption that what has happened before will happen again, or that what existed in the past has relevance to what exists now' (Chorley, 1965, pp. 60–61). The most important of the models that have been used in deserts are the 'aeolian' model, Davis' model of landform evolution in arid climates and its successors, allied schemes for semi-arid and savanna climates, and contemporary models based on the analysis of systems.

(a) Extravagant 'eolation'. In the nineteenth century there grew up an idea that desert landforms were produced largely by mechanical weathering and wind action. James Geikie's declaration that in rainless and desiccated regions insolation and deflation play the most important role, and that rain and running water play only a very subordinate part in forming the topography, reflected the concensus view of deserts at the turn of the century (Geikie, 1898), although evidence to the contrary had been accumulating for some years. Foremost amongst the advocates of aeolian activity were Albrecht Penck (1905), Passarge (1904*b*), Walther (1924) and Keyes (1912).

The concept that wind activity was dominant in deserts probably arose simply from the observation of the common coexistence of dry, vegetation-free surfaces with loose materials and strong winds; direct observations of the work of debris-charged winds on many an expedition may have helped to impress the efficacy of wind action on explorers.

Evidence used to support the view that aeolian processes are dominant was almost entirely drawn from qualitative observations of landforms, rather than from study of the processes themselves. Polished rocks, shallow enclosed depressions, rounded sand, and loess deposits were frequently cited as evidence of wind action. On a larger scale, extensive plains, and in particular the presence of isolated mountains surrounded by plains—the *Inselberglandschaft*—were considered to be peculiarly desertic phenomena allied to the wind-erosion process.

Most persistent and outspoken of the Aeolianists was Keyes (e.g. 1908; 1910*a* and *b*; 1911; 1912; 1932). In a series of long and rather indigestible essays he asserted repeatedly that 'recent investigations in arid and semi-arid countries appear to demonstrate beyond all shadow of doubt that as a denuding, transportive and depositional power the wind is not only fully competent to perform such work, but that it is comparable in every way to water action in a moist climate' (Keyes, 1912, p. 541). Wind erosion ('eolation') was said to be much less vigorous than normal water erosion on resistant rocks (which generally formed the mountains), but very much more powerful on weak rocks, which usually underlay the plains (Keyes, 1910*a*); it effected erosion in the deserts themselves, and permanent deposition occurred only beyond the desert boundaries. A cornerstone of the Aeolianists' theory was the concept of desert-levelling by wind action. This process differed from river action in being unrelated to conventional base-levels of erosion and restricted only by ground-water levels. The wind was supposed to attack mountains, as does the sea an exposed coast, carving out bedrock plains and causing mountain fronts to retreat.

Keyes (1912) drafted his own 'deflative scheme of the geographic cycle in an arid climate'; it was a response to the model of W. M. Davis, and embodied most of the extreme views expressed by the proponents of wind action. In youth, relief increases rapidly by the excavation of broad, flat-bottomed troughs bounded by mountains; some material is washed into the troughs from the mountains by flowing water, but it is soon removed by the wind. In maturity, plains are more extensive and relief of the mountains increases as the troughs continue to be preferentially exploited; marginal drainage increases slightly in importance. As maturity passes into old age, the mountains are reduced, streams disappear, inequalities of relief due to lithological differences are removed, and eventually a simple extensive plain is formed.

The views of the Aeolianists were never widely accepted. The reasons for this are numerous. Firstly, although both Davis and Keyes based their conflicting ideas on very few data, Davis was persuasive and influential, Keyes was not. Secondly, detailed evidence to support the association of planar surfaces with wind activity has never been forthcoming, for although there are very extensive hardrock areas of the Sahara with no signs of fluvial activity they have never received the attention they deserve, and the huge parallel arcuate grooves around Tibesti have come to light only on recent aerial photographs. Thirdly, wind-eroded features which were authenticated—such as faceted pebbles and pedestal rocks—were shown to be only minor embellishments of forms produced by other processes. Fourthly, the Aeolianists assiduously avoided collection and analysis of wind data and the notion of pervasive, high-velocity winds has been largely discredited. Finally, the cyclic system depended on protracted periods of aridity, and such periods have not been substantiated by investigations of climatic change in deserts. But Aeolianist views were very much a product of their times, reflecting a widespread ignorance of deserts, a specification of process through observation of form, and a fashionable attention to the evolution of landforms in the post-Darwinian era. They represented a significant attempt to counterbalance the influential views of Davis. More importantly, they focused attention on a critical problem, a problem still quantitatively unresolved—the relative importance of water and wind in deserts.

(b) Water and wind. W. M. Davis (1954) adapted his 'geographical cycle of erosion in humid temperate climates' to accommodate information on landforms from areas of arid climate. In this new scheme, he drew heavily upon his experience in the south-western United States and on Passarge's work in southern Africa. Davis's deductions are well known and need only brief mention here. Initial basins of deformation comprise independent centripetal drainage systems. In youth, relief slowly diminishes by removal of waste from the watershed zone and its deposition on basin floors by streams and surface floods; wind action plays a secondary role in exporting debris from deserts and in reducing desert relief. The stage of maturity is reached when erosion of divides, accumulation in basins and headward erosion lead to coalescence of basins and to the transport of debris from higher into lower basins; eventually graded, rock-floored plains sloping towards debris-floored basins will be formed. The activity of water decreases as slopes decline and maturity progresses, but the work of wind is hardly altered and thus it becomes relatively more important. In old age, deflation hollows become more common, runoff is further reduced and drainage becomes disintegrated. Extensive plains are formed and as these are lowered by wind and water, resistant rocks remain as upstanding *Inselberge*.

Davis' cycle was based on very little precise information and a great deal of speculation. In this respect it was identical to the less well received cycle of Keyes. A major weakness of the scheme was that its deductions were based on evidence drawn from areas as widely different in their tectonic and climatic histories as the south-western United States and southern Africa. There is no reason to believe that the basin-range topography of the former will ever be converted into a landscape similar to that of the latter or even, for that matter, to one similar to the adjacent Colorado Plateau. The notion of basin integration has never been adequately corroborated, and in areas where it might seem to be acceptable, such as in parts of the Atacama and Mojave deserts, landform evolution is apparently more dependent upon the complex commingling of tectonic and climatic changes in the recent past than upon a single, uninterrupted sequence of development. A change in the relative importance of wind and water with time remains a completely unverified conjecture. The concept of slope reduction is a matter for debate. A final criticism of the cycle is that not only is so much of it unsubstantiated, it is unsubstantiatable; for this reason the cycle remains of value only to those who have the faith to believe.

Despite these criticisms, Davis' cycle has provided a framework within which many desert geomorphologists have worked, and it has been considerably modified and extended. In Europe, many geomorphologists have explored the field of 'climatic geomorphology', in which associations are sought between landforms, landform evolution and climate. Notable contributions include those of Tricart and Cailleux (e.g. 1965), Birot (1968) and Büdel (e.g. 1963). Birot, for instance, described an arid-climate cycle in which wind dominates over wash, linear erosion is unable to develop an organized drainage net, and relief modification occurs in hydro-aeolian basins through deepening by wind and widening by wash. Cotton (1942), Davis' main disciple, accepted the cycle of arid erosion, and illustrated it with examples from many deserts.

(c) Semi-arid and savanna cycles. Cotton's (1942) semi-arid cycle differs from the arid cycle in having base-level control by rivers flowing to the sea. Scant vegetation results in mountain sculpture similar to that under arid conditions. Beyond the mountains, accumulation forms such as alluvial fans and bajadas develop in early stages of the cycle, and soon divides are overtopped by alluvium. As the supply of material dwindles with the reduction of relief, erosion, mainly by lateral planation, extends into the aggradation zone. Ultimately a semi-arid peneplain is produced. Birot (1968) also sketched a semi-arid erosion cycle characterized by parallel retreat of slopes and pediment development. Of critical importance to both schemes is the development of pediments at the expense of mountains; the processes by which pediments are formed has been a matter of controversy and we shall consider them in Part 3. The semi-arid cycles may be criticized on the same grounds as Davis' arid cycle.

An important variation on the cyclic theme, which owes much of its initial inspiration to semi-arid environments and which also places emphasis on pediments, was provocatively propounded by L. C. King (1962). He described the 'standard epigene cycle of erosion'. A cycle of erosion is initiated by scarp formation. The scarp attains a gradient appropriate to its environment and then proceeds to retreat parallel to itself, leaving a pediment at its base. Pediments coalesce to form pediplains. Successive cycles of erosion are represented in all continents by extensive pediplains at several levels. The main erosion processes are rill and gully action on the scarps, and linear and sheetflow of water on the plains; at the stage of senility, soil creep may become dominant. The pediplanation process is most efficient in semi-arid environments, but 'the pediment . . . is the fundamental form to which most, if not all, subaerial landscapes tend to be reduced, the world over' (King, 1962, p. 144). A significant contrast between Davis's and King's schemes is that in the former the whole landscape is said to become progressively older, whereas in the latter the landscape, at any one time, is 'young' at the scarp, and the pediplain is older with distance from the scarp. King's backwearing concept owes much to the earlier views of Walter Penck (1953) and others.

One difficulty with the application of the backwearing concept is that 'if extensive areas are exposed in this manner, then one is requiring increasing work, in terms of distance of sediment transport, of the streams draining a mountain scarp through time, whereas the catchment areas for these streams must diminish through the same time interval' (Lustig, 1968, p. 102). King's ideas tend to come from sitting 'passively upon hills just letting the scenery "soak-in"' (King, 1962, p.v.) and his scheme, although bolstered by a wealth of detail, is almost as deductive and speculative as its predecessors.

A savanna climate, according to Cotton (1942), is characterized by an alternation of wet with rather long dry seasons in which high temperatures occur, so that vegetation is scanty except along watercourses. The landscape is often composed of plains punctuated by inselbergs. Cotton suggested that the scarps are worn back by weathering and local pedimentation, while the surrounding plains are extended by lateral river planation. In recent years several alternative models of savanna-landscape evolution have been formulated, and all seem to suffer from the failings of Cotton's scheme—they are rather qualitative and speculative; and they tend to gloss over details of the processes upon which their veracity depends. Büdel (1957; Cotton, 1961), concerned especially with the origin of inselbergs in Africa, suggested a two-cycle model. In the first cycle, a surface suffers surfacewash and marginal pedimentation, and is deeply weathered; in the second cycle, following rejuvenation of drainage, much of the weathered material is removed and rocks resistant to subsurface weathering are exposed as inselbergs. Evidence which tends to support this hypothesis, generally if not in detail, was described by Ollier (1960) in Uganda, by Thomas (1966) in Nigeria and by Twidale (1962) in southern Australia: these authors minimize the role of hard-rock pedimentation, and emphasize the role of structurally-controlled deep weathering and subsequent stripping of weathered debris. Louis (1959; Cotton, 1962) argued that some gently undulating plains, such as those in East Africa and Thailand, may be produced under humid-savanna or monsoonal rainforest conditions by slopewash and weak linear erosion, and that as these surfaces are worn down, inselbergs gradually emerge.

There are many conflicts among the models of King, Davis, the climatic geomorphologists, and others. Holmes (1955) has attempted to resolve some of those arising from apparent contrasts between humid, semi-arid and arid landscapes. He suggested that slopes everywhere are composed of only two fundamental types—*wash slopes* (or graded sur-surfaces of sediment transport) and *gravity slopes* (which supply the sediment for transport). Differences of topography and slope evolution in different areas represent differences in the number, size and distribution of the two fundamental types. In arid and semi-arid regions the two types occur in large units which dominate the landscape. In more humid regions, vegetation cover has the effect of breaking up the elements into smaller units, decreasing the contrast in their characteristic slope angles, and mingling the two types in units on slopes of intermediate steepness. Gravity slopes are said to retreat at their characteristic angles everywhere. In semi-arid and arid areas, gravity slopes persist and the 'back-wearing' hypothesis is applicable; in more humid areas the gravity slopes are progressively eliminated, wash slopes are integrated, and the overall effect is of decreasing valleyside slopes and 'downwearing'.

(d) Contemporary models. For many reasons there has been growing dissatisfaction in recent years with the grand, comprehensive evolutionary models. Too often the schemes appear to be based on deduction and guesswork and on gentlemanly 'views from the tops of hills'. Although fascinating and imaginative sequences of change are described, many of their details are at present not susceptible of proof. The fact that some assumptions and assertions have been shown to be inaccurate and unjustified has tended to jeopardize the remaining unverified comment. In addition, as we shall discuss shortly, they are schemes that apply only to the cyclic scale of Schumm and Lichty (1965) and not to the practical details of measurable forms and processes in contemporary systems.

As a result, geomorphologists have tended to move away from these models towards generalizations based more on induction and empirical research, and on the solving of smaller-scale and more tangible problems. There is now a concern with actual working systems, composed of various forces and resistances and having a distinct structure as a result of energy flows. There is interest, for instance, in weathering systems, drainage basins and aeolian systems. More precise methods of recording and analysing data are being employed. Already work in this field has yielded interesting generalizations concerning, for instance, 'dynamic equilibrium' conditions in systems (e.g. Strahler, 1950), the structure and functioning of river systems (e.g. Leopold, Wolman and Miller, 1964) and a small-scale semi-arid cycle of erosion in valley floors (Schumm and

Hadley, 1957).

1.4.2 THE GEOMORPHOLOGICAL SYSTEM IN DESERTS

A general model is clearly desirable in any introduction to desert geomorphology. It is needed as a means of categorizing the literature in a meaningful way, as a source of hypotheses that can be tested by observation, to define the field and the relationships of its component parts, and to help in identifying areas about which little is known, or areas in which knowledge is vital to an hypothesis. Our review of the approaches to desert geomorphology has shown that the form of the model adopted depends very much on the questions that need to be answered. This has meant that work in deserts has been focused on themes that have been common throughout geomorphology in the early twentieth century, namely on evolutionary sequences or on climatic interpretations, or on both. These interests are of course viable academic themes, and the field is far from exhausted, but they have been unfruitful in certain respects. We prefer to adopt a different frame of reference and a different emphasis.

Our starting point is an interest in the functioning of geomorphological systems. In order to proceed we must first briefly review more general geomorphological principles.

(a) Features of the system. The system of earth sculpture functions because there is an input of energy which meets and responds to resistances. It is a dynamic, open system (Chorley, 1962) in which work is performed, energy is dissipated and entropy increases (Leopold and Langbein, 1962); and which has a measurable morphological expression.

Energy comes from outside the system itself. It comes from solar sources and from the interplay of tectonics and gravity. Tectonic forces perform work against gravity and provide surfaces upon which solar energy and gravity can operate. Solar energy is dissipated along a number of paths on reaching the earth: some is returned immediately to space; a small proportion is used in performing work on or just beneath the incident surface; and the remainder powers the 'atmospheric heat-engine'. We are interested mainly in two expressions of one subsystem within the atmosphere: the rain and the wind. Water vapour is lifted, translocated and deposited as rain and this rainwater provides potential energy in the fluvial system by reason of the gravitational force. Atmospheric energy is redistributed over the earth's surface and creates pressure differences which result in winds; in deserts, these are an important subsidiary source of energy in landscape sculpture.

(b) Scales of operation. In order to discuss the nature of the geomorphological system itself, it is necessary to consider the issue of scale. Many of the problems that have confused geomorphologists working in deserts are considerably clarified if their scale setting is defined.

Any discussion of scale in geomorphology must consider the work of Schumm and Lichty (1965) who argued that, as the scale at which a problem is to be studied is increased, either spatially or temporally, so the number of variables that need to be considered is also increased, and so changes occur in the nature of these variables, in their relationship to one another, and even in their modes of operation. They identified three characteristic scales at which geomorphological processes can be seen to act: the steady, the graded and the cyclic. The distinctions between these scales are, of course, arbitrary. The desert situation seems to call for three rather different divisions of the scale range. From our review of the literature, and our experience of desert research, we have identified three useful groupings: *the local*, *the basin* and *the regional*. The local scale is the scale of a single slope unit within a drainage basin; the basin scale encompasses drainage basins from first order to the whole of a basin-range unit; and the regional scale is the scale of a whole desert, a concept almost identical to the 'cyclic' scale of Schumm and Lichty.

It needs to be emphasized that the scale range is a continuum. These units represent an arbitrary division of the range, and have been chosen because they enable us to study the landscape and to review the literature. Each involves quite different perspectives of and insights into the desert landscape and, more important, requires different methods of study. The regional scale has been the province of the denudation chronologist who sought answers to problems in geological perspective; the basin scale has been the context within which much of the geomorphological work in the earlier part of this century was tackled, when the questions related to peculiar features such as pediments and where the methods employed were often based on wide-ranging exercises in mapping; the local scale is the context of more recent work where the questions have been more practical and immediate. There is a nested quality in this system: slope within basin, and basin with region. It is our belief that the problems of desert geomorphology should be attacked from the better known at the local scale towards the less known at the regional scale.

(c) The system at the local scale. At the scale of a single slope, the desert geomorphologist is able to exercise some observational techniques on processes actually in progress. Measurements (such as of soil erosion,

soil salinization, pavement formation and so on) can be made in a reproducible and closely controlled experimental framework.

We can define the basic unit at the local scale as the 'facet' (e.g. Gregory and Brown, 1966). In this unit the principal source of energy is the potential energy of the weight of rainwater acted on by gravity over the height and length of the facet. A series of subsystems can be identified, each of which has its own set of resistances, paths of energy flow, and resulting morphology: rock weathering and soil formation; slope processes such as wash and creep; and channel erosion and deposition.

(i) The soil and weathering system. Rocks at the surface are altered by a set of atmospheric forces. A small proportion of the energy involved derives directly from incoming solar radiation. Some of this is engaged directly in rock breakdown or in the movement to the surface of aqueous solutions where the resistances include the thermal and reflecting characteristics of the surface. Direct solar radiation is probably more important to the system when it is channelled through biota, although a very small percentage of the total incident energy at the desert surface is actually used in this way. In Arizona it has been estimated that only some 0·06 per cent of the annual available radiant energy is used by a *Larrea tridentata* community (Chew and Chew, 1965). This energy, once fixed by primary producers, is channelled in various ways into the soil and weathering system: organic acids provide sources of hydrogen ions for hydrolysis, and this liberates bases into the soil solution; roots loosen and break up the soil; and animals eat the plants and burrow and trample the soil surface. These forces meet resistances in the chemical and mechanical properties of the rock.

The relative importance of these processes is probably not the same in deserts as it is in more humid climates. More of the solar radiation may reach the ground surface directly, but on the other hand desert surfaces usually have a higher albedo, and so more of it is reflected. Indeed the high albedo of desert sands means that even a few centimetres beneath the surface there are only very small annual fluctuations in temperature (e.g. Williams, 1954). Desert vegetation is relatively much less efficient at energy fixation than temperate vegetation. For example, the sparse *Larrea* community mentioned above is only between three and ten per cent as efficient as forest stands in England (Chew and Chew, 1965). As we have seen, more of the primary production in deserts is involved in root formation within the soil and more is in circulation at any one time.

Water is another important source of energy within the soil because of its potential energy under gravity. As we have seen, the fraction of the rainfall that is involved in inflow to the soil is probably less in deserts than in wetter climates. This downward and laterally moving water carries away the products of weathering and leads to soil horizonation. This process meets with resistances such as the permeability and wettability of the soil, the swelling and adsorbing characteristics of the clay fraction and so on.

All things considered, however, weathering and soil formation are probably not nearly as active in deserts as they are in wetter climates and this means two things: firstly desert-weathered mantles are thin, and secondly the processes taking place in them are always primary rather than secondary breakdown processes, since the latter never have time to operate because of the complementary activity of the slope processes.

(ii) The slope subsystem. The main source of energy on slopes is the potential energy of rainwater; a secondary source is the potential energy of loose soil. The potential energy of the rainwater acts either on the surface as overland flow or just beneath it as throughflow. These forces meet a great range of resistances that are related to the chemical, sedimentary and physical properties of the parent rock and the weathered mantle, such as permeability, wettability, cohesion, strength, cementation, and so on. The slope subsystem has a number of internally regulated properties such as slope angle, surface roughness and the like (e.g. Emmett, 1970), and it is often dependent on the characteristics of the channel at the base of the slope which removes the sediment produced on its surface.

(iii) Channel subsystems. The source of energy in channels is almost exclusively the potential energy of water. Channels have a number of internal characteristics that can be regulated such as the grain-size of their load, their bed-roughness, their slopes, their cross-sectional shapes and areas, and their patterns. The regulation is in response to a number of resistances such as given slopes, grain-sizes of sediment input, bank and bed permeability, water discharge and bank cohesion (Leopold, Wolman and Miller, 1964).

(iv) The last stage in the transport of water and sediment in desert basins is often *the playa.* This in many ways can be regarded as a separate subsystem at the local scale. Here the potential energy of the water is no longer available, and it is the incoming incident solar radiation that is the most important source of energy. Water is drawn to the surface where salts may crystallize and shrinkage-cracks result. The source of energy in the wind may come into play and remove salts to higher parts of the basin.

(d) The system at the basin scale. We can define a basin as any unit within which several of the subsystems described at the local scale operate. Thus a basin is an essemblage of local-scale subsystems. Basins are functional areas with fairly well-defined structural properties and distinct boundaries. The investigator uses maps and plans whereas at the local scale he often uses only profiles and sections.

Studies of basins differ between those that deal with small first- or second-order basins and those that deal with a whole basin/range assemblage. The quantitative studies of the last two decades have seldom reached beyond the small drainage basin. One might cite the seminal work by Melton (1957 and 1958) in this connection. The sources of energy and the resistances it meets here are the same as those at the local scale, but an integrated study allows the analysis of morphological variation, of sediment transport, and of the adjustment of form to certain controls.

At a larger scale the basin-and-range framework has been a popular one within which to study geomorphology in deserts, and indeed it is very appropriate to the endoreic conditions of many deserts. These studies have been very fruitful in the basin-and-range province of the south-western United States and northern Mexico, from whence we get many of the terms used to define the parts of the system. It is in the basin framework that mountain fronts, pediments, and alluvial fans begin to have meaning.

The basin model is clearly also appropriate for studies in other basin-and-range deserts as in Iran and South America. It can be extended for use in the basins between the cymatogenous domes of the older shield deserts in Australia and Africa, where the basins are shallower and wider and are the result of warping rather than of faulting (e.g. King, 1962). As in the basin-range system, cymatogeny produces endoreic basins whose denudation history has an internal coherence. Many of the warped basins, however, are very large and some of their landforms could be said to have a 'regional' rather than a 'basin' character (e.g. Mabbutt, 1955*b*).

The study of basins in desert geomorphology has a useful analogue in the basin analysis of sedimentologists (Potter and Pettijohn, 1967). Basins in this sense are used as units within which to study the production, translocation and deposition of sediments, and they can vary from very small to very large units. The framework is said to have predictive and theoretical advantages. Palaeoenvironmental aspects of desert basins have been studied in this framework (e.g. Opdyke, 1961).

An appreciation of the basin context of a local-scale study helps in the understanding of the latter. For example, the characteristics of the soil-slope-channel pattern change as one moves from an upland across a piedmont area to a basin centre. On the upper slopes there are thin eroded soils on steep slopes, whereas in the piedmont many slopes are aggradational and the soils developed on the low-angle landsurfaces are deep and distinctly horizonated.

(e) The system at the regional scale. As we have shown, the concern with regional landscape evolution has dominated many works on desert geomorphology. These studies have seldom yielded general statements about the development of desert forms, since their main concerns have been with the development of denudati on chronologies that do not differ significantly from similar chronologies in other areas, and with the establishment of answers to geological (and especially structural) questions. Endogenetic forces become of very great importance at this scale, and these do not conform to climatic boundaries. For these reasons we shall be referring to such studies very little in the following pages.

(f) Aeolian systems. The obvious and widespread effects of the wind are characteristic of deserts. The wind is seldom as effective in denudation as running water, but minor wind erosion is evident in many areas and reaches major importance in some restricted localities; and depositional features created by the wind cover up to as much as 25 per cent of some deserts. The aeolian system can only be said to interact with the fluvial one in a very restricted way, so that we need a fresh definition of the framework within which to discuss the work of the wind.

Aeolian systems involve a different form of energy and a new set of resistances. The energy of the wind meets resistances in the drag of a surface and in its yield of loose particles. Aeolian forms can also be seen within a different kind of scale framework: there are at least three different and distinct size groupings of phenomena corresponding to ripples, dunes, and larger features known in north Africa as *draa*. Each of these depositional features has an erosional analogue and each obeys similar laws, but their super-imposition leads to more complex forms as the scale increases. As with fluvial forms, the increase in areal scale must be matched by an increase in temporal scale: ripples respond to daily changes, for example, and *draa* only to climatic changes. The basin scale in the fluvial model corresponds to a whole sand sea in the aeolian model. It is by looking at the assemblage of forms in a sand sea that we can get ideas about trends in the wind, grain-size patterns, and sediment flow. There is no meaningful unit at the regional scale in the aeolian system that corresponds to the regional unit of the fluvial system.

These schemes can be regarded for the moment merely as a convenient framework within which to discuss the geomorphology of deserts. Before any more acceptable framework can be constructed we shall have to know very much more about the flow and forms of energy in the landscape, and about the movements of sediments. Once more is known in these fields a new scheme will undoubtedly emerge, since the present model, like the models before it, must be regarded as merely a fashionable expression of our state of knowledge.

In Part 2 of this book we are concerned with geomorphological phenomena at the local scale in the soil surface system. Part 3 is concerned with the integrated patterns of change on slopes and in channels, and with the larger features within basins such as pediments, fans and playas. Part 4 deals with aeolian systems.

PART 2

Desert Surface Conditions

2.1 Introduction

'Desert surface conditions' is a comprehensive phrase which covers the nature of surface and shallow subsurface materials, the landform features associated with them, and the processes that operate on them. Important specific features considered in this context include soils and weathering mantles, landforms produced largely by weathering, stone pavements, patterned ground, and surface crusts. We are especially concerned with processes of rock disintegration, chemical transformation, soil horizonation, particle sorting and concentration near the surface, and patterned-ground development.

The desert literature includes a wealth of references to surface conditions and features, but much of it relates only to individual occurrences of outstanding phenomena; the general context and relative importance of the phenomena, especially in terms of their distribution, have been neglected. There are a few exceptions. For example, Mabbutt (1969 and 1971) classified the general surface conditions of Australian deserts (Fig. 2.1) and calculated the percentage of the arid zone covered by each type (Table 2.1). Glennie (1970) has mapped the distribution of sedimentary environments in parts of the Gulf States and Libya. Seligman, Tadmor and Raz (1962) have cal-

TABLE 2.1 *Extent of Desert Surface Conditions in Australia*

	Percentage of arid zone
Mountain and piedmont desert	17·5
Riverine desert	4·0
Shield desert	22·5
Desert clay plains	13·0
Stony desert	12·0
Sand desert	31·0
	100·0

SOURCE: Mabbutt, 1971.

culated the area covered by various surface types in the Negev Desert. And Clements *et al.* (1957) classified desert surface characteristics into 'militarily significant groups' and calculated the areas covered by these groups in four areas (Table 2.2).

The problem with many such classifications is that the categories are not always mutually exclusive or collectively comprehensive, nor are the distinctions between them clear. For example, in Table 2.2, 'desert flats' could presumably include 'playas'.

TABLE 2.2 *Comparison of Areas of Desert Surface Types*

Group	*South-western U.S.*	*Sahara*	*Libyan Desert*	*Arabia*
Playas	1·1	1	1	1
Bedrock fields (inc. hammadas)	0·7	10	6	1
Desert flats	20·5	10	18	16
Regions bordering throughflowing rivers	1·2	1	3	1
Fans and bajadas	31·4	1	1	4
Dunes	0·6	28	22	26
Dry washes	3·6	1	1	1
Badlands and subdued badlands	2·6	2	8	1
Volcanic cones and fields	0·2	3	1	2
Desert mountains	38·1	43	39	47
	100·0	100·0	100·0	100·0

SOURCE: Clements *et al.*, 1957. Figures are percentages of area.

As a preface to this section, we have constructed a speculative figure showing the simplest and most frequently recurring desert profile, which is composed of a mountain flanked by plains (Fig. 2.2). On it we have illustrated diagrammatically the occurrence and relative importance of various surface conditions with respect to this profile. (The maximum development of all features is shown by the same bandwidth.)

At the outset of this section it is important to emphasize a cardinal principle: it should not be assumed that the occurrence of the same landform in different areas justifies the conclusion that the same processes were responsible for it. Meckelein's (1965) observations on

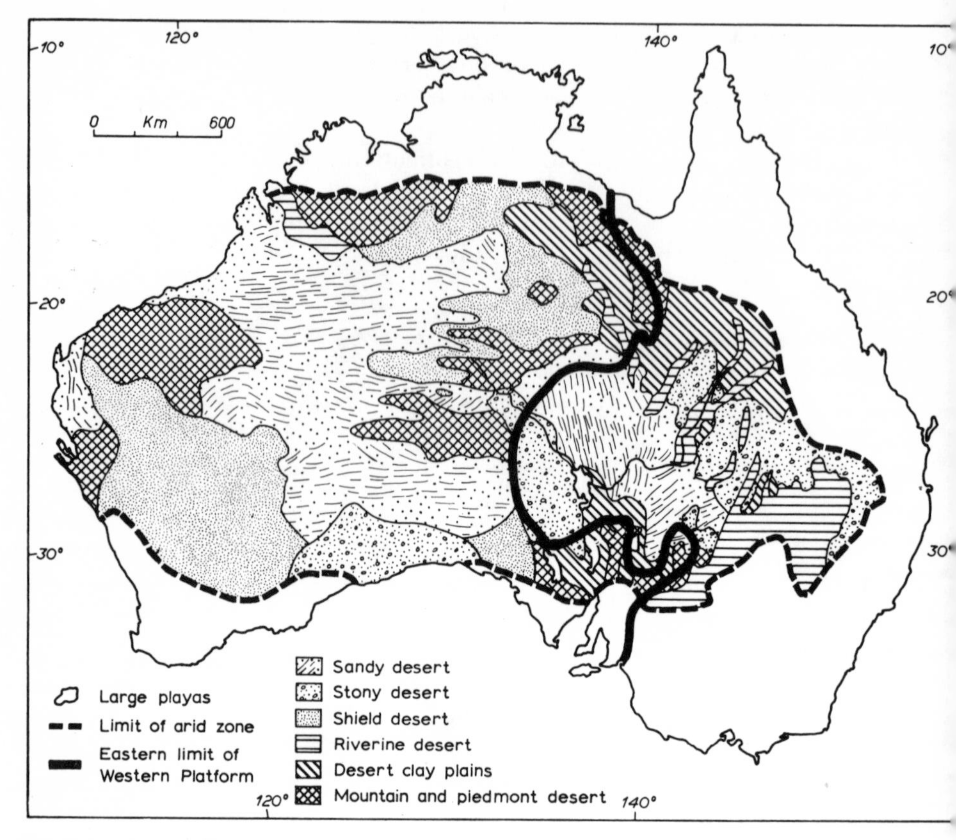

2.1 Classification of physiographic types in the Australian arid zone (after Mabbutt, 1969).

geomorphological similarities between polar and hot deserts, for example, make it clear that such an assumption is unjustified. Indeed, it is dangerous, if not impossible, to determine processes responsible for a feature merely by examination of the feature itself. We shall find repeatedly that similar features may be formed in contrasted areas by different processes or combinations of processes operating within the same or different systems.

2.2 Weathering Forms and Mechanical Weathering Processes

2.2.1 WEATHERING FORMS

In discussions of desert weathering features there is often confusion over the processes responsible. This confusion usually arises because the

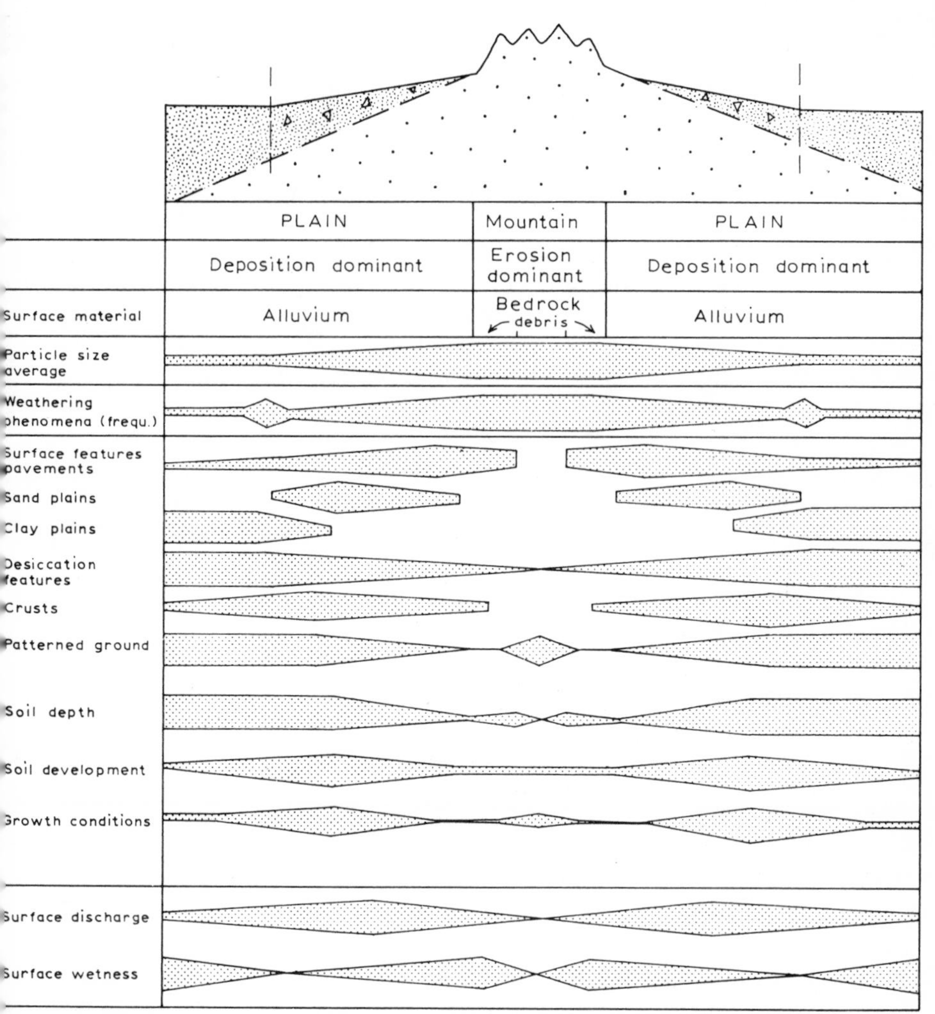

.2 A characteristic desert profile, and the common relations of various surface conditions to it. This is a speculative rather than an authoritative summary diagram.

features alone may not reveal the processes which produced them. Thus, for instance, one cannot merely look at a split rock and accurately deduce that it was formed by a mechanical weathering process, even though there may be no apparent chemical alteration of the rock; and still less can one decree that a specific process, such as frost action, was the cause. Most weathering forms in deserts have been ascribed to many different agents. In this section we shall describe some of the more common weathering phenomena and mention briefly some of their various explanations. The mechanical processes will be considered in the next section.

Desert weathering, as Mabbutt (1969) emphasized, has two immanent characteristics: superficiality and selectivity. Its superficiality is indicated by the shallowness of soil profiles and weathering mantles, and the preponderance of surface crusts and patinas. It results largely from the fact that the zone of water penetration and temperature change tends to be shallow in deserts. The selectivity of desert weathering is demonstrated by the localization of such features as weathering pits and caverns. Selectivity results mainly from two causes. Firstly, as the proportion of exposed rock is fairly high, structural and lithological properties of rocks impose strong differentiation on the effectiveness of superficial weathering processes. In particular, joints, faults, crystal boundaries and other lines of weakness may be exposed, and they tend to be loci of exploitation. Secondly, extreme local variations of temperature and humidity across the surface are an important cause of selective weathering. For instance, shaded crevices may be weathered more rapidly than exposed, dry surfaces.

Rock outcrops and superficial debris are commonly affected by *flaking*, *spalling*, *splitting*, and *granular disintegration*. Flaking, spalling, and splitting (the *Kernsprung* of German authors) all involve the loosening of fragments from parent particles, often without any apparent chemical alteration. Significant generalization about these features is difficult. For example, examination of split, flake and spall phenomena affecting superficial boulders in Chile and California (Cooke, 1970*b*) suggested that: they occur on most rock types, but less commonly on coarse-grained igneous rocks; the particles affected vary in their size and shape, and in the extent of their immersion in finer material; and the orientation of split planes to the surface, the width of spaces between splits, and the number of splits also vary greatly. A distinctive form of splitting, with the cleavage of a fragment along two or more subparallel fractures roughly normal to the groundsurface, is illustrated in Plates 2.2 and 2.3. Such split rocks have been described in most deserts (e.g. Brown, 1924; Cooke, 1970*b*; Ollier, 1963; Tricart and Cailleux, 1969; Walther, 1924; Yaalon, 1970; Young, 1964). Sometimes salt crystals are present along split planes (Yaalon, 1970). Debris resulting from spalling, splitting and flaking is characteristically angular and coarse, and it often provides a large proportion of the detritus on the desert surface, especially where there has been little transport of material (e.g. Jutson, 1934, part 16). A number of mechanical weathering processes have been invoked by various authors to explain these phenomena; and many chemical weathering processes have also been implicated. One point should be emphasized—the splitting of rocks, especially those embedded in superficial detritus, may involve two phases: the establishment of the split, and its enlargement; the same process is not necessarily responsible for both phases.

A distinctive form of splitting commonly affecting granitic rocks is the exfoliation of rock sheets along curved planes. Such exfoliation can occur on a large scale, perhaps involving sheets several metres thick (Plate 2.4). This phenomenon has been attributed to expansion by pressure release along joint planes in granite following erosion of superincumbent rocks, and to the creation of curved joint planes under conditions of radial compressive stress (e.g. Twidale, 1964). The small-scale exfoliation of individual silicate-rock boulders, sometimes called 'onion-skin weathering', was qualitatively analyzed in a classic paper by Blackwelder (1925). He argued convincingly that the peeling of skins involved expansion and that theoretically this could be effected by temperature changes (caused by fire, lightning or insolation), expansion of foreign substances (e.g. ice, salt, rootlets), or chemical alterations (absorption of water by colloids; hydration or oxidation of silicates). He favoured an hypothesis that involved the establishment of cracks (by undefined processes) that may act as channels for percolating water; chemical alteration is then effective along the cracks, even when the exposed surface of the boulder is dry and, as a result, the undersurface of the plate expands more than the exposed surface, and the plate is warped. The cracks might be established along boundaries between concentric zones of rock surrounding dense cores which solidified at an early stage in the consolidation of the magma (Schattner, 1961).

In non-siliceous rocks an alternative explanation of small-scale exfoliation to that of Blackwelder is required. LaMarche (1967) observed a contrast between the angular weathering of unmetamorphosed dolomite and the spheroidal weathering of metamorphosed dolomite marble in the White Mountains, California. The unmetamorphosed dolomite weathers to polycrystalline fragments and fine debris, whereas the marble weathers by granular disintegration into the constituent mineral grains of the rock. The reason suggested for this contrast is that the marble has developed intergranular stresses as the result of the large drop in temperature since crystallization, and stress relief, for example by fracturing along grain boundaries, may promote corrosion. Exfoliation of spheroidally weathered rocks may depend on the fact that porosity increases near the margins of the blocks because of intergranular corrosion. An expansion mechanism related to the porosity differences would explain concentric fractures: in the White Mountains, repeated freezing and thawing of pore-water provides such a mechanism.

Granular disintegration—the breakdown of rocks into individual crystals or clusters of crystals—is chiefly found in coarse-textured igneous and metamorphic rocks. Plate 2.5 shows an example where a rounded, granitic boulder has been disintegrated until it is flush with the desert surface. Such weathering has been attributed to both mechanical and chemical processes, and different explanations may be correct under different circumstances. In our experience, granular disintegration usually affects rocks in which at least some minerals have been chemically altered.

Two more localized weathering phenomena which have attracted some attention are *tafoni* and *gnammas*. Neither can be regarded as an exclusively desert phenomenon.

Tafoni has come to be used as a general term for cavernous weathering features. (*Alveoles*, features of honeycomb weathering, are similar to tafoni in many respects, and will not be considered separately here.) Typically, tafoni have arch-shaped entrances, convex walls, overhanging upper margins, and fairly smooth, gently sloping floors (Plates 2.6–2.8). Diameters of tafoni vary from a few centimetres to several metres. Their shapes are equally variable, as a series of cross-sections of tafoni in granitic and gneissic rocks in South Australia shows (Fig. 2.3). Tafoni usually occur in groups, prompting an analogy between the

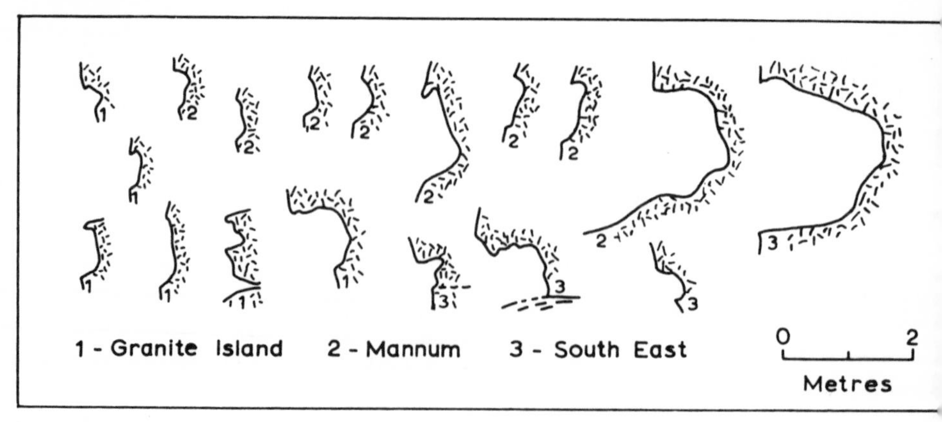

2.3 Vertical sections of a sample of tafoni in granitic and gneissic rocks in South Australia (after Dragovich, 1969).

rock outcrops and Swiss cheese. They are common in many different climates, but they have been described most frequently in deserts, particularly foggy coastal deserts. Some significant studies include those of Blackwelder (1929) in the Mojave Desert; Bryan (1925) in the Sonoran Desert; Calkin and Cailleux (1962) in Antarctica; Grenier (1968), and Segerstrom and Henriquez (1964) in the Atacama Desert; Dragovich (1967 and 1969) in South Australia; and Wilhelmy (1964) in Corsica, Uruguay and elsewhere. Two examples, one developed on porphyry north of Vallenar in the Atacama Desert, the other on granite in the Mojave Desert, are shown in Plates 2.6 and 2.8. Several generalizations about tafoni appear to be justified. Firstly, they occur in many different rock types. (Examples in the literature include argillite, porphyry, rhyolite, conglomerate, lava, limestone, gneiss and granite.) Secondly, their formation includes at least two phases: disintegration of the walls, and removal of debris from the floors. Thirdly, once they are established, the air and surfaces of the shaded caverns are often cooler and moister and the ranges of temperature and relative humidity tend to be smaller than those outside, so that the caverns may be loci of chemical weathering. Fourthly, cavern walls are generally crumbly, often breaking up by flaking or granular disintegration, and impregnations of salts are common. Finally, tafoni may be either fossil or active. The former may be preserved by lichen covers, desert varnish or case-hardening. Activity in the latter may sometimes be demonstrated by the disintegration of carved names (e.g. Bryan, 1925*a*) and cave drawings, and by loss of a varnished surface, etc. Most of the published descriptions appear to be of active caverns, although Grenier (1968) recorded 'dead' tafoni in Chile, and mentioned similar features in the heart of the Sahara.

The origin of tafoni has not been studied in detail. Often the features appear to have been initiated along pre-existing joints and other rock structures, in randomly distributed cavities resulting from general surface weathering, or at points of mineralogical weakness, especially where water naturally accumulates. Schattner (1961) suggested that in Sinai some tafoni on granite may have originated by the detachment of 'core' boulders. In South Australia, Dragovich (1969) established a correlation between tafoni and present ground-level. She suggested that cavern development accompanies 'ground-level weathering' in which there is a disparity between the rates of disintegration above and below the surface. Above-surface weathering is more rapid because changes of temperature and moisture conditions are more pronounced, and this is especially true on the downslope side of boulders and rock outcrops. Twidale (in Jennings and Mabbutt, 1967) indicated that the base of cliffs in the Flinders Ranges, South Australia, were very susceptible to cavern initiation, for these are sites of relatively rapid rock weathering.

Many authors consider that the disintegration of tafoni walls proceeds by hydration, promoted perhaps by the relative dampness of the cavity environment. Other processes which have been invoked include: wind-scouring, insolation weathering, frost action, burrowing animals, solution and salt weathering (Segerstrom and Henriquez, 1964). Any one or any combination of these processes might be responsible for tafoni in particular circumstances. For example, along the foggy littoral of the Atacama Desert, where salts compose the condensation nuclei of the fogs, salt weathering may be a particularly significant process. Removal of debris from the floor of tafoni may be by wind, soil creep, rainwash or channelled flow of water.

Forms similar to tafoni but of different origin are the flared lower slopes of some granite inselbergs in South Australia and elsewhere (Twidale, 1971). Such steepened, sometimes overhanging slopes which sweep smoothly upward and outward are often found in incipient form beneath the alluvial mantle fringing the inselberg. The flaring is best explained by subsurface weathering around the margins of inselbergs, where runoff is concentrated, followed by erosion of debris to expose the slopes.

Weathering pits, which are sometimes known by the Australian aboriginal word *gnammas*, occur widely in relatively bare, horizontal or gently sloping surfaces of granite, sandstone, quartz porphyry and other cohernet and massive rock types (Plate 2.9). They are by no means confined to deserts, but they are most common as unvegetated surfaces. Gnammas may have diameters of up to about 15 m, and depths of up to some 4 m; shapes in plan are varied, sometimes reflecting detailed structural patterns; but the pits are often elliptical or circular on horizontal surfaces, and asymmetrical on sloping surfaces. In cross-section, gnammas are usually flat-floored or semicircular, and they occasionally have overhanging sidewalls. The pits appear to be caused by differential weathering of the rock and subsequent removal of the resulting debris. In a study of granite weathering pits in South Australia, Twidale and Corbin (1963) concluded that the depressions originate by differential disintegration which is concentrated at points of physical weakness or along joints, by flaking of granite surfaces, or by lichen disintegration; once established, they become the sites of water accumulation, and chemical alteration—especially hydration—becomes important; material is lost from the pits by periodic flushing or deflation. The possibility of granular disintegration occurring beneath a case-hardened rock surface, especially where the pits flare out beyond a small entrance, has been suggested by Ollier (1969). Overhanging sidewalls might also be due to the fact that water remains longer on the lower part of the pit (Twidale and Corbin, 1963).

2.2.2 MECHANICAL WEATHERING PROCESSES

Recognition of weathering debris which has apparently been produced without chemical alteration—such as cracked and split rocks, and sharp, angular material—has led to the conclusion that mechanical weathering processes are active in deserts. Indeed, mechanical processes have been said to be more important than chemical and biological processes in arid and semi-arid lands because the former are enhanced by the direct relationship between rock surfaces and the atmosphere, whereas the latter arc reduced by limited plant cover and low humidities. In this section we review some of the main mechanical weathering processes which may be at work in deserts, and may be responsible for some of the weathering features.

(a) Insolation weathering. The concept of 'insolation weathering' involves the idea that rocks and minerals are ruptured as the result of large diurnal and/or seasonal temperature changes, temperature gradients from the surface into the rock, and the different coefficients of thermal expansion of various rocks and minerals.

The idea is an old one, often defended by reports of fusillades of heard but unseen rocks cracking in the desert night. It was particularly cherished because it evidently provided a mechanism for initiating a split, whereas many other processes were thought to be able only to exploit established lines of weakness; it theoretically provided a mechanism of instant rupture (although stresses may be cumulative), and it had the appeal of simplicity and an apparent sympathy with the desert environment: its more recent advocates include Tricart and Cailleux (1969) and Ollier (1963 and 1965).

Many workers have discredited the idea (Birot, 1951; Blackwelder, 1925 and 1933; Reiche, 1945; Schattner, 1961). But neither the advocates nor the adversaries have paid great attention to the two key groups of variables that determine its efficacy: the magnitude, frequency and rate of diurnal and seasonal temperature change, and the coefficients of rock and mineral expansion. Data are indeed sparse. Blackwelder (1925) subjected sound igneous rocks to sudden changes of temperature from 15°C to 210°C by plunging the cold rock into boiling oil, and produced no spalling, cracking or even weakening of the samples. In later experiments, Blackwelder (1927 and 1933) several times heated rocks to certain temperatures and then cooled them by immersion in water. He showed that most igneous rocks withstand repeated *rapid* heating and cooling through a temperature range of 200°C, but begin to fail under such treatment between 300°C and 375°C; that many rocks endure *slow* changes through a 400°C–600°C range of temperature; and that the capacity of different igneous rocks to withstand temperature changes varies greatly. Griggs (1936) subjected polished granite to rapid temperature changes through 110°C for a period equivalent to 244 years without causing weathering; the rock quickly began to disintegrate, however, when it was cooled with water. Laboratory experiments by Birot (1951) suggested that thermal variations were more important in increasing rock permeability—and thus providing avenues for the penetration of chemically active water—than they were in actually causing rupture. Birot also demonstrated that the crystalline rocks that break down most rapidly in dry climates are those which absorb water most quickly, such as coarse-grained granites. All these experiments point towards the conclusion that variations in insolation are inadequate to cause rock breakage.

Relevant data concerning field conditions are also sparse. In the Tibesti Massif it has been maintained that micro-climatic differences have marked effects on mechanical breakdown. Narrow valleys have less marked temperature contrasts and therefore slower breakdown of rock than wider valleys (Capot-Rey, 1965). Daily temperature ranges are high in many deserts (see Table 1.2). Monthly means of daily *air* temperature ranges may be as much as 29°C (Meckelein, 1959), and individual instances of daily ranges as high as 54°C have been recorded (Keller, 1946). The daily range of surface temperature is often much higher than that in the air or a little below the surface, as Meckelein's (1959) and Hörner's (1936) measurements show (Fig. 2.4). Such ranges are extreme; whether rocks and minerals react sufficiently to cause disintegration remains a matter of speculation. But two points deserve emphasis: firstly, the fabric of a rock may be weakened by numerous small volume-changes over protracted periods, rather than by spectacular daily changes, and 'rock fatigue' may be important; secondly, as Smalley (1966) emphasized in discussing the origin of quartz sand, some rocks such as granite may already have considerable *internal* stresses caused by crystallization which are an important supplement to external forces of rock disintegration.

Roth (1965) described an eight-month series of field measurements on temperature and humidity conditions on and within a quartz monzonite boulder in the Mojave Desert, California. Briefly, his conclusions were: *(i)* the maximum difference of temperature, measured by thermisters, at any particular time between any two points in the rock was 24°C; *(ii)* the maximum diurnal temperature range at any one point was 24°C; *(iii)* the range was greater in summer than in winter; *(iv)* temperatures on and in the rock were generally higher than the temperatures of the adjacent atmosphere, especially during the day. The average coefficient

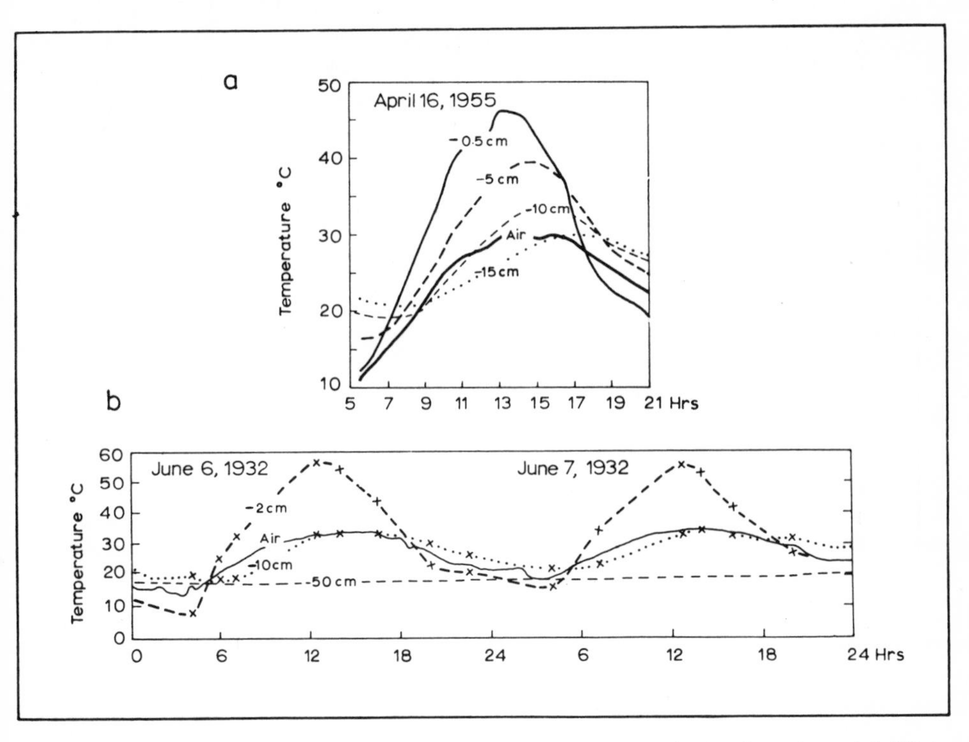

2.4 Diurnal variations of air and soil temperature at two desert locations (*a*) Wau el-Kebir, Libya (after Meckelein, 1959) and (*b*) central Asia (after Hörner, 1936). Centimetre measurements refer to recording depths below the surface.

of volumetric expansion of the rock was estimated to be 0·0028 per cent, and the maximum volumetric change in 24 hours was calculated as being 0·00672 per cent. Such changes would appear to be considerable, although Roth argued that since the figure is a maximum, temperature gradients are fairly uniform, and there is little time-lag in the temperature response of the rock as a whole, the insolation weathering process is probably ineffective. Roth also measured moisture through the block, found it to be variable but appreciable, and preferred Blackwelder's hypothesis that chemical alteration is likely to be more important than temperature changes in causing rock disintegration under such desert conditions.

(b) Frost weathering. Frost weathering—the result of pressures exerted by the volumetric expansion of water when it is frozen in confined spaces or by crystal growth—has received little attention in deserts,

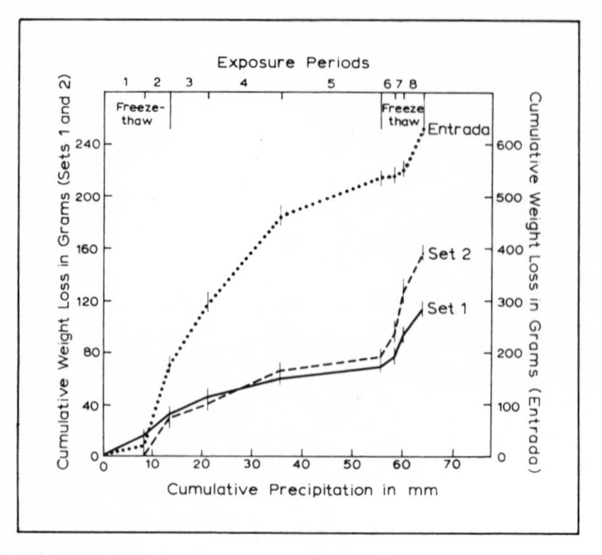

2.5 Double mass curve of cumulative weight loss plotted against cumulative precipitation for each of eight periods in which sandstone samples (Set 1, Set 2, and Entrada Sandstone) were exposed to precipitation at Denver, Colorado. The greater slope of curves in periods 1, 2, 6, 7 and 8 indicates greater rate of weathering at those times, which coincide with periods of freeze-thaw (after Schumm and Chorley, 1966).

although it was mentioned by Tricart and Cailleux (1969), emphasized by Coque (1962), experimentally investigated by Schumm and Chorley (1966) on sandstones in the Colorado plateaux (Fig. 2.5), and invoked by Melton (1965*a*) as a mechanism which produced detritus during colder phases at higher elevations in Arizona. Laboratory studies of frost weathering (Tricart, 1956; Wiman, 1963) indicate a need for detailed climatological data on moisture conditions, and the frequency and extent of ground and soil temperature fluctuations about freezing point; but such data are rarely available.

In the Mojave Desert, California, the average number of days per month over a 10-year period in which screen air temperatures fall below 0°C has been calculated, and plotted against station altitude (Fig. 2.6). Although temperature conditions appear to be suitable for frost action, water is also required, and precise data on this aspect of the problem are

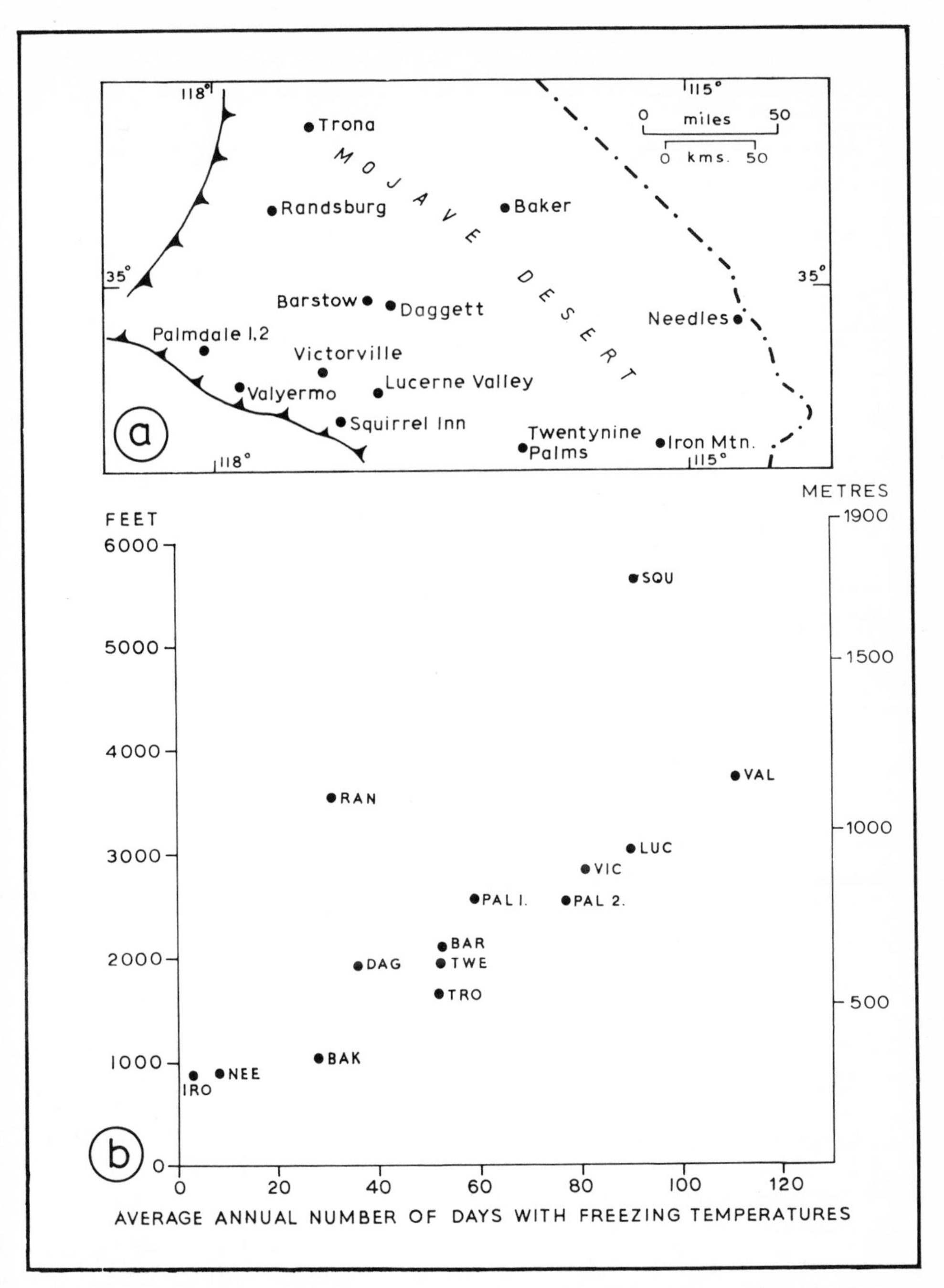

2.6 Average number of days with freezing air temperatures for stations in the Mojave Desert, California in a 10-year period. (*a*) Shows the locations of stations and (*b*) the frequency for each station plotted against altitude of the station (after Cooke, 1970*b*).

largely absent. However, freezing temperatures occur most frequently in the winter months of December, January, February and March, when 47 per cent of all precipitation occurs. Furthermore, humidities are highest, and frost and dew are commonly formed at this time of year. This limited evidence gives some support to the view that frost action may be effective in the Mojave Desert; if the process is at work today, it is likely to have been effective during the several cooler, moister periods of the Quaternary era. If this argument is true for the Mojave Desert, it is probably also true for high-altitude areas in many deserts, and generally for deserts with cold winters and seasonal occurrence of ice (Tricart and Cailleux, 1969). Lustig (1966), however, has pointed out that the differences between the process at present and in the past are difficult to specify and may not be very important in terms of the amount of debris produced.

(c) Salt weathering. Cooke and Smalley (1968) identified three main mechanisms by which salts may lead to rock disintegration: *(i)* stresses exerted by the expansion of many salts in confined spaces as they are heated, *(ii)* stresses caused by hydration of certain salts in confined spaces and *(iii)* stresses caused by crystal growth from salt solutions in confined spaces.

(i) The extent to which salts may expand when they are heated depends on their thermal characteristics and the temperature ranges to which they may be subjected. Many of the salts that commonly occur in deserts have coefficients of volumetric expansion that are high, and are higher than those of many common rocks, such as granite (Fig. 2.7). Fig. 2.7 shows the thermal expansion characteristics of five of these salts over a range of about 300°C. The temperature/volume relationships are approximately linear in each case and it is assumed that this linearity extends to atmospheric temperatures, and that the percentage expansion which the salts undergo in natural conditions can be estimated from this diagram. The potential of salt expansion was also noted by Hunt and Washburn (1966) who stated that the thermal coefficient of linear expansion of rock salt per 1°C at 40°C is $40{\cdot}4 \times 10^{-6}$, which approaches that of ice, $47{\cdot}4 \times 10^{-6}$.

The nature of diurnal temperature ranges in deserts has been outlined above. The effect of such temperature changes on salts precipitated from solution in crevices and pores in rocks near the surface may be considerable. The stress-causing system in rock associated with salt in a small crack is efficient in that the stresses caused by thermal expansion of the salt produce tensile stresses in the rock which are concentrated at the tip of the crack, the point at which further failure must occur. If the

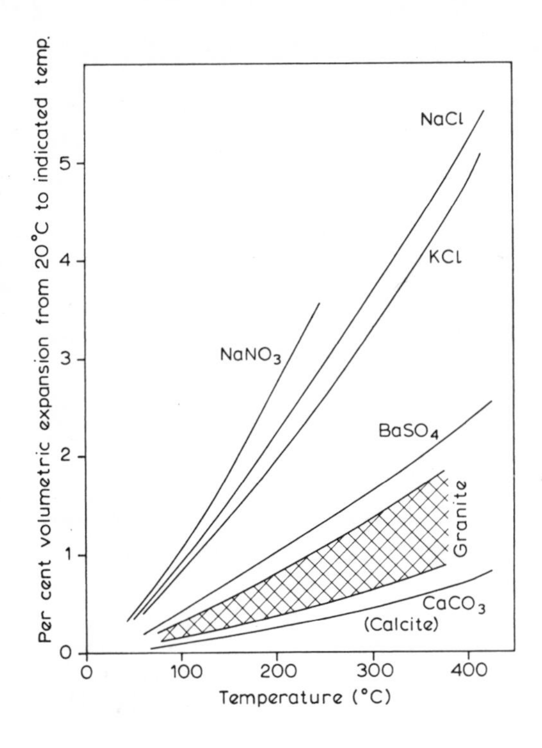

2.7 Relationship between temperature and the volumetric expansion of various salts and granite (after Cooke and Smalley, 1968).

rock is granite, with its considerable internal stresses caused by crystallization (Smalley, 1966), then the combination of these internal stresses with the stresses caused by salt expansion may produce considerable fracturing. Of course, the cracking produced by salt expansion provides avenues for the further penetration of saline solutions, and the process may continue until the rock disintegrates.

(ii) Disruptive stresses may also be exerted by anhydrous salts, dehydrated in high desert temperatures, which became hydrated from time to time (e.g. Mortensen, 1950; Winkler and Wilhelm, 1970). The presence of both anhydrite and gypsum in deserts suggests that some conversion of one to the other may occur and, in fact, the presence of other salts tends to promote the dehydration reaction. There is certainly a volume change associated with the hydration of $CaSO_4$; when a test-tube is partly filled with plaster of Paris and water and allowed to set, the tube is often broken by the apparent expansion, yet there is an overall diminution in volume of about seven per cent when the reaction represented by the equation:

$$2(CaSO_4.\tfrac{1}{2}H_2O)+3H_2O = 2(CaSO_4.2H_2O) \quad \ldots (2.1)$$

Salt crystallisation

proceeds to completion. It seems that, although there is a theoretical decrease in volume of seven per cent during the hardening of a $2(CaSO_4.\frac{1}{2}H_2O)-H_2O$ paste, there is in fact a 0·5 per cent expansion (Chatterji and Jeffery, 1963). Thus emplaced salts which have become even partially dehydrated could be expected to exert a force on the walls of the containing fissure when they are wetted. The various forms of hydrated and anhydrous $CaSO_4$ have widespread occurrence in deserts, and the weathering mechanism proposed involves deposition of a hydrated form in a rock fissure, followed by slow hydration in high desert temperatures and in the presence of dehydration-assisting salts, which is followed by a relatively sudden wetting, causing rapid hydration with consequent tensile stresses on the fissure walls.

(iii) It has been demonstrated many times by workers in various fields that rock disintegration may be caused by salt crystallization in confined spaces (Evans, 1970); indeed this is probably the most important salt weathering process. It may assume considerable significance in deserts (Wellman and Wilson, 1965), especially where salts are concentrated (e.g. in and around playas, along channels, and in moister areas such as the underside of boulders) and where wetting and drying is most common (e.g. in humid, coastal deserts such as the Atacama, where fogs are frequent, and salts are available from fog nuclei, the sea and chemical weathering). Evidence of the effectiveness of this weathering process has been reported from many deserts such as those in Iran (Beaumont, 1968), the western United States (Blackwelder, 1940), western Australia (Jutson, 1918) and northern Chile (Wright and Urzúa, 1963). Rock splitting seems to be one of the commonest manifestations of salt crystallization (Yaalon, 1970; Young, 1964); and granular disintegration of granite can be accomplished by this process, as Kwaad (1970) demonstrated experimentally.

Experimental studies by Goudie, Cooke and Evans (1970), in which rock specimens were immersed in saturated salt solutions and subjected to a temperature change comparable to that in soils of the central Sahara* daily for 40 days, demonstrated four significant facts. Firstly, salts vary in their ability to disintegrate a rock by crystal growth (Fig. 2.8), a fact also demonstrated by Kwaad (1970). Secondly, the effect of the same salt on different rocks varies according to the type of rock (Fig. 2.9). Thirdly, a significant factor in determining rock susceptibility is rate of water absorption. (Rank correlation of weight loss after 40 daily cycles with water absorption rates for nine rock types gave a coefficient, r_s, of 0·9, which was significant at the one per cent level.)

*60°C for six hours, and 30°C for the remainder of each 24-hour cycle.

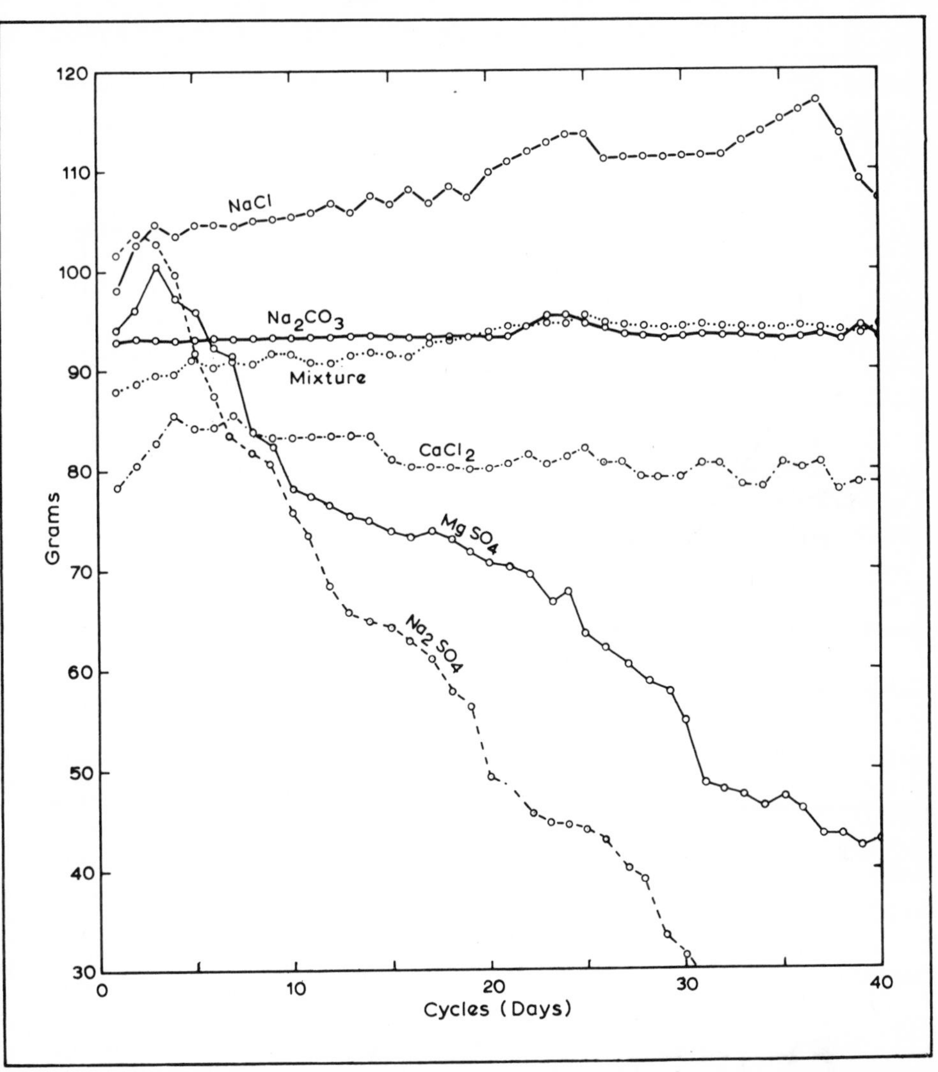

2.8 Change in weight of Arden Sandstone samples on treatment with different salts (after Goudie, Cooke and Evans, 1970).

Finally, in the experiments using a sandstone, both granular disaggregation *and* splitting occurred, and Na_2SO_4 was most effective in both cases.

This form of salt weathering is often compared with frost action but, although the weathering products may be similar, the processes are distinct for several reasons: salt crystallization involves both a salt and a

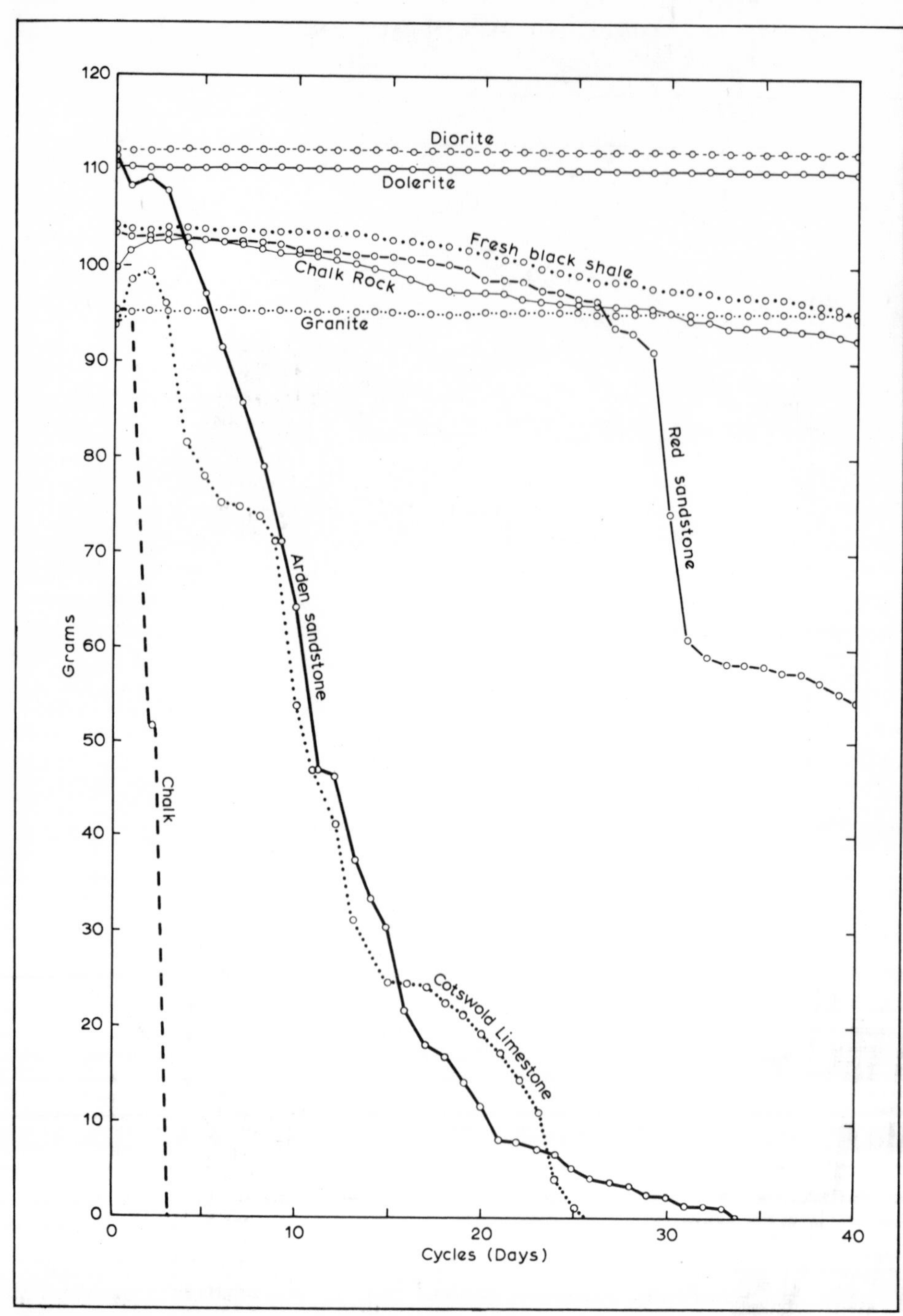

2.9 Change in weight of different rock types on treatment with sodium sulphate (after Goudie, Cooke and Evans, 1970).

solvent and only the salt enters a solid phase; volume-increases of the system are smaller in salt crystallization; and where salt crystallization results from evaporation, the system is open and pressures cannot be exerted by volume-change (Evans, 1970).

The force is one of crystal growth under pressure. Stress is transmitted through a thin film of supersaturated solution, and crystallization continues until the stress reaches a magnitude determined directly by the degree of supersaturation and inversely by the stress coefficient of solubility of the salt. A small supersaturation may be attained by rapid evaporation and may produce a force in excess of rock tensile strength (Evans, 1970).

(d) Wetting and drying weathering. In the experiments quoted above, Goudie, Cooke and Evans (1970) showed that wetting and drying of fresh black shale in distilled water caused the rock to disintegrate. Ollier (1969) reported unpublished experiments by Condon in which fine-grained rocks such as shale disintegrated by flaking and splitting when repeatedly wetted and dried. A possible explanation of this process is in terms of 'ordered-water' molecular pressure. Ollier noted that, as water is a 'polar' liquid with two positively charged hydrogen atoms at one end of a molecule and a negatively charged atom at the other, the positively charged end is attracted to the negatively charged surface of clay or other materials, and other molecules line up to form an ordered water layer. Wetting and drying may allow the molecules to become increasingly ordered and thence perhaps to exert an expansive force that thrusts apart the confining walls.

In the desert, the surface may be wetted by rainfall or dew. In some areas wetting by dew may be most important, permitting at times a daily wetting and drying cycle. In Israel, Yaalon has noted that wetting by dew can extend up to several millimetres below a bare surface and up to several centimetres under trees and bushes. Since the dew in Israel is saline (Yaalon and Ganor, 1968), weathering of surface material could be promoted there by both wetting and drying and salt.

(e) Weathering by lichens. Lichens grow on many rock-surfaces in deserts. They are firmly attached to the rock, and their thallial tissues contain a large proportion of gelatinous and mucilaginous substances. These substances will contract on withdrawal of water, a pressure will be exerted on the rock, and under appropriate conditions a small rock fragment may be broken off, in much the same way as drying gelatin will chip flakes from the surface of a glass plate. Disintegration might also be produced by the wetting of thalli lodged in confined spaces, which swell

and may exert pressure on the rock causing it to fracture. In a series of laboratory experiments, Fry (1924 and 1927) demonstrated the nature and effectiveness of these mechanical processes on various rock types. Their efficacy in deserts has not been analyzed, but the presence of both lichens and wetting and drying conditions in lichen tissue could be responsible for a small amount of surface flaking. We might speculate that other micro-organisms may also play a role in breaking up rock-surfaces in deserts.

2.3 Desert Soils and Weathered Mantles

Soil and weathering mantles are important components of the system of desert surface degradation at the local scale. The role of these mantles will be outlined in Part 3: it is, briefly, that the weathered and 'pedo-morphosed' mantle supplies material to the fluvial or the aeolian sub-systems.

The balance between the mantle subsystem on the one hand and the erosional subsystem on the other is vital to the understanding of the soils themselves. Slopes can be distinguished as either erosion-controlled or weathering-controlled (Culling, 1963). Where erosion predominates, soils are thin or absent; where weathering predominates, soils are deeper. In many deserts it is erosion which generally 'wins' this battle. Soils are often thin or non-existent, and the surface details are those determined by weathering or by a set of characteristic surface processes. Only on more stable surfaces such as those of low-angle alluvial fans and pedi-ments does weathering 'win' and can true soils be found.

In discussing desert soils, pedologists have been drawn ineluctably into the discussion of environmental change, because of the important and complex evidence of the soil profiles themselves. The complexity of real profiles, however, does not mean that a statement about desert soil processes is undesirable or impossible. In the discussion that follows, we concentrate on the part that soils play in the landscape and attempt to describe the contemporary processes of soil formation in deserts. We shall later outline briefly some of the evidence that desert soils have suffered environmental change.

To understand the distinctiveness of desert soils we can first isolate the factors which control their formation, and then examine the processes within the profile that lead to their distinctive features. In doing this we use two models: Jenny's (1941) view of the factors of soil formation and Simonson's (1959) generalization about soil-profile processes.

2.3.1 FACTORS OF DESERT SOIL FORMATION

Soil characteristics can be seen as functions of several factors, principally climate, vegetation, parent material, relief and time (Jenny, 1941).

(a) Desert climate as a soil-forming factor. Climatic influences are, of course, the primary contributors to the distinctiveness of desert soils, and as Jenny discovered (1941), there are several soil characteristics that can be closely correlated with climatic parameters. The clay contents of desert soils are lower than those of humid soils (Fig. 2. 10a; Barshad, 1962; Harradine and Jenny, 1958; Jenny, 1941). For this reason, most desert soils have low exchange capacities (van der Merwe and Heystek, 1955; Rodin and Bazilevich, 1965; Scott, 1962), but these are usually saturated with cations. In South Africa and in Arizona, base-saturation is nearly 100 per cent in desert soils (van der Merwe, 1954; Whitaker *et al.*, 1968). In East Africa, complete base-saturation persists as the rainfall rises from very low figures to 51 cm, but then there is a very rapid fall-off in saturation percentages (Fig. 2.10b; Scott, 1962). Australian desert soils, on the other hand, are more acid, a fact which has been given palaeoclimatic significance (Stephens, 1961; Wild, 1958).

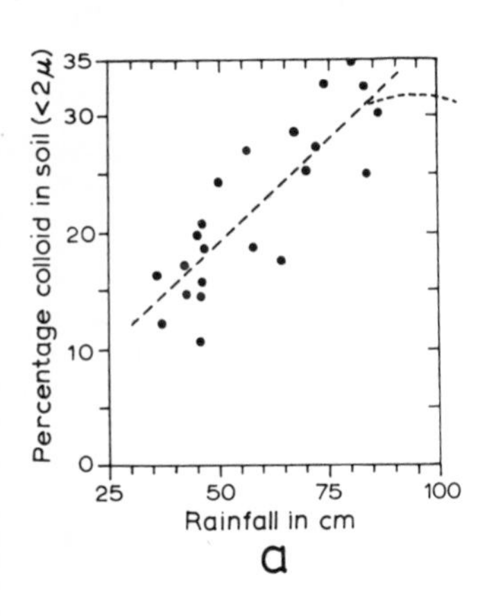

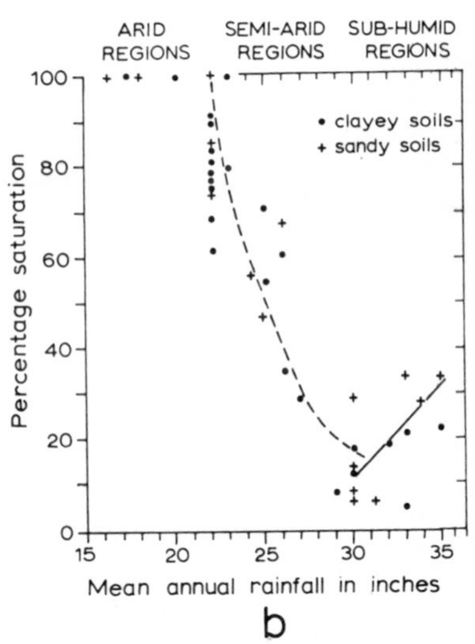

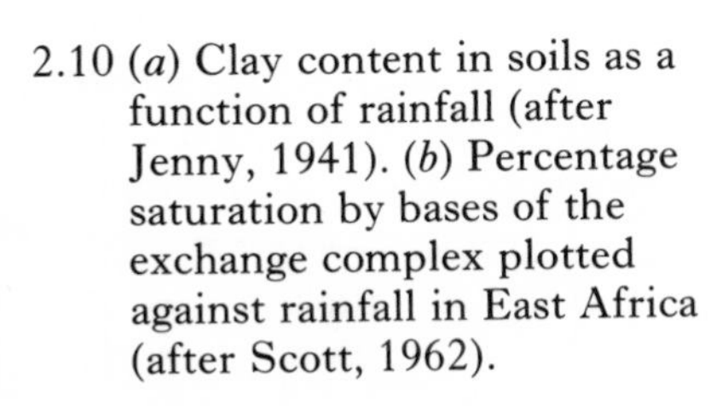
2.10 (*a*) Clay content in soils as a function of rainfall (after Jenny, 1941). (*b*) Percentage saturation by bases of the exchange complex plotted against rainfall in East Africa (after Scott, 1962).

The exchange capacities of desert soils may also be linked to their clay-mineral suites. Several workers have claimed that desert soils have distinctive mineral suites (Barshad, 1955 and 1962; Ismail, 1970; Jackson, 1959; van der Merwe and Heystek, 1955; Weinert, 1965). In a South African study, for example, it was found that the Karoo Dolerite was broken down principally by mechanical weathering in dry areas to give hydrous micas, whereas beyond a distinct climatic boundary, the same rock was chemically altered to different clay- minerals (Weinert, 1965). Brown and Drosoff (1940), Ismail (1970), Stankhe *et al.*, (1969) and Walker (1967) are amongst many who claim that montmorillonite and hydrous micas are characteristic of desert soils. In spite of these findings, it can be shown that a very wide range of clay-mineral suites can occur in desert soils; this has been explained by the slow breakdown of a variety of inherited clay minerals (Buol, 1965; Elgabaly and Khadr, 1962).

We discuss later some of the processes that are thought to give distinctive mineralogy in desert soils and also the forces which lead to the shallowness of desert soils—a distinctive characteristic linked closely to climate.

(b) Desert vegetation and desert soils. Because vegetation and other biota are in such close association with the soil and because vegetation is itself closely controlled by climate, it has been maintained that soils and vegetation may usefully be considered as a single system (Jenny, 1961).

Desert vegetation has low biomass and low productivity. For example, wormwood and saltwort desert in central Asia may have a biomass of 40–140 kg/ha which is at least 10 times less than in temperate forests and several times less than in steppes (Rodin and Basilevich, 1965). In the hotter and drier deserts, biomass is often zero. In Arizona, the productivity of desert vegetation is less than 200–500 $kg/m^2/yr$. However productivity and vegetation pattern can vary widely, and slightly moister areas can have surprisingly high productivity (Greig-Smith and Chadwick, 1965; Robin and Bazilevich, 1965; Whitaker *et al.*, 1968).

Low productivity and biomass does not necessarily mean low litter-fall. On the contrary, it appears that in the Soviet Union litter-fall is higher in some deserts than in coniferous forests (Rodin and Bazilevich, 1965); this may be due to the higher proportion of annuals and ephemerals amongst desert plants. It means that a large proportion of the biomass is in circulation. Much of the litter (frequently 60 per cent) is subsurface, since roots are well developed in desert species. The shallow depth at which most roots of desert species occur means that much of this root litter is added to the near-surface soil (Cloudsley-Thompson and

Chadwick, 1965; Rodin and Bazilevich, 1965). In some Sudanese soils, however, moisture-retaining, fine-textured subsurface horizons have higher organic matter contents than might be expected because, it is thought, roots seek out these horizons and add organic matter to them when they decay (Jewitt, 1952).

These characteristics of desert vegetation are very important to desert soils. It has been observed that some soils have redder hues in drier climates. This is in part due to the masking in wetter climates of the redness by organic matter (e.g. Jenny, 1929 and 1941; Ruhe, 1967; Whitaker *et al.*, 1968). Few desert soils in central Asia have organic matter contents greater than two per cent and figures of less than one per cent are common. In hotter deserts, organic matter is usually less than one per cent in *A* horizons and is frequently zero (Rodin and Bazilevich, 1965). We discuss the red coloration in section 2.3.2.*c.ii*.

The litter of desert vegetation has other characteristics of importance to soils. Desert vegetation, which in places contains a high proportion of leguminous plants, accumulates nitrogen more than humid vegetation as a response to the harsh environment (Rodin and Bazilevich, 1965; Whitaker *et al.*, 1968). This is one of the reasons for the relatively low carbon/nitrogen ratios of desert soils (Jenny, 1929). Pottassium and calcium are also accumulated more by desert than humid vegetation, and halophytes accumulate salts which, as litter, may contribute to the surface salinity of some soils (Coque, 1962; Rodin and Bazilevich, 1965). Silica is an important accumulation in desert grasses and is often the most important element in circulation (Rodin and Bazilevich, 1965). This may be important in the mobility and concentration of silica in desert soils. For example, leaf analysis of some Australian arid-zone plants showed that they accumulated silica preferentially (Hallsworth and Waring, 1964; section 2.3.2.*d.vi*). It has been noted too that some arid-zone plants accumulate sulphur preferentially and this has been suggested as a mechanism of gypsum formation (Glennie and Evamy, 1968).

Phreatophytes with deep-rooting systems may be important in keeping water-tables low in many desert soils. Some species in fact transpire large amounts of water as an adaptation to the harsh desert environment (Cloudsley-Thompson and Chadwick, 1965). Removal of such species on desert margins can lead to rises in the water-table. Such a situation is said to have occurred in northern Nigeria and in Western Australia after the agricultural colonization of semi-arid sandy soils (Bettenay *et al.*, 1964; Jones, 1957), and cutting of vegetation is seriously contemplated in the arid United States to raise water-tables (Miller *et al.*, 1961). In some situations deep-rooting species prevent salinization by keeping water-tables low (Bettenay *et al.*, 1964). Some species add colloids to the soil and help to stabilize it by adding salts and $CaCO_3$ by root secretion (Gimmingham, 1955; Ravikovitch, *et al.*, 1958; Yabukov and Bespalova, 1961). Yet others have the stability, like marram grass on temperate coastal dunes, to grow upward through accumulating silt or sand and thus contribute to a marked micro-relief (e.g. Gile, 1966; Gimmingham, 1955). Some arid soils, especially sands, are 'water-repellent' because of organic-matter additions (Bond, 1968).

Algae and sometimes lichens are very important in some desert soils. In central Asia, algae often form the sole cover of takyr (clay soils) (Glasovskaya, 1968; Lobova, 1967). They may form a resistant crust, protecting the soil from raindrops but increasing runoff (Bolyshev, 1964; Shields *et al.*, 1957). Algae, too, tend to fix nitrogen and thus are another contributor to the high carbon/nitrogen ratios of desert soils (Fuller *et al.*, 1960; Mayland and McIntosh, 1963; Rodin and Bazilevich, 1965; Shields *et al.*, 1957). Certain bacteria are also important nitrogen fixers and others are thought to be instrumental in concentrating gypsum in some desert soils (Martin and Fletcher, 1943). Others have been accredited with sulphate reduction and the formation of Na_2CO_3 in alkali soils (Kelley, 1951), but in general, bacteria, fungi and insects have a smaller role to play in desert soils than elsewhere. Termites become less and less obvious in soils as the rainfall declines in West Africa from the savannas northward into the desert; north of the 200 mm isohyet they are unimportant (Tricart, 1958), although still evident in valley soils (Bernard, 1954). Ants are also confined to moister environments. In semi-arid central African soils, they may however play an important translocational role (Bocquier, 1968). Small mammals are only of very localized significance as burrowers into the soil.

(c) Soil parent materials. Although the range of bedrock parent materials available for soil formation in deserts is probably as great as in any other climatic regime, there are distinctive desert substrates. Few desert soils develop on the types of weathered mantle which characterize temperate areas because in deserts the rate of erosion usually exceeds the rate of weathering and the surface has only a thin, coarse, weathered mantle. In some situations, however, deeply weathered mantles (usually attributed to more humid phases in the past) do provide parent materials for modern soils (section 2.3.2.*f*). But it is young sediments that are the most important soil substrates in most deserts. The characteristic soils of many arid and semi-arid areas are developed on alluvial surfaces composed of coarse cobbles, boulders and sands (e.g. Coque, 1962; Ruhe, 1967) and the saline and alkali soils of most deserts are

found on the alluvium of large flood-plains such as the Indus or the Amu Darya where sediments are usually fine-grained and deep (Hunting Technical Services, 1965; Lobova, 1967).

In semi-arid areas, former aeolian sands that are more or less fixed by vegetation are very common and have distinctive soil profiles (Crocker, 1946; Grove, 1969; Grove and Warren, 1968; Hillel and Tadmor, 1962; Raychaudri and Sen, 1952; Suslov, 1961; Warren, 1970; White, 1971). In these and in more arid areas the sands have higher densities of vegetation than nearby heavier soils because of their permeability and thermal insulation properties (Bagnold, 1954*a*; Branson *et al.*, 1961; Cloudsley-Thompson and Chadwick, 1965; Hillel and Tadmor, 1962; Prill, 1968; Smith, 1949; Yang, 1956). Williams (1954) and Cvijanovich (1953) found that about 0·2 to 0·3 m beneath a sandy soil there were hardly any diurnal temperature or humidity fluctuations. Water can only penetrate to a depth of about eight times the immediate precipitation and is therefore held in the soil (Bagnold, 1954*a*).

(d) The relief factor. The contrasts between soils on different parts of a slope are often more pronounced in dry areas than in humid ones because of three principal factors. Firstly, above a certain slope angle (which depends *inter alia* on rock type), the erosion/weathering balance is tipped in favour of erosion. Below this critical angle, slopes are more stable and soils can develop. It is probable that this is a distinct breakpoint: above it there are no soils, below it there are soils. Slope can also affect the proportion of runoff as opposed to infiltrating water. In the Rio Grande Valley, New Mexico, for example, salty soils are often associated with slightly higher, more steeply sloping sites, where there is less infiltration than in hollows (Wiegland *et al.*, 1966).

The second differentiating set of processes concerns the water-table. Where this is deep it has little effect on soils, but where it rises above a certain well-defined critical depth (section 2.3.2.*d.ii*), it can affect soil properties by capillary rise, and saline or alkaline soils are the result. It is interesting that in both the erosional and capillary processes there are indications of a precise threshold in deserts between processes that lead to soils with very different characteristics.

Water-movements have other influences on soil characteristics in addition to capillary rise from a water-table. On the slopes of many desert inselbergs in crystalline rocks, water flows off the higher-angle slopes and accumulates in the coarse alluvium of the plain, where it leads to locally deeper weathering (Bocquier, 1968; Twidale, 1962) or it may even seep out at the surface, depositing fine sediment which it has collected in its throughflow in soils upslope. It is said that flow off the upper slopes removes weathering products to lower slope areas. One-to-one clay minerals and leached soils are associated with the former and two-to-one clay minerals and sometimes alkalinity with the latter (Goss and Allen, 1968; Ruxton, 1958; Ruxton and Berry, 1961).

Thirdly, these two characteristic processes mean that soils on different slopes are often of different ages. In New Mexico, for example, steeper slopes have younger soils, for it is here that there is either faster erosion or faster deposition; lower-angle slopes have older soils (Gile, 1967; Ruhe, 1967). The relationships may be more complex on alluvial surfaces that have suffered repeated phases of erosion. Some soils are then developed on the pre-weathered material eroded from older soils (Gile and Hawley, 1968).

(e) Time in desert soils. The effects of time on desert soils are best examined by studying soil horizons and profiles, and it is to these that we now turn.

2.3.2 SOIL PROFILE DEVELOPMENT

In the preceding discussion we have considered soils as a whole. But one of the distinctive features of soils is that their development is marked by the appearance of a profile of differentiated horizons. Since soil horizons, particularly subsurface ones, are usually more persistent features of soils than some of the properties we have so far considered, they are particularly likely to survive changes in the soil environment, and to persist, say, from a wetter past into a drier present. This makes the discussion of profile development under desert conditions hazardous, in that much of the evidence can be given a palaeoclimatic interpretation, and very important (geomorphologically) in that the horizons have differential erosion-resistances. In the first part of this subsection we shall persist in trying to describe processes peculiar to desert soils before discussing palaeosols in a later section.

The processes of horizon differentiation can be divided into four groups: additions, removals, transformations and translocations (Simonson, 1959).

(a) Additions. Additions have a more important part to play in the development of desert soils than humid soils. Aridity creates a number of conditions which are favourable to this. In the first place, dry loose surfaces, unprotected by vegetation, are particularly liable to erosion by wind or water. The resulting sediments are deposited on soil surfaces elsewhere. Secondly, upward movement of water by capillarity through profiles means that salts may be added to the soil surface; moreover the

salt crystals are themselves mobile in the wind and they tend by crystallization to rework and loosen surface sediments and render them more liable to removal.

Because of their limited mobility in runoff and seepage after deposition, oceanic salts added by the wind are more obvious in arid than in humid soils. This question was reviewed at length by Eriksson (1958). In Israel, more salt is deposited on the more humid northern coasts than on the drier southern ones, but salt accumulates in the drier soils much more than in the wetter northern soils (Yaalon, 1963). The same is true in the Soviet Union (Tsyganenko, 1968). In some zones, oceanic water near to arid coasts can be expected to yield more salts than less saline waters near humid coasts (Eriksson, 1958).

Oceanic salt additions do indeed appear to be of widespread importance to arid soils. In Australia, the coastal semi-arid lands have much more saline soils than those in the interior (Jackson, 1957). In South Australia it has been estimated that 168·96 kg/ha/yr of NaCl and 14·28 kg/ha/yr of gypsum are added to the soils of the York Peninsula, whereas in Western Australia 125 to 750 kg/ha/yr of all salts are deposited near the coast and 12·5 kg/ha/yr further inland (figures quoted by Jennings, 1955). Eriksson (1958) has mapped this fall-off from about 150 kg/ha on the coast to about 5 kg/ha in the interior. In Israel, Yaalon and Lomas (1970) have plotted salt fall-off curves with some accuracy. Hallsworth and Waring (1964) suggested that the slow addition of oceanic salt was important even far from the Australian coast; and it is estimated that some 3 kg/ha are deposited on the inland 'steppe' regions of West Africa (Eriksson, 1958). In central Asia, 60–90 mg/cm^2 of salts are received annually in the rainfall (Tsyganenko, 1968). In Israel as a whole, 100,000 tons of NaCl are estimated to fall each year: NaCl predominates, especially in the wetter areas; sulphates become more important in more arid areas (Yaalon, 1963).

Most airborne salts, however, are probably of more local origin. In Victoria, calcium (in higher percentages than in sea-water) becomes increasingly important inland (Hutton and Leslie, 1958). Calcium greatly increases as a percentage and sodium declines away from the coast into inland Australia (Hutton, 1968). Sodium is usually associated near the coast with chloride and nitrate. In central Asia, sulphates and carbonates are the most important land-derived salts (Tsyganenko, 1968).

In many areas the airborne journey is merely a part of a 'salt cycle' on a basin scale. Salts are slowly dissolved from rocks and soil in the upper basin, and are then transported haltingly in solution across the intervening zones to depositional areas near base-level. Here they are concentrated near the surface by evaporation, from whence their crystals can be transported back again by the wind to higher parts of the basin. In pre-Saharan Tunisia such a cycle is envisaged for gypsum (Coque, 1962). Gypsum is slowly dissolved from Cretaceous and Tertiary outcrops in the higher parts of the region; it is carried down to the great chotts and crystallized by evaporation, and then blown back to the slopes again to form the *croûte gypseûse* of the area. Chlorine may have a local cycle from sebkha to lunette and back near Oran, in Algeria (Boulaine, 1954).

Carbonates in arid soils are another important addition often thought to have originated by aeolian deposition (section 2.3.2.*d.v*).

More inert, less weathered airborne material is also added to many desert soilsurfaces. Direct evidence of this is abundant, although it shows that the desert is more often the source of dust added to semi-arid lands; for example, dust is added to the surfaces of many North American soils, and is most marked in semi-arid areas; anything up to 83·7 kg/ha is added in some months of the year (Smith *et al.*, 1970). Clay particles appear to be important in aeolian transport in arid areas. This has been observed in Israel (Dan *et al.*, 1968) and in North America (Smith *et al.*, 1970). In south-eastern and in Western Australia there is a widespread clay deposit known as *parna*, which is thought to be of aeolian origin (Butler, 1956; Butler and Hutton, 1956). Dust is often trapped in the 'dead' zones around small shrubs, and so forms a very uneven surface of accumulation (Coque, 1962; Gile, 1966*a*; Petrov, 1962). Very fine particles in the form of 'aerosolic dust' have been reported to be important, for example in the West Indies, where, it is said, the dust is derived from the Sahara (Syers *et al.*, 1969).

Coarser particles can be locally important as additions to arid soils. The basin slopes around the Carson Desert in Utah, for example, receive regular additions of sand blown up from the basin floor (Morrison, 1964). In the semi-arid Sudan, clay soils with an addition of red sand to the surface horizons are known locally as *qardud*. These soils retain the patterns of ancient dunes which have now been eroded away (Warren, 1970).

Airborne additions occur at different rates. Most commonly the rate of addition is slow and material is incorporated into pre-existing soils. Inert particles are worked down in cracks by 'pedoturbation' and soluble particles are dissolved and redeposited lower down the soil profile. In some situations, additions are so heavy that a completely new solum develops on the deposited material, or a new horizon appears to give a complex 'bi-sequal' profile. Several of these complex profiles have been described, for example, in dunes, parna and riverine alluvium in southeastern Australia, in the Sudan and in Israel (Butler and Hutton, 1956; Churchward, 1961 and 1963; Dan *et al.*, 1968; Karmeli and Ravina, 1968; Williams, 1968).

Fluvially-borne additions are usually of a different character. Surface wash probably effects an important redistribution of fines, for example on bare rock-surfaces where weathered fines accumulate in hollows, but fluvial activity is more episodic and violent than that of the wind, so that additions from rivers and floods tend to produce new sola and horizons, rather than slow surface additions. Complex soil profile patterns can result where there is local erosion in one place and deposition in another on the surface of alluvial fans (Gile and Hawley, 1966).

Additions to soils need not come from above. Ground-water movements slowly transport salts which can be deposited in the soils when the water evaporates. Many ground-waters are saline in arid zones, partly because their salts have been concentrated by evaporation (Coque, 1962; Shotton, 1954; Yaalon, 1963). These saline waters, even if they initially have very low salt concentrations, are undoubtedly the principal source of salts in saline soils (Kelley, 1951). An important source in some areas may be infiltrating river waters. In Arabia these appear to have cemented their immediate beds and banks with $CaCO_3$ so that subsequent erosion of the surrounding uncemented areas has left anastomosing and meandering channels as upstanding ridges (Glennie, 1970). In New South Wales, Butler (1950) and Hallsworth and Waring (1964) speculated that rivers have injected salts into local soils. Salts are undoubtedly added to certain soils by flooding river water, for example along the Nile (Buursink, 1971) and along the lower Rio Grande (Horn *et al.*, 1964).

Organic matter additions have been discussed in the earlier part of this subsection. It is important to remember that they are surprisingly heavy and a high proportion is in root form, that they are quickly broken down, and that in some situations, notably with halophytic vegetation, the litter may be rich in salts and so materially add to salt concentration in the surface horizon. Dead organic matter is probably only a small contributor to airborne salts in Australia (Hutton, 1968); it is thought to be important in central Asia (Tsyganenko, 1968).

(b) Removals. The removal of material from desert soils is relatively less important than its addition. Surface removals are to some extent the obverse of additions, but many of the important airborne additions are removed from 'not-soil' areas in dunes or playas, so that, in general, soils *sensu stricto* receive more than they lose. Surfacewash may be

important on some slopes, and is important to the development of surface carapaces and stone pavements (section 2.4.1). More drastic removals of material by surface erosion have the profound effect of exposing subsurface horizons; this is an important part of the soil-geomorphological system which will be discussed later.

Removal of salts and other substances by leaching is much less important in desert than in humid soils. Indeed it is the imbalance between addition and removals which accounts for the accumulation of substances such as salts and silica and which is so characteristic of desert soil profiles. Most salts are removed only from a shallow surface horizon down to the depth of maximum penetration by rainwater. Only the most mobile of ions can leave arid and semi-arid soils by leaching. An example is chlorine (e.g. Bettenay and Hingston, 1964; and Boulaine, 1954).

(c) Transformation. (i) The processes of chemical transformation of parent rock minerals into soil materials in deserts are similar to those in other environments but there are differences in the rate, and in the relative importance of the various reactions. Water is the principal reactant in hydrolysis, the most important of the transformational processes, and it is also fundamental as a solvent in the leaching process (e.g. Barshad, 1962; Keller, 1962). Organic matter both as humus and living plant and animal tissue is another important source of hydrogen ions for the hydrolysis reaction. Plant roots exchange hydrogen for nutrients with rock and clay minerals; the H^+ ions then insinuate themselves into mineral crystal lattices, disrupting them and releasing bases. Since both water and organic matter are scarce in deserts, one would expect that weathering might be slower and might give rise to different end-products.

There is evidence from desert soils and weathering mantles that this is the case, but the inheritance of weathering products, not only from soils formed in different climatic environments in the recent past, but also from older sediments, can complicate the story. Montmorillonites and hydrous micas are common in desert soils (section 2.3.1.*a*) although they are not peculiar to this environment. Their formation might be explained partly by the greater solubility of silica at high *p*Hs. Silica is thus mobilized in desert soils and can then recombine with alumina to form two-to-one clay-minerals such as the montmorillonites. The very slow leaching rate in arid soils means too that silica cannot be so easily removed in drainage-water as in humid areas.

The transformation of biotite to a clay mineral has been followed in the laboratory (Ismail, 1969 and 1970), and in actual desert soil samples the transformation of hornblende to iron-rich montmorillonite has been observed with an electron probe (Walker *et al.*, 1967). In acid conditions biotite is weathered to vermiculite by the loss of octahedral ions, but in alkaline conditions there is a loss of surface charge, and montmorillonite forms. Hornblende appears to be altered gradually to iron-montmorillonite, while iron oxide is also formed to coat the surrounding large particles. The mobility of iron is discussed in section 2.3.2.*c.ii*.

Another piece of evidence that throws light on the nature of desert transformations is the weathering of limestone. Limestones weather little in very arid conditions due as much to the absence of water as to the absence of organic matter. In North Africa there seems to be little limestone weathering in areas with rainfalls less than 150 mm (Capot-Rey, 1939; Cornet and Pinaud, 1957; Estorges, 1959). Similar observations have been made in Syria (Wirth, 1958), in Israel (Nir, 1964) and in Australia (e.g. Jennings, 1967). In California, calcium and magnesium losses increase in wetter areas, probably because of increased soil carbon dioxide due to increased organic activity (Feth, 1961*b*).

Since rates of erosion are low in very arid deserts and since the only weathering mantles appear to be palaeosols (section 2.3.2.*f*), one must conclude that weathering is a very slow process, characterized only by very small scale translocations (such as the formation of iron oxide coloration), rather than by major losses of materials. In semi-arid areas, on the other hand, a greater range of weathering intensities is found. There is a loss of silica from the upslope soils and an accumulation of silica in soils downslope. The upslope soils have kaolinite or other one-to-one clay-minerals; the downslope soils (often clayey vertisols) are characterized by two-to-one clay-minerals such as the montmorillonites. As one proceeds into the wetter tropical areas, red soils with one-to-one minerals extend over more and more of the landscape as more and more silica is lost from the soil landscape. On the wetter temperate margins of deserts greater horizonation appears as ions and clay-minerals become more and more distinctly separated within the soil profile itself.

(*ii*) *Red coloration.* The redness of some soils and sediments, and particularly of aeolian sands, is an obvious feature of some desert surfaces. Since many of these deposits can be shown to be contemporary or very recent in origin, it would seem natural to attribute the colour to a contemporary process. It would seem that the hematite and other iron oxide coatings which are responsible for the colours are yet another manifestation of the mobility and precipitation properties of iron in the desert environment that results also in desert varnish (section 2.3.2.*d.i*). Unfortunately, the huge and acrimonious literature on 'red beds' has confused rather than clarified these simple observations.

Much of the literature on the red coloration of sediments concerns their palaeoclimatic interpretation, mostly in ancient deposits, and it has often paid scant regard to the findings of pedologists or even of geochemists. Because many of the processes that have been invoked probably take place deep beneath the surface during diagenesis, the discussion of much of this literature is of little relevance here. Moreover, this literature has recently been the subject of several reviews, notably by van Houten (1964), Norris (1969) and Glennie (1970). Glennie has constructed a useful table that summarizes the positions adopted by the principal contestants in the red-bed debate.

What is relevant here is the red coloration of certain surface sediments and soils since this is a part of the total picture of weathering processes in deserts. By no means all desert surface materials are red, as several authors have noted (e.g. Glennie, 1970). In the arid parts of the Sudan, for instance, Buursink (1971) found very few red or reddish soils. In Australia, Litchfield (1962 and 1963) also found a number of soils with drab colours. Nevertheless reddish colours have been widely reported. In the semi-arid Sudan, Warren (1970) reported the frequent occurrence of 5YR 5/6 colours and some much redder colours.* White (1971) noted colours of 5YR 5/6–8 and 2·5YR 4/6–8 in similar fixed aeolian sands in the Republic of Niger, and Worral (1969) in summarizing much of the work on the sands in the southern Sahara, noted widespread colours in the red, reddish-yellow and reddish-brown ranges. In arid Kenya, sandy soils have colours in the 5YR 4/6–4 range (Hemming and Trapnell, 1957). Glennie (1970) reported red aeolian sands from the Trucial Coast. Litchfield (1962 and 1963) noted colours in the 10R range amongst central and western Australian sands and clayey sands. In the Sonoran Desert colours are commonly 2·5YR 6/4, and elsewhere in Mexico colours of 10R 6/4 are reported (Walker, 1967).

The brightest colours in the soil are often in the *B* horizon. This can be attributed to more advanced weathering, to iron illuviation (section 2.3.2.*d*) or to the fact that red colours may not be masked by organic matter in this position. It is also often observed that the red coloration does not extend beneath the solum, presumably because of the limited penetration by moisture (e.g. Walker (1967) and Walker and Honea, (1969) for the Sonoran Desert; Edmonds (1942) and Warren (1970) for the Sudan; Poldervaart (1957) for the Kalahari).

The red coloration is due to coatings of oxides of iron, sometimes as hematite, on the coarser grains in the soil. A brilliant red colour can be achieved with only 0·01 per cent of free iron (Walker and Honea, 1969). The clay fraction may contain much more iron, as can the unweathered rock minerals, but these combined forms of iron do not contribute to the

* all colours are quoted in the Munsell Colour Chart terms.

colour. The free iron is released by the weathering of these minerals, notably from iron-rich montmorillonite in the clay fraction and from hornblende in the rock-mineral fraction (Schmalz, 1968; Walker, 1967; Walker *et al.*, 1967). As we shall note in the case of desert varnish (section 2.3.2.*d.i*), the levels of the hydrogen-ion-concentration and the oxygen-reduction-potential in desert soils particularly favour this kind of weathering. The iron appears to be transformed to an amorphous ferric oxi-hydroxide which is then transformed to goethite. These intermediary forms are generally yellow or reddish-yellow in colour. They are then dehydrated to hematite or to another red iron oxide. The goethite to hematite reaction is virtually irreversible (Schmalz, 1968). Crystalline hematite could not be detected in many of the red soils of the Sonoran Desert (although it is characteristic of ancient red beds), so that it is thought that much of the red colour of more recent soils must be attributed to oxides other than hematite (Walker, 1967).

It is clear then that the geochemistry of these reactions requires both moisture to release and transport the iron, and a dry period in which to dehydrate it. Accordingly a climate with wet and dry seasons has usually been postulated as characteristic of red bed palaeoenvironments (e.g. Schmalz, 1968): the savanna climate, in which there is, in addition, a higher temperature to speed the reaction, has been suggested by many workers. Red colours are indeed very characteristic of the savanna-zone sands in west Africa (e.g. Worral, 1969), and it has been observed here that redness usually becomes more intense towards the wetter areas (e.g. White, 1971). However, many desert soils, and perhaps sands especially, experience similar alterations of wetness and dryness. We have already mentioned (section 2.3.1.*c*) the evidence that many sands retain moisture for long periods (e.g. Cvijanovich, 1953). It seems reasonable to assign to this wetting, and the subsequent drying, the processes that lead to redness of modern desert sediments and soils. Walker (1967 and 1968) has strongly advocated this point of view. He has noted in addition that desert soils are usually fairly alkaline and tend to have low water-tables, both factors favouring red-bed formation. He has followed, with the aid of an electron probe, the transition from hornblende to iron-rich montmorillonite in desert soils and has noted the associated occurrence of red iron oxide coatings (Walker *et al.*, 1967). Norris (1969) has accumulated more evidence that red coloration is common in desert sands. There may be more to the distribution of redness in sands than weathering at the surface for it is probable that in the ripple-and-dune forming process the heavy iron-bearing minerals are sorted out to accumulate in certain zones and lend their redness to these (Rim, 1951).

The palaeoclimatological interpretation of redness must clearly depend on the acceptance of the arguments outlined here. Nevertheless several authorities have hazarded opinions on the subject. In Africa it is very widely held that redness is associated in deserts with wetter climates in the past (e.g. Kubiena, 1955, Meckelein, 1957, Monod, 1958 and Tricart and Cailleux, 1964 in the Sahara; and Cooke, 1961, in the Kalahari).

Redness is more characteristic of aeolian sands than of other sediments. In the Sahara and its sahellian margins, Gautier (1928) and Tricart and Brochu (1955) are among many who have noted that the larger, evidently more stable dunes are red, whereas the smaller, putatively more recently active dunes tend to be yellow. Norris (1969) has drawn together a range of evidence, mostly from the United States and Australia, which also attests this relationship. It is substantiated for the Trucial States by Glennie (1970). It seems that the more stable dunes have given reddening sufficient time to develop, whereas the moving sands are either too mobile for the process or are subject to abrasion that wears off the pigment.

The reddening of soils and sediments does generally appear to deepen with time, but it is as difficult to assign a precise age to redness as it is to assign an age to desert varnish. Walker (1967) suggested that the development of full redness required several hundreds of thousands of years, but this seems too long for the Saharan and sub-saharan sands. Many of the larger dunes in the Saharan ergs can be no more than a few thousand years old (Wilson, 1970), and in the Sudanese fixed dunes it is probable that stability has lasted only some 9,000 years (Warren, 1970). It is likely that redness develops at different rates in different situations.

(d) Translocation. Translocations of soluble salts, ions, organic matter and clay particles lead to soil-horizon differentiation. The vertical component in the movement of materials is usually considered to be most important, although a strong horizontal component in the movement may be important in certain cases. The process of horizonation takes place over considerable periods of time, and it is in this section therefore that we return to the time factor in soil formation.

Desert soil profiles, like soil profiles in other climatic zones, are seldom simple. Changing conditions, additions of new horizons and stripping of old ones, have made most soils very complex. However, distinctive classes of soil must be recognized as the basis for soil classification systems. The soils in these classes are often seen as combinations of 'type' horizons, and in some classificatory schemes it is the horizons alone that are classified (e.g. Fitzpatrick, 1969). In this section the characteristics of some type horizons are described, and then their conventional combination into some soil classes is briefly discussed.

In the discussion that follows we shall first mention surface features assignable to translocational processes and then we shall consider subsurface translocations within the soil itself.

(i) Desert varnish. The blackened appearance of rock surfaces is one of the most striking features of many desert landscapes. It is the result of a thin mineralized patina. Merrill in 1898 aptly called this desert varnish.

The literature on desert varnish has been adequately reviewed by Engel and Sharp (1958), and their conclusions have been substantiated in a study using an electron-probe by Hooke, Yang and Weiblen (1968). The brief discussion below is based largely on the work of these authors. Information on varnish in northern Africa comes from Tricart and Cailleux (1964).

Desert varnish forms a dark layer of very variable thickness over individual stones as well as over large bedrock exposures (Plates 2.19 and 2.20). The thickness varies from locality to locality and also over the surface of single stones. Engel and Sharp illustrated samples of varnish that are commonly 0·06 mm thick becoming 0·1 or 0·2 mm thick in hollows. The varnish described by Hooke *et al.*, (1968) is between 20 and 50 μ thick. The patinas described from the sahel zone of West Africa by Tricart and Cailleux (1964) are often thicker: they described patinas at Mopti that are commonly 0·5 to 7 cm thick, and patinas at Goundam that are 0·2 to 0·4 cm thick.

The outer surface of the varnish varies greatly in its appearance, from lustrous and black to dull and dark red. Engel and Sharp (1958) found that below the immediate surface layer the varnish was layered and usually redder. Hooke *et al.* (1968) distinguished between an outer or 'main' layer that is more enriched in manganese and iron, and an inner 'subordinate' layer that is more silica-enriched. The outer layers are generally amorphous and non-crystalline. The varnish itself is underlain by a weathered 'rind' of altered rock, usually about 0·5 mm thick, but sometimes as much as 3 mm thick. Tricart and Cailleux (1964) also described the weathered rind beneath the sahellian patinas that is commonly 1 cm thick. Beneath the soil surface the patina on a pebble becomes abruptly and markedly thinner, softer and redder (Plates 2.21 and 2.23).

The thickness and character of the varnish can change sharply even within one locality. It appears to be better developed on isolated boulders than on large bedrock outcrops. There is almost always a thickened layer near the ground surface on pebbles, known as the 'groundline band'. Hooke *et al.* (1968) found that these thicker varnishes also occur on the sides of stones and in cracks between stones. Tricart and Cailleux (1964) illustrated a situation in which the thick varnish in the cracks of a stone is preserved after the stone has itself been weathered away. Plate 2.22 shows a pebble that has lost its core of softened sandstone to leave only the hardened patina as a cup-like shell.

In some deserts virtually all rock-surfaces are covered with varnish; in others, varnish is absent. Although many stable surfaces tend to have a varnish, there is a correlation between varnish thickness and rock type. Vein quartz usually has a rather thin varnish. Sandstones commonly have a well-developed varnish, and the thickest varnishes are found in fine-grained sedimentaries and volcanics with a rough, porous surface (Daveau, 1966; Engel and Sharp, 1968; Tricart and Cailleux, 1964). It is commonly accepted that limestones and similar rocks have surfaces that are too unstable for the varnish to be preserved. The presence of varnish is indeed as much a matter of preservation as of

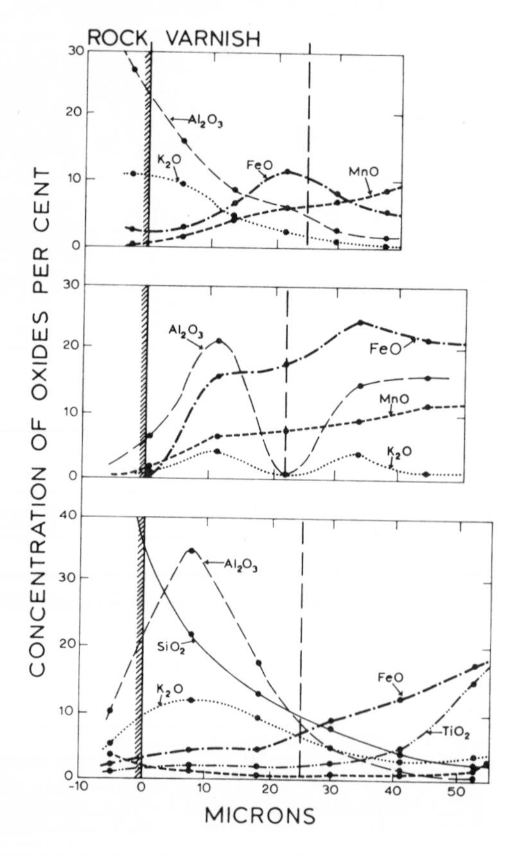

2.11 The concentrations of various oxides in thin sections of parent rocks and desert varnish. 'Microns' refers to distance from surface (after Hooke *et al.*, 1969).

formation (Engel and Sharp, 1958).

The varnish itself is almost always highly enriched in iron and manganese when it is compared with the underlying rock. Ratios of between two and six between the Fe_2O_3 content of the varnish and of the rock, and of between 66 and 292 between the MnO contents, were reported by Engel and Sharp (1958). Some of the graphs of Hooke *et al.* (1968) are illustrated in Fig. 2.11. These authors found that in most cases FeO and MnO contents increase from the rock to the varnish, whereas SiO_2, Al_2O_3, and K_2O contents decrease. There is a distinct point between the main and the subordinate layers where SiO_2 contents fall away sharply and FeO and MnO contents increase very noticeably. MnO contents increase more rapidly outwards through the varnish than FeO contents. MnO content is higher not only in the outer layer of the varnish, but also in the groundline bands and in hollows and cracks. The reddened undersides of rocks beneath the soilsurface are relatively depleted in MnO, but enriched in FeO.

There is some disagreement about the environmental associations of desert varnish. Engel and Sharp (1958) thought that a rainfall below 13 cm and a mean annual temperature of 16 to 21°C were the optimum for varnish formation. Hunt and Mabey (1966), and Tricart and Cailleux (1964), to the contrary, maintained that conditions needed to be much wetter. Tricart and Cailleux (1964) concluded that a markedly seasonal pattern of rainfall, with annual totals of about 500 mm, is the optimum. Varnish thicknesses, according to them, decrease towards the drier parts of the southern Sahara. We might note that this is very much at variance with other findings. The desert varnish of Tricart and Cailleux is distinguished by Daveau (1966) from true desert varnish. She draws a line between the redder sahellian patinas and the blacker varnish of the desert in the Baten d'Adrar of Mauritania. Hooke *et al.* (1968) suggested that there may be no clearly defined environmental limits. They cited evidence of similar coatings from Antarctica and from positions near to glaciers in Colorado.

French workers on desert varnish (Tricart and Cailleux, 1964) and the early American work led to a conclusion that desert varnish was the result of weathering of the underlying rock and the migration in solution of iron and manganese from the weathered areas to the outer surface. This was held to be due to the outward capillary movement of moisture after rain or heavy dew. Several authorities favoured the idea that the precipitation was helped by organic agencies (e.g. Blackwelder, 1948; Hunt and Mabey, 1966). In Russia, Glasovskaya (1968) advanced this explanation, suggesting that the resulting organic complexes are very stable.

The chemical data of Engel and Sharp (1958) and of Hooke *et al.* (1968) make this explanation difficult to accept in its entirety. Hooke *et al.* found that some of the constituents of varnish decrease in concentration outwards from the rock and can therefore be attributed to the rock, whereas others increase outwards and are therefore likely to be from an external source. For example, Al_2O_3 and K_2O decrease outwards when they are abundant in the parent rock, but they actually increase outwards if they are scarce in the rock. When Al_2O_3 and K_2O contents decrease outwards, in other words when it appears that they come from the rock, there are indeed distinct relationships between rock contents and varnish contents; and when FeO content decreases outwards (when it can be inferred that the FeO is from the rock) there is also a correlation between rock and varnish FeO contents. Conversely, there is no correlation between rock and varnish contents for substances that increase in concentration outwards. This seems to suggest that varnish constituents may come from both internal and external sources.

If this is accepted, the nature of the sources must still be decided. Engel and Sharp (1958) concluded that windborne material might well supply some of the elements, especially some of the trace elements such as molybdenum, but that wind could not be taken as the only source.

It seems more likely that the source of much of the material is in weathered soil and debris surrounding the varnished surfaces, and that the peculiar constitution of the varnish must be attributed to fractionation processes that occur during the movement of solutions to the rock-surfaces.

The relative mobility of various elements will be discussed in section 2.3.2.*d.ii.* Of particular interest here are the relative mobilities of iron and manganese. 'Under the Eh-*p*H conditions likely to be found in desert soils, iron is much less soluble than manganese and iron may be precipitated during the early stages of evaporation before the solution reaches the rock surface on which the varnish is forming' (Hooke *et al.*, 1968, p. 285). There are probably several different processes that act in desert as in other soils to make Mn more mobile than iron, and a number of complementary processes acting on the rock surfaces that help to precipitate it (Engel and Sharp, 1958). A similar conclusion about the greater mobility of Mn has been reached in Russian studies (Fig. 2.12; Glasovskaya, 1968). The prior deposition of Fe, particularly on the undersides of stones embedded in the soil, accounts not only for the reddened undersides, but also for the relative richness of Mn in the groundline band, since the solutions are seen as moving preferentially through the weathered rind, being fractionated as they go. Fractiona-

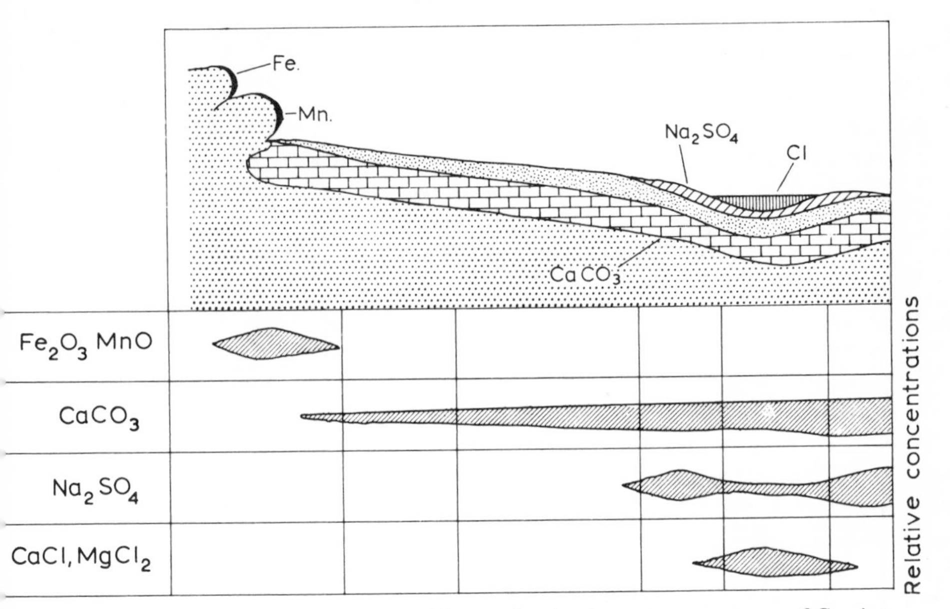

2.12 The concentrations of various oxides and salts in toposequences of Soviet deserts. The relative positions of the substances illustrate a mobility sequence (after Glasovskaya, 1968).

tion probably also occurs at the molecular level in thin films of water moving over rock-surfaces. This probably accounts for redistribution of Mn and Fe in the varnish giving heavier Mn concentrations in hollows and in the outermost layers (Hooke *et al.*, 1968). The fractionation explanation, however, is less useful, as Engel and Sharp noted, when one is discussing larger surfaces.

The age of desert varnish is a point of some disagreement. Blackwelder (1948) claimed that a complete varnish needed some 20,000 to 50,000 years in which to form, while Engel and Sharp (1958) claimed much shorter formation times. Hooke *et al.* (1968), on the other hand, claimed that pebbles very recently inverted on a playa surface were developing a Mn-rich varnish.

Desert varnish has not escaped palaeoclimatic interpretation. Tricart and Cailleux (1964) referring to work in the Sahara, and Capot-Rey (1965) in Tibesti were of the opinion that a much wetter climate than the present is needed for the formation of varnish. It is therefore relict, and Hunt and Mabey (1966) attributed it to a wetter climate some 2,000 years B.P. Daveau (1966) suggested that the varnish in the Baten d'Adrar in Mauritania is active, but that in north Africa it is probably relict.

Desert varnish cover is an indicator of the relative activity of contemporary erosion. Where patinas cover large rock-surfaces, it often is only disturbed at the sites of recent rock-falls. Elsewhere, the patina may be undercut by contemporary erosion that must be faster than the rate of patina formation (Plate 2.20). In the Baten d'Adrar, for example, the hardest rocks are thickly patinated, whereas softer, more readily eroded rocks are not (Daveau, 1966). Tricart and Cailleux (1964) invoked a veritable cycle of patination. Patinas are formed, the rock beneath being softened in the region of the rind. When erosion cuts into this it is quickly eroded away down to the fresh rock. This new surface, being more stable, develops a patina, and the cycle is repeated.

(ii) Upper horizons. In discussing soil upper horizons, it is convenient to discuss first those horizons that are the result of predominantly upward movements of material. Such horizons are confined to arid climates (of extreme cold as well as of extreme heat) because of the shortage of moisture for downward leaching. They are, nevertheless, of limited extent and are confined to the lower ends of toposequences and generally to fine alluvial parent materials which are geologically young (Kelley, 1964).

There is a very extensive literature on these horizons (e.g. Duchaufour, 1965; Kelley, 1951; Richards, 1954), since they are often associated with agricultural production. Details of their chemistry, agricultural potential and reclamation are well known; they are of interest to us here only from the point of view of their place in the landscape system.

Horizons of this type are characterized by certain ions in the exchange complex and by combinations of these ions in certain salts. The cations are usually dominated by sodium; calcium and magnesium are generally more common in arid soils than in humid soils and may occasionally dominate (Kelley, 1951; Makin, *et al.*, 1969); potassium may be important in places; and boron is very occasionally found in toxic quantities. The anions concerned are predominantly chloride, sulphate, bicarbonate and carbonate; nitrates are occasionally important. The characteristics of the soil depend very much on which ions predominate in the solution, on the nature of the clay adsorption complex, and on which salts crystallize out.

The salts are often concentrated near the surface, particularly in 'young' soils (Fig. 2.13). In many Australian soils the salt ratio between the 0–5 cm and 5–15 cm layer is commonly three and sometimes six. The ratios for sodium are commonly 1·5 and sometimes as high as 10 (Hutton, 1968).

Where NaCl is found in the surface horizon, as is the case in many soils, the soil is known as saline, white alkali, saline-alkali or *solonchak.*

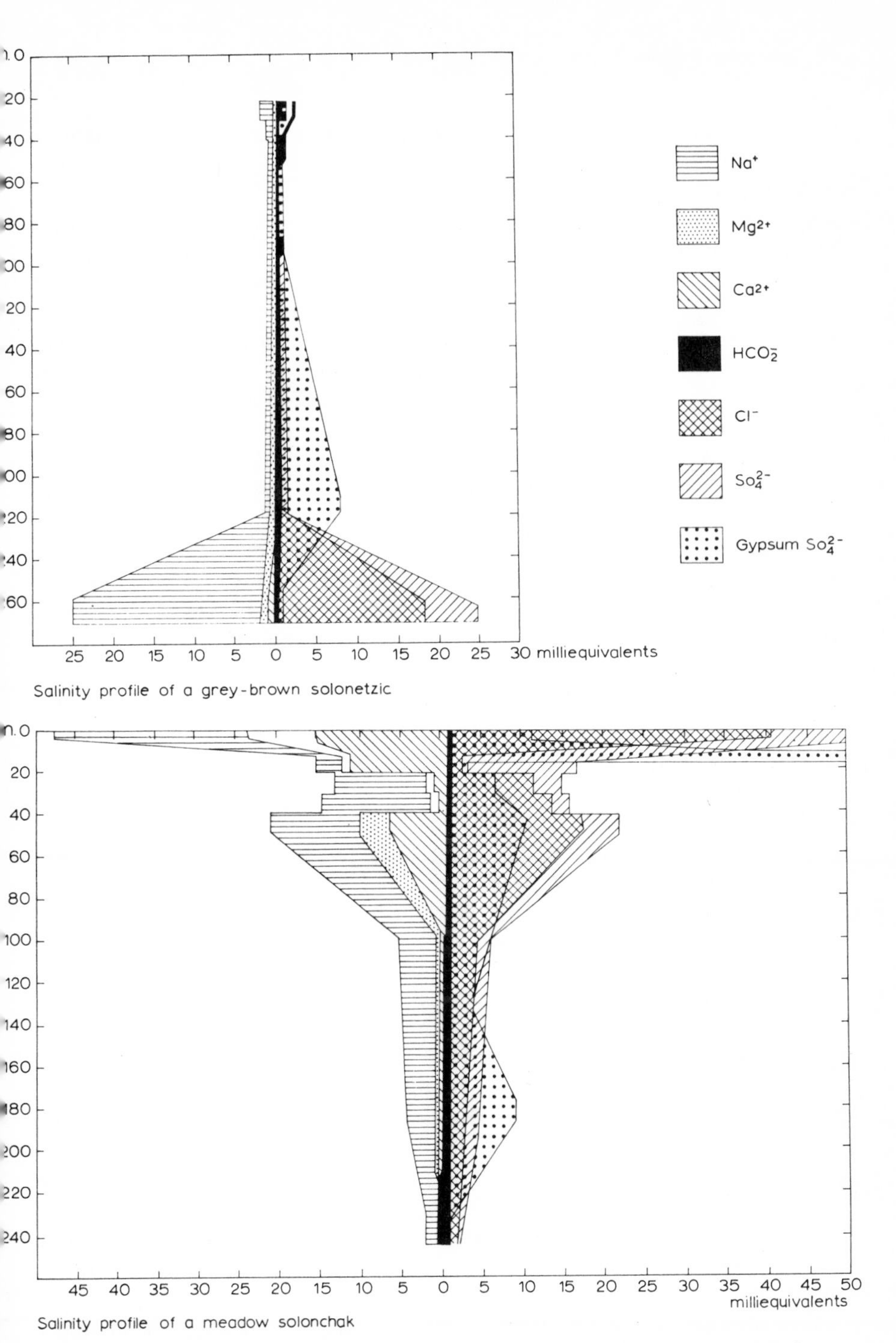

2.13 Salinity profiles in desert saline and alkaline soils of the U.S.S.R. (after Lobova, 1967).

The surface is often whitish, with large, glistening salt crystals which have broken the surface into a puffy mulch. When dry, the structure is granulated. High osmotic pressure and concentrations of Na^+ eliminate plants and other soil life (Kelley, 1951). Deliquescent salts such as calcium and magnesium chlorides in the upper horizons give the surface a firm, smooth, damp appearance, even in the driest weather.

If Na_2CO_3 is present in the surface horizons or if much Na^+ is held in the exchange complex, the *p*H is high and the soil may have a dark appearance, due to the solubility of organic matter in alkaline solutions of sodium. These are black alkali soils or *solenetz*. Because of the high hydrated radius of sodium and the weakness of the sodium monovalent bonding, the clay in these soils is highly dispersed when wet and the soils can be very impermeable as a result. Mg^{++} ions, in spite of their divalent charge, tend to have the same effect as Na^+, but K^+ ions do not. Appreciable quantities of magnesium are often associated with sodium in alkali soils (Horn *et al.*, 1964). In some alkaline soils distinctive types of clay minerals such as analcite may be found (Kelley, 1961 and 1964). But the main variations amongst solonetz soils must be attributed to the kinds of ions in the soil solution.

The nature of the ions in the soil solution and the salts that may be deposited within the soil depend firstly on the availability of ions in ground-water, surface-water, parent material or in airborne additions, and secondly on the ionic balance. For example, it has been said that sodium ions will only accumulate on the exchange complex of clays if the exchangeable $Na^+/Ca^{++}+Mg^+$ ratio is greater than two (Kelley, 1951). If added irrigation or floodwater has a high bicarbonate content, $Ca(HCO_3)_2$ precipitates and sodium becomes the dominant ion on the exchange complex. Waters high in carbonate or silicate anions are said to have the same effect (Allison, 1964). On the other hand the calcium may not be mobile at high *p*H (it will not ionize) and therefore a high total calcium content need not affect the Na^+ status of the soil (White and Papendick, 1961). Yet another factor that may affect the kind of salt in the soil is bacteria which may be important to the formation of, for example, Na_2CO_3 (Kelley, 1951).

The ions and salts in saline and alkali soils are not often *autochthonous*. They are derived ultimately either from sea salt, or from rock weathering elsewhere in the desert basin. In the United States there appear to be distinct associations within desert basins between saline soils and shales, and between non-saline alkali soils and granites, and accumulations of boron and nitrates can often be traced to local rocks containing these ions (Kelley, 1951). The ions are usually leached from the upper basin and transported in the ground-water to lower areas. In the process, filtration may increase the proportion of some ions (such as Cl^-) at the expense of others (Bettenay *et al.*, 1964; Boulaine, 1954). In some leached solonetz soils in the United States, the sodium does seem to be autochthonous (e.g. White and Papendick, 1961).

The salts in saline soils are concentrated in the soil by evaporation at or near the surface. Water is drawn up through the soil in thin capillary pathways from the water-table, but some salts may even diffuse through thin water-films (Kelley, 1951). The capillary process occurs only below a certain critical depth depending on soil texture and structure. Collis-George and Evans (1964) found in one locality that there was a very sharp fall-off in soil salinity when the water-table retreated below 91 cm. Yaalon (personal communication) maintained that in sandy soils the critical depth is one metre and in clayey soils it is three metres. The occurrence of a distinct critical depth explains the patchiness and precision of boundaries in soil salinity patterns, since a very slight difference in depth to the water-table can make the difference between saline or salt-free soils (Kelley, 1951). Where irrigation has raised the water-table above the critical level, there has been extensive salinization (e.g. Hunting Technical Services, 1961). Other sources of salt concentrations may be from the litter-fall from certain halophytes (Hutton, 1968; Rodin and Bazilevich, 1965; Walter, 1961), and more importantly, from air-borne salts (section 2.3.2.*a*).

Because Cl^- is so mobile, it is easily leached downward in drainage waters, and because Na^+ is unstable in the pesence of Ca^{++}, NaCl tends to be an unstable salt in the upper soil. If there is downward movement of water after saline soil (solonchak) formation they therefore tend to be transformed by leaching into *solonetz* soils or into *solodized-solonetz* or *solod* soils in which Na^+ has been leached to lower horizons and Cl^- has been totally removed.

Since Na^+ tends to disperse clays, these tend to be easily leached down the soil profile and out of the upper horizons, leaving them depleted in clay, and added to the lower horizons which become enriched in clay.

Where the water regime allows it, therefore, the upper horizons of mature desert soils are commonly leached. In these *A* horizons there are signs of the predominantly downward movements either of weathered material or of chemically unstable additions. These are the horizons within which the surface processes referred to in section 2.5.1.*a* take place. The evidence of weathering processes and movements that might be taking place in the *A* horizons of desert soils is usually indirect. Most of the evidence of movement comes from accumulations in the lower horizons (section 2.3.2.*d.iii*), but by calculating ratios between highly weatherable CaO and very stable ZnO_2 in desert *A* horizons, Smith and Buol (1968) used direct evidence to conclude that most weathering took place there, rather than in the *B*. Similar calculations will be referred to later.

Some *A* horizons show clear evidence of eluviation. These are the *A* horizons of the non-saline alkali (e.g. *solodized-solonetz*) soils and their putative derivatives, solodic *A* horizons. Clays are mobilized in alkali soils because of their dispersal in the Na^+ environment. They are washed down leaving a clay-poor and even acid upper horizon. Solodization (or solonization) is a widely recognized phenomenon (section 2.3.2.*d.iv*).

The most distinctive features of desert *A* horizons in general are a low content of organic matter and the biogenic restructuring by roots and animals (Rodin and Bazilevich, 1965). In cooler deserts, fulvic acids from litter breakdown may chelate iron and manganese and move them down the profile (Glasovskaya, 1968; Rodin and Bazilevich, 1965). More commonly, organic litter is rapidly oxidized in the upper *A* horizon because of extreme surface temperatures. *A* horizons may also develop a vesicular structure in deserts because the creation of a superficial crust (section 2.4.2) traps air beneath the surface, and vesicles are formed due to air expansion when the soil is heated during drying (Yaalon, in preparation).

(*iii*) *Illuviation in general.* The products of weathering in *A* horizons, as well as chemically unstable additions to the soil surface, are passed downwards to illuvial horizons where many of them accumulate. Indeed, little material passes beyond the illuvial horizon in desert soils.

The mode of deposition in illuvial horizons has been discussed by many soil scientists. In deserts, as in humid soils, it is likely that there are a number of processes often acting in concert, but one process appears to be both characteristic of semi-arid illuvial horizons and at the same time very important, if not predominant in them: it appears, not surprisingly, that dissolved products are deposited at the depth to which percolating rainwater normally penetrates. This view is subscribed to by a number of authorities (Arkley, 1963; Bocquier, 1968; Crocker, 1946; Gardner and Brooks, 1957; Gile *et al.*, 1965; Kusnetzova, 1958; Rodin and Bazilevich, 1965; Yaalon, 1965). Once in this lower horizon 'the adsorption and condensation of [some materials are] greatly enhanced by drying' (Barshad, 1955, p. 42).

The mode of operation of the illuviation process can be viewed theoretically and has been confirmed experimentally (Yaalon, 1965). One can imagine the soil profile as a chromatograph—substances pass down the profile at different rates depending on their chemical characteristics. There will be a depth of maximum concentration of a substance and a tail of lessening concentration above and below this point (Fig. 2.14). If $\bar{X}$ is the depth of maximum concentration, then:

$$\bar{X} = \frac{P_{ef} R_m}{W_a} \qquad \ldots (2.2)$$

where P_{ef} is the effective precipitation which carries down the material, R_m is the relative mobility of the material, depending on its properties of adsorption to the soil complex and its solubility etc., and W_a is the proportion of the soil occupied by mobile moisture (more or less the field capacity of the soil). R_m can be redefined as the ratio between the distance travelled by penetrating moisture and the distance travelled by the substance.

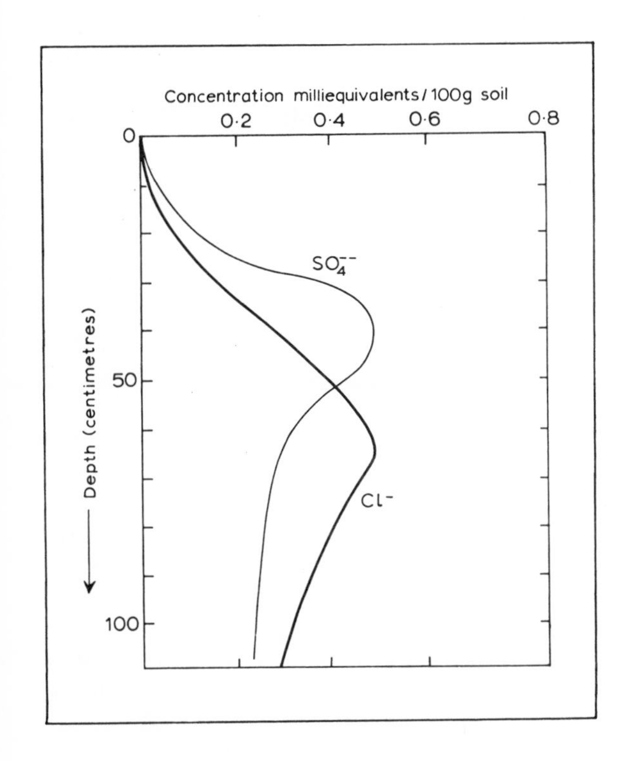

2.14 The relative positions of chloride and sulphate concentrations in an experimental soil column after leaching. The soil remained dry beneath the depth shown (after Yaalon, 1965).

Thus the salts that are found in arid soils should each have their own characteristic depth of concentration. The distribution about this point might, however, be normal or skewed, depending on a number of factors, such as the power of their adsorption on the exchange complex. Yaalon noted that chlorides, for example, move at about the speed of the percolating water, sulphates at about two-thirds of this rate. A similar differential was noted by Hallsworth *et al* (1952) in the leaching of irrigated soils in New South Wales. Yaalon confirmed this model in some experiments and in an examination of Israeli and Sudanese salt profiles in soils. He envisaged that his formula would apply to the mobile salts of his experimental work as well as to less mobile, slower moving soil salts. Yaalon's view is upheld by the work of Miller and Ratzlaff (1961), who used sodium (because of its greater mobility and replaceability compared to calcium) as a measure of the maximum penetration of rainwater. Hutton's (1968) figures for downward movements in desert soils also confirm this basic model.

Another approach to the general problem of illuviation is through the calculation of water movement through the soil. If one can calculate from the known texture of a soil its available moisture capacity (the amount of water between field capacity and the wilting percentage), and if one knows the rainfall and evaporation at the site, one can, making some assumptions, calculate the amount of water available over a set period to move materials down through the soil (Arkley, 1963). Assuming that calcium, as an example, is present in the original soil and that none is being added (probably a gross assumption, as we shall see in section 2.3.2.*d.v*) one can then calculate how long it would take to move the 'calcium front' down through the soil to its present position. Arkley's calculations of the time needed to move calcium carbonate to its present position in late Wisconsin soils fit quite well with the supposed age of the landsurfaces involved.

(iv) B horizons. Below the *A* horizon in many desert soils there are redder horizons known as *cambic* that are enriched in clay. They often have distinct structural development, carbonates have been redistributed, and the original bedding in sediments has been eliminated. In many cases these characteristics are also associated with silica enrichment and this will be dealt with in a succeeding section. In cool deserts organic acids may also accumulate in the *B* (Robin and Bazilevich, 1965).

A distinction can be made between *B* horizons that are enriched in sodium and those that are not. When sodium reaches over about 15 per cent of the ions in the exchange complex the morphology of the *B* is distinctive of the type found in the soils known as *solonetz, solodized-solonetz,* and *solod* (Fig. 2.8). The illuvial horizon in these soils is often enriched in clay as well as sodium, and it has a very characteristic columnar structure, the columns having a polygonal arrangement in plan. Often the upper part of the *B* is enriched in silica (Hallsworth and Waring, 1964). The classic view of these soils is that they have developed from an original saline solonchak soil since the dispersal of the clay in the *A* horizon has meant the loss of clay to the *B*. Another part of the classic concept of these soils is that they are low in exchangeable calcium (Kelley, 1951).

More thorough research into these soils reveals that a different sequence of soil formation must often be postulated. Many people have noted that solodized-solonetz soils occur on quite undulating topography in positions too far above the water-table for the initial development of a saline soil. In addition, White (1961) noted that the parent material on which this kind of soil developed in South Dakota was quite calcium-rich. Local conditions of parent material or of water relations must be invoked. In New South Wales and in Western Australia it appears that there has been very slow addition of sodium to the land-surface over many thousands of years, and that this has been redistributed by drainage and throughflow (Bettenay *et al.*, 1964; Hallsworth and Waring, 1964). A very similar genesis is proposed for Tchadian solodized-solonetz soils (Bocquier, 1968). In South Dakota, and in semi-arid Kenya it appears that the parent materials of the solodized-solonetz soils are rich in sodium (Makin *et al.*, 1969; White and Papendick, 1961). In the South Dakota case, calcium in the bedrock does not seem to affect early profile genesis because of its low solubility at high *p*H. It is released only later as the *p*H falls. In their review of the literature on these soils, White and Papendick (1961) noted that a number of American authorities had been maintaining for several years that saline soils were not necessarily precursors of solonetzic soils. The same opinion is now held in Russia (Tyurin *et al.*, 1960). The distributions of ions and salts in these *B* horizons can be explained with Yaalon's model. Such an explanation was used by Mueller (1963) to explain the distribution of salts in the Chilean nitrate fields, where Cl^- ions were the lowest in the profile. The clay-enriched *B* horizon of solonetz and solod soils is perhaps easier to explain than clay enrichment in other soils, but its origin must still be in doubt, as we shall now see in our discussion of clayey *B* horizons not enriched in sodium.

Considerable argument still surrounds the clayey *B* horizons that are found in desert soils. Nikiforoff (1937), in his classic introduction to desert soils, reasoned deductively that since there was not enough moisture to move clays downward in desert soils, and since the high base status of most of the soils meant the clays could not normally be dispersed, illuviation was impossible. Dispersal under sodium enrichment was a special case.Barshad (1955, p. 40) also maintained that 'low rainfall, poor plant growth and active plant decomposition increase the electrolyte content, induce flocculation and hinder clay migration'. Nikiforoff maintained that clayey *B* horizons could be explained if there was a zone beneath the surface in which moisture was retained more easily than at the surface, and that here one would have more weathering and clay formation. We have already pointed to the evidence that, in sandy soils at least, there is considerable evidence that subsurface horizons do stay wet longer than the surface (e.g. Bagnold, 1954; Cvijanovich, 1953; Williams, 1954). Buol (1965), while acknowledging such evidence, pointed out that there is no evidence that this wetter horizon corresponds with the zone of clay enrichment. Indeed he cited evidence for a wet horizon above of the clayier *B*. Another approach is to calculate gains and losses from the parent material in each horizon. Barshad (quoted by Buol, 1965) found one profile which seemed to have suffered illuviation while another had not. Oertel (1968) more recently found that the bulk density of the *B* in Australian soils was lower than that of the *A*, a finding that could not have been made had there been inwashing of clay into the *B*. Studies of the distribution in the profile of very slightly soluble elements, such as zircon, confirmed Oertel in his opinion that there was no evidence for illuviation even in a solodized-solonetz in New South Wales. Ruhe (1967) made some similar calculations for soils in New Mexico and came to the same conclusion.

The argument has not been helped by the definition of an argillic horizon in the U.S. Department of Agriculture 7th Approximation Soil Classification (1960). Clay skins, as seen in this section, are there taken as an indication of clay illuviation. But Buol and Hole (1961) showed that clay skins and orientated clay could very easily be disturbed by 'pedoturbation' in soils and that, where montmorillonites (typical of desert soils) were found, shrinkage and swelling would probably easily destroy clay orientations. Nettleton *et al.*, (1969) have confirmed this, especially for fine soils with swelling potentials greater than four per cent.

Not surprisingly, the answer to the problem of clayey *B* genesis probably lies somewhere between the ideas of the two opposing schools. In some Arizona soils it has been shown that there is evidence in the same profile for both clay illuviation and weathering *in situ*. CaO/ZnO_2 ratios show there has been maximum weathering at the surface in the profiles, and thin sections show some evidence of downwashing, but there has also been weathering at depth (Smith and Buol, 1968).

Whatever the answer to the problem of contemporary *B* horizon formation, it is likely that many of the more distinctive *B* horizons were formed in climates that were wetter than the present. This is maintained by Gile (1967 and 1968) for a series of soils in New Mexico and by Brewer (1956) in Australia. Gile and Hawley (1969) also maintained that there is no evidence for the formation of argillic *B* horizons in contemporary calcareous soils in deserts.

(v) Horizons of carbonate accumulation. In the soils of the semi-arid margins of the deserts there are commonly horizons below the *B* horizon that are enriched in carbonates, particularly calcium carbonate (Plates 2.10 and 2.11). The literature on these horizons is enormous, and the number of local names quite alarming, as the extensive discussion of the topic by Goudie (1970) showed. It is perhaps more surprising that descriptions which are widely separated spatially, temporally and by linguistic barriers should often be so essentially similar. There are close correspondences, often without cross-reference, between photographs, sketches, analyses and even colour names. A widely accepted terminology is, however, lacking; in the descriptions which follow we shall adopt the terms proposed by Gile and his co-workers (1965), and shown in Fig. 2.15. The terminology depends on the recognition of a '*K*-fabric' defined as 'fine grained carbonate which occurs as an essentially continuous medium', siliceous particles appearing to be suspended in and separated by this medium. The definition is independent of carbonate content; the lower limit, at which the continuity of the *K*-fabric is broken, occurs anywhere between 15 and 40 per cent $CaCO_3$, depending on the texture of the parent material (Gile *et al.*, 1965).

Horizons above and below the *K* horizons, in which there is less than 50 per cent *K*-fabric, are known as either the B_{ca} or the C_{ca}. Here the carbonate content is usually below 20 per cent and carbonate occurs in forms varying from thin, soft filaments to sizeable nodules. Below the B_{ca} horizon, in a type-profile, the *K* horizon is divided into three zones: an upper transitional *K*1 horizon, the carbonate horizon proper, or *K*2, and a lower horizon transitional to the C_{ca}, the *K*3. Very hard and indurated, massive *K*2 horizons are designated *K2m*. A very distinctive banded horizon often occurs above the *K2m*, and is known as the *K*21*m*; it appears to be very similar to the *croûte zonée* or *dalles* of North Africa described by Boulaine (1961), Coque (1962), Estorges (1959) and Ruellan (1968).

The *K*2 horizons of New Mexico were found to have carbonate contents ranging between 23 and 60 per cent (Fig. 2.16; Gile *et al.*, 1966). The *K*21*m* commonly has contents of 75 per cent in both New Mexico and in Tunisia (Coque, 1962; Gile *et al.*, 1966). In Texas, Reeves

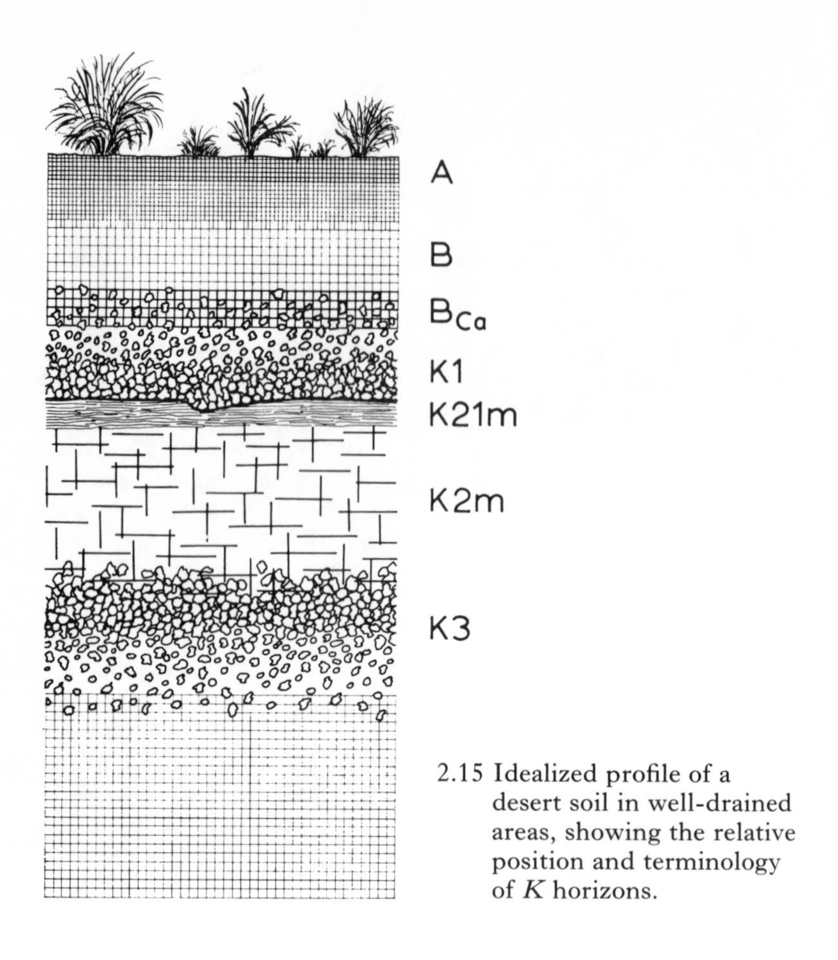

2.15 Idealized profile of a desert soil in well-drained areas, showing the relative position and terminology of *K* horizons.

and Suggs (1964) found some concentrations of up to 91 per cent. Colours vary, but Brown (1956) and Coque (1962) described independently the *K*2 as 'salmon'-coloured. An extensive summary of the literature on chemical content and mechanical properties of carbonate horizons, particularly in southern Africa, can be found in Goudie (1970).

The close similarity in the descriptions of the morphology of *K* horizons is not paralleled by agreement on genesis, even within one region. The various theories have been discussed by Brown (1956) in North America and Coque (1962) in north Africa. However the last decade has seen the crystallization of opinion, and most authorities agree that, for the most part, *K* horizons have been formed by the downward eluviation and deposition of carbonates in an essentially pedological process; but some major puzzles remain to be solved.

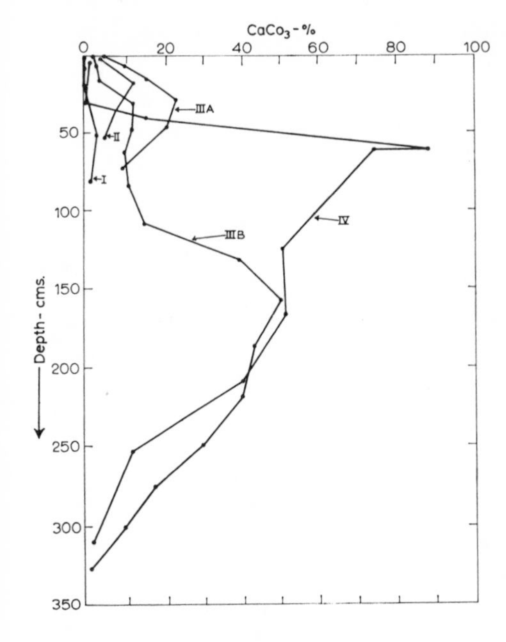

2.16 Calcium carbonate concentration with depth in some New Mexico soils. Roman numerals refer to stages of *K*-horizon development as explained in the text (after Gile *et al.*, 1966).

The strongest support for the eluviation hypothesis comes from the position of carbonate horizons in soils. They usually occur at nearly constant depths in the profile, despite undulations in the landscape (Coque, 1962; Gile *et al.*, 1966; Ruhe, 1967; Stuart *et al.*, 1961). In most cases the correspondence between the relief of the *K* horizon and the surface relief is closer than it would be if the carbonate has been deposited at or near a water-table.

Further support for the eluviation hypothesis comes from the association of apparent sequences of *K*-horizon development with datable sequences of geomorphic surfaces. In New Mexico a flight of piedmont surfaces provides an excellent example of such a sequence. In the lower horizons of soils on the younger surfaces there are only thin carbonate coatings on pebbles; as the surfaces become older, *K* horizons become evident, and they thicken, become more massive and are finally capped by a *K*21*m* laminar horizon. The growth of the *K*21*m* may even lead to the forcing upward of the overlying soil (Gile *et al.*, 1965; Gile, 1967; Ruhe, 1967). A more precisely dated sequence from New Mexico is discussed in section 2.3.2.*d.vii*. Similar sequences have been described in Texas (Hawker, 1927), in Morocco (Ruellan, 1968), in the Sudan (Buursink, 1971), and in South Africa (Goudie, 1970). The general sequence is illustrated in Fig. 2.16. The *K* horizon grows upward into the *B* horizon after the lower horizons have been made impermeable by carbonate plugging. The evidence for this process is the high silicate clay content of the clay included in the upper *K*2 which appears to be the remnant of a former *B* horizon (Gile *et al.*, 1966; Gile, 1967). Later stages in New Mexico include the development of multiple *K*21*m* horizons.

Climatic associations also provide evidence to support the eluviation hypothesis. *K* horizons appear to be associated with semi-arid areas. It is in semi-arid rather than in drier areas that organic acids are available to facilitate carbonate mobility (Rodin and Bazilevich, 1965). In southern Africa, a limit to the occurrence of *K* horizons can be found at the 50·8 cm rainfall and 127 cm free surface evaporation isolines (Goudie, 1970). In the Sudan, only soils with rainfall less than 750 mm have carbonate horizons (Buursink, 1971). In Tunisia, a rainfall between 250 mm and 600 mm appears to be the optimum for calcrete development (Coque, 1962). In Syria, *K* horizons are characteristic of the 600–100 mm rainfall range (Wirth, 1958). In Israel, *K*-horizons appear to be optimally developed at 350 mm precipitation (Yaalon, personal communication). However *K* horizons are frequently found in drier areas and this is seen to be the result of climatic change (Bryan and Albritton, 1943; Capot-Rey, 1943; Coque, 1962; Daveau, 1966; Estorges, 1959; Meckelein, 1957; Ruellan, 1968; Ruhe, 1967). In Texas most of the *K* horizons are seen as having been developed in the Tertiary (Brown, 1956).

The climatic evidence is further supported by the fact that *K* horizons become thicker and more diffuse and occur lower in the profile as rainfall increases (e.g. Arkley, 1963; Jenny, 1929; Ruhe, 1967). This supports the widely-held view that carbonates are deposited at the normal maximum depth of percolating rainwater (section 2.3.2.*d*). The invocation of other depositional mechanisms such as down-profile changes in bulk density, CO_2 pressure, pore-water pressure, or *p*H, does not invalidate the eluvial hypothesis.

There are problems in accepting the eluviation hypothesis without reservations. Radiocarbon dating of the carbon in the *K* horizon, for example, does not always show a consistent picture. In one Arizona profile, the B_{ca} horizon had a date of 2,300 B.P., whereas an underlying *K* horizon (beneath a distinct discontinuity) had dates ranging from about 9,800 B.P. at the top to 32,000 B.P. at the base (Buol and Yessilsoy, 1964). This was interpreted to mean that, whereas the B_{ca} may be pedological, the *K* is not connected to the present landsurface and may even be considered as a geological stratum (Buol, 1965). Some datings in New Mexico, on the other hand, lend support to the eluviation hypothesis. The radiocarbon age of the carbonate in one profile becomes younger from the *K2m* upward to the uppermost lamina of the *K21m* laminar horizon. Other datings, however, produced a more confused pattern, which could only be explained if deposition in each horizon was supposed to vary in age (Gile *et al.*, 1966). Ruhe (1967) was sceptical of all radiocarbon datings of *K* horizons. Some of Ruhe's dates were, he said, 'nonsense'. Many *K* horizons have been seen as being of different ages in different zones of the profile. Coque (1962) and Daveau (1966) in north Africa join Buol and Yessilsoy (1964) in the United States in this opinion.

A more serious problem is the source of the carbonate itself. One seems to have to postulate an impossible thickness of overlying leached horizons to yield the amounts of carbonate in many *K* horizons. Moreover, many *K* horizons occur in parent materials that are quite acid (Brown, 1956; Coque, 1962; Crocker, 1946; Gile *et al.*, 1966). It has also been pointed out that the amount of weathering necessary to release the carbonates would have released appreciable amounts of clays at the same time, a fact for which there is no evidence (Ruhe, 1967). Solutions to the problem in some cases can be sought in downslope movements of carbonate (e.g. Gigout, 1960), or in the addition of carbonates from percolating or flooding streams (e.g. Glennie, 1970; Motts, 1958), but these are not general explanations.

A more popular explanation is the addition to the soilsurface of carbonate-rich dust which, being unstable in the *A* horizon, is translocated to the *K* (Bettenay and Hingston, 1964; Brown, 1956; Coque, 1962; Gile *et al.*, 1966; Hutton, 1968; Jessup, 1960*a* and *b*; Ruhe, 1967; Tsyganenko, 1968). Direst evidence that carbonate dust in sufficient quantity is being added to soilsurfaces is not hard to find (e.g. Ruhe, 1967). High percentages of carbonate were also found in dusts in Victoria (Hutton and Leslie, 1958). It is less easy to see from whence this carbonate dust is derived. In southern Australia it is thought to be from the Tertiary limestone on the Nullarbar Plain or from earlier carbonate soils (Jessup, 1960*a* and *b*); in Victoria it is certainly terrestrial, because it far exceeds the proportion of carbonate in sea-water (Hutton and Leslie, 1958); in Tunisia, Coque (1962) favoured the dust hypothesis. Organic sources in leaf-litter may be a minor contributor (Rodin and Bazilevich, 1965; Teakle, 1937).

Another puzzling feature of *K* horizons is their frequent multi-sequal nature and their great thickness (up to 50 m). In New Mexico, where the *K* horizons are probably no older than mid-Pleistocene, and in Texas, where they are probably Tertiary in age, an explanation of multi-sequal profiles is that older horizons have been buried and new horizons have developed on the overlying new soil (Brown, 1956; Ruhe, 1967).

Sequences of this kind on Australian dunes can be seen in terms of repeated phases of dune growth (Churchward, 1961). Some bisequal profiles however are seen as the result of two different processes—there having been downward eluviation to the upper horizon, and emplacement from a water-table to the lower (Buol and Yessilsoy, 1964; Buol, 1965; Coque, 1962; Hunt and Mabey, 1966; Ruhe, 1967). In New Mexico multiple *K*21*m* horizons appear only in older soils (Gile and Hawley, 1969).

The character of the laminar horizon is another problem. In west Texas many workers were of the opinion that this was an algal limestone (e.g. Reeves, 1968). Although in some places evidence clearly supports a freshwater source from former lakes (Motts, 1965; Williams, 1966), this seems unlikely over large areas. Other earlier explanations included surface-water flushing to induce travertine-like deposition (e.g. Estorges, 1959). In central Australia some travertines have undoubtedly been deposited around former artesian springs (Wopfner and Twidale, 1969).

Brown (1956) from his experiences in Texas and Coque (1962) in Tunisia were both of the opinion that the *K*21*m* was an horizon of aeolian deposition. They cited its blanket-like occurrence, the well-sorted nature of the included siliceous matter and the banding. Brown (1956) extended his explanation to the whole of the *K* horizon, explaining the difference in structure by different rates of lithification: fast deposition and lithification would give the *K*21*m* character. Some surface horizons of carbonates in South Australia have been seen as the result of aeolian addition (Jessup, 1961*a* and *b*). On the other hand, Gile and his co-workers (1965 and 1966) are of the opinion that plugging of the subsoil by *K*2*m* growth forces percolating water to migrate laterally and deposition from this leads to lamination. A similar explanation has been evolved independently by Ruellan (1968) in Morocco.

The chief alternative to the downward eluviation hypothesis is the *per ascensum* hypothesis in which carbonates are seen as having been deposited by evaporation from a water-table. Although less popular, this hypothesis still has supporters, and it may indeed be applicable in some areas. In Tunisia the *encroûtements* or thicker, less dense *K* horizons are attributed to such a source (Coque, 1962). Hunt and Mabey (1966) used this hypothesis in Death Valley; Ruhe (1967) has used it in New Mexico; Wirth (1958) used this explanation in Syria; and Netterberg (quoted by Goudie, 1970) discussed the idea as applied to South African *K* horizons. Calcreted and upstanding river channels in Arabia probably owe their $CaCO_3$ content to river water (Glennie, 1970).

Post-depositional lithification processes are evidently not disputed.

Percolating rainwater and dew redissolves and redistributes the carbonate (Brown, 1956; Coque, 1962; Gigout, 1960; Gile *et al.*, 1966; Ruhe, 1967).

(vi) Other illuvial horizons. Gypsum ($CaSO_4.2H_2O$) is another salt which forms soil horizons in semi-arid areas. Jenny (1929) noted that gypsum characterizes more arid areas than calcium carbonate, and this is substantiated by Coque (1962) in Tunisia. Gypsum is also more restricted in its occurrence in toposequences, being confined to such places as the low areas around the oases and chotts south of the eastern Atlas Mountains in north Africa, and to basins and 'slick-spots' in the United States. Widespread 'gypcrete' horizons are found south-west of Lake Eyre in South Australia (Jessup, 1961; Wopfner and Twidale, 1967) and in the northern part of the Great Eastern Erg in Algeria. Bellair (1954) has described a zone of some 10,000 km^2 where 'gypcrete' horizons occur at thicknesses between 30 cm and 1 cm. Gypsum can form mechanically strong crystals. It is found therefore as sand-sized particles in some arid areas and as such it can be blown to accumulate as dunes (Bettenay, 1962; McKee, 1966; Trichet, 1963).

In north Africa, gypsum soil horizons occur in the same form as carbonate horizons. Around the oases, particularly of El Souf, gypsum crystallizes in the sand dunes as a massive horizon of beautifully interlocking twin crystals or sand-roses (Bellair, 1954). The deposit is known locally as *deb-deb*. It is generally agreed that *deb-deb* is emplaced from a gypsum-rich ground-water (Coque, 1962), whereas purer gypsum with virtually no sand included is associated with ancient lake deposits (Bellair, 1954).

A *croûte gypseuse zonée* occurs, and it is analogous to the *croûte calcaire zonée*. It does not occur in the same areas as the *deb-deb*, but as a blanket that covers small *nebkha* (section 4.3.3.*g*) on the northern margins of the Great Eastern Erg and neighbouring pediments and fans. Its most obvious distributional association is with the great chotts around which it occurs in an annular zone. The mechanism which is invoked to explain this *croûte* is similar to that for the *K*21*m* (Coque, 1962). Gypsum seems to be attributable to the chotts and is blown on to to an undulating surface. As it accumulates, vegetation is eliminated, until a balance is achieved between sparse vegetation and the additions and removals of gypsum dust from the surface. Because it is less soluble than other salts, gypsum in crusts is more stable, but there is probably a slow *cycle de gypse* back to the chotts.

Extensive gypsum horizons are also found in northern Iraq, Arabia and northern India (Glennie and Evamy, 1968; Smith and Robinson, 1962) and in many other arid-zone soils. It is thought that in many of these areas sulphur is accumulated in plant roots and that this accumulation helps gypsum formation, leaving a characteristic pattern of root-channels (known as *dikaka*) in the resulting formation (Glennie and Evamy, 1968). These patterns have also been noted in the *deb-deb* of North Africa (Coque, 1962) as in calcareous horizons in Algeria (Boulaine, 1961).

Silica-enriched horizons are another peculiarity of some desert soils. They are reported from arid and semi-arid areas (e.g. Flach *et al.*, 1969; Hallsworth and Waring, 1964), but they have not received the attention paid to carbonate or even to gypsum enrichment.

Silica enrichment takes a variety of forms. At one extreme there may simply be enrichment of an horizon with very small silica particles (Hallsworth and Waring, 1964). At the other extreme there may be complete cementation of an horizon with a 'sponge-like' matrix of silica, or a complete matrix surrounding other particles similar to the *K*-fabric (e.g. Mabbutt, 1967). The cementing silica may be, by weight, a very small percentage of the horizon (Flach *et al.*, 1969). The silica is often described as chalcedonic or opaline. Sometimes silica cements part of a calcrete horizon, and may even have replaced the carbonate in some cases (e.g. Goudie, 1970). Some of these forms occur in contemporary soils whereas others, and perhaps all the thicker silcretes, are thought to be palaeosols (e.g. Langford-Smith and Dury, 1965; Mabbutt, 1967). We shall discuss the palaeosols in section 2.3.2.*f*.

The association of silcrete with arid environments seems to be agreed but the genetic meaning of this association is not entirely clear. Several authorities have noted that in alkaline environments, particularly where the exchange complex is saturated with sodium, silica should be especially mobile. However, the form and occurrence of the horizons seems to indicate that there is more to it than this. In the upper parts of the *B* horizons of solodized-solonetz soils in arid New South Wales the silica occurs as well-rounded silt-sized particles that are remarkably similar in appearance to the form in which silica is found in the leaves of local grasses. Because of this and because of the high silica contents in the live parts of these grasses, Hallsworth and Waring (1964) have suggested that the grasses must be the origin of the silica. It is widely observed that desert grasses do have exceptionally high silica contents and that silica is commonly the most important element in circulation in the desert ecosystem (e.g. Rodin and Bazilevich, 1965). The association of silica enrichment with sodium in the soil seems widespread, and silica gels may in fact lead to sodium retention, though the causal chain is more commonly thought to be first alkalinity and then silica movement (e.g. Bettenay and Hingston, 1964; Lewis and White, 1964). The occurrence of silica cement in some of the horizons of soils on alluvial fans and other formations in Western Australia suggested to Litchfield and Mabbutt (1962) that there might be an association between sheet-flooding and the silcretes. There appears also to be an association between silcretes and certain parent materials. In the western United States volcanic rocks seem to have soils with well-developed silica-cemented horizons even after a relatively short period of time, whereas on alluvial soils derived from granitic rocks, silica cementation seems to take much longer to develop (Flach *et al.*, 1969). Parent material association has also been noted in the case of supposedly ancient silcretes in central Australia (Mabbutt, 1967). The depth to hardpan was observed in Western Australia to be greater beneath more permeable soils (Litchfield and Mabbutt, 1962). They also observed that since the horizons were not very deep below the surface of aggrading alluvium, they must be forming at a similar rate to aggradation.

(vii) Soil profiles and soil-forming factors. The profile characteristics of desert soils have commonly been linked to climatic factors. The early Russian pedologists noted that the *A* horizons of desert soils were redder in drier areas and attributed this to decreasing amounts of organic matter. Marbut and the staff of the United States Department of Agriculture (1935) made colour the chief criterion for the classification of the soils of arid areas: they distinguished a range of soils from 'red desert' in arid areas (with 'serozems' in cooler deserts), to 'chocolate brown' and 'chestnut brown soils' on the semi-desert margins. Arkley (1963), and Jenny (1941) noticed that the horizon of carbonate concretions became deeper towards the wetter margins of desert areas (Fig. 2.18a).

Some circularity of argument surrounded the association of soil types with climatic parameters. Stephens (1946) noted that there were fairly distinct rainfall isohyets bounding various mapped soil types in Australia. Prescott (1949) plotted the positions of soil types as recognized in Australia against two climatic parameters (rainfall *P*, and saturation deficit *E*, a measure of evaporation). He found there were distinct patterns, with soils which had been described as 'desert soils' falling into particular groups on the diagram (in areas where *E* was between 0·20 and 0·60, and *P* was between 10 and 45 cm). This may show no more than the soil surveyors' awareness of their environment when they described the soils. A slightly more rigorous test, expanding on Prescott's ideas of water movement through the soil, has been made by Arkley (1967). Arkley used the soil profile data taken from a generalized soil map of the western United States, and the climatic figures for several stations in this area. For each station he computed three synthetic climatic

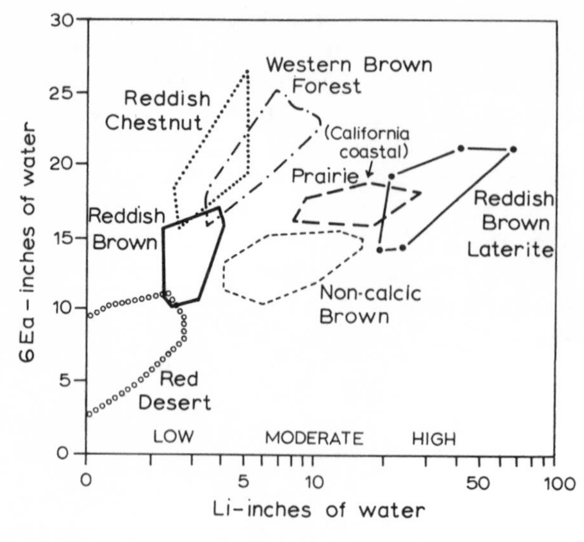

2.17 The climatic relationships of desert soil types. 6Ea is a measure of the evaporation from the soil assuming that it holds six inches of water, and Li is a leaching-effectiveness index or measure of the amount of water entering the soil (after Arkley, 1967).

parameters: L_i, a leaching effectiveness index which is a measure of the amount of water penetrating the soil surface; $6E_a$, an estimate of the amount of evaporation from a soil assuming that it can hold 6 inches of rainfall; and T, the mean annual temperature. Some of Arkley's results are shown in Fig. 2.17. For the three temperature ranges the soils fall within quite well-defined areas on the plots of $6E$ against L_i. One could conclude from these studies and from the rather similar study by Fränzle (1965) that a very general correspondence has been shown to exist between the described characteristics of desert soil profiles and climatic controls.

Some of the characteristics of salt and ion accumulation in soils are linked to climate, as we have seen. For example, gypsum characterizes drier areas than calcium carbonate. In the U.S.S.R. it is recognized that exchangeable sodium enrichment in the clay complex is characteristic of areas with 300–500 mm rainfall whereas horizons with sulphate and chloride salts are more characteristic of semi-arid areas (Janitzky, 1957, quoted by Buursink, 1971).

The degree of horizon development in desert soils is also related to time. Over the years the characteristics of the sediments or the simple weathered mantle that forms the parent material disappear and are replaced by a set of distinctly pedological features. For soils on well-drained sites we can base our discussion of this development on a study by Gile and Hawley (1969) who worked in southern New Mexico where the rainfall is between 200 and 300 mm. They were working in an area where the sequence of geomorphic surfaces is fairly well understood and were fortunate to find a site where alluvium in various surfaces could be dated by radiocarbon techniques. They extended the sequence back to the mid-Pleistocene by comparing the local surfaces with the surfaces in the surrounding area of the Rio Grande Valley. The results of their study are summarized in Table 2.3. It can be seen that there is an orderly development of soil horizons. It is thought by these authors that clay-enriched *B* horizon development is not taking place in the modern environment and that it must be assigned, where it is found, to a wetter period in the Pleistocene. However, using similar evidence in the same area it has been calculated that *B* horizons take some 9,500 years to develop (Ruhe, 1967). An hypothetical sequence of soil development is sketched in Fig. 2.18b.

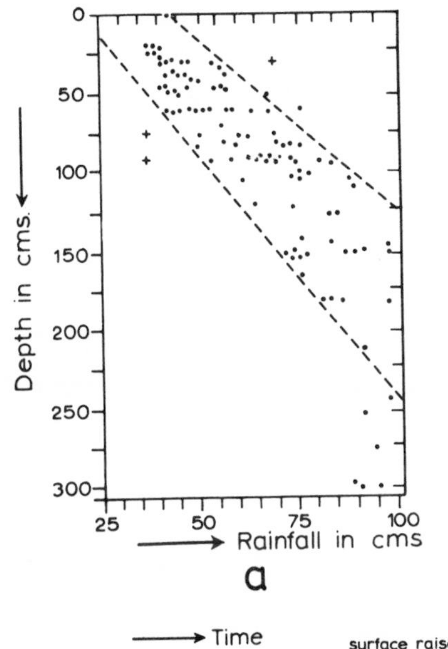

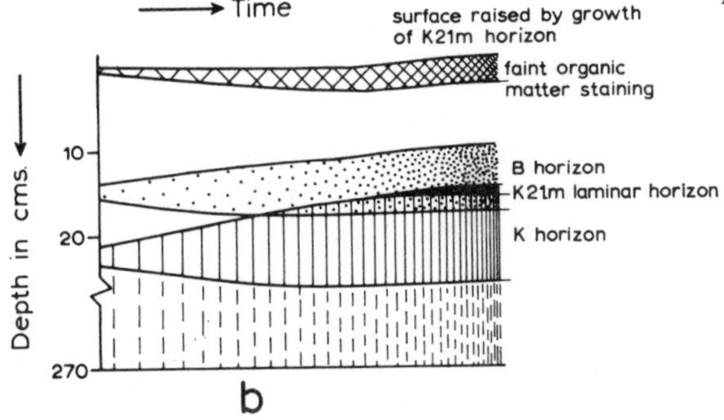

2.18 (*a*) The depth to zone of calcium carbonate accumulation against rainfall for some United States soils (after Jenny, 1941). (*b*) Generalized sequence of desert soil-profile development.

TABLE 2.3 *Soil Ages and Characteristics in New Mexico*

Soil age in years (soils developed in materials of ages)	*Features of soil development*
100 years (?)	Thin grey *A* horizon, vesicular in places; slight organic matter accumulation; original sedimentary bedding still present.
100 (?) to 1,000 years	Very slight carbonate accumulation; most of the original bedding has been destroyed, and only thicker beds of fine earth remain.
1,100 to 2,100 years	No distinct signs of carbonate accumulation in soils with little gravel, but in gravelly soils a weak but distinct horizon of carbonate accumulation in the form chiefly of pebble coatings.
2,200 to 4,600 years	In non-gravel soils a weak carbonate horizon has developed in the form of coatings on structure units, and a distinct structure has developed.
Late-Pleistocene	Orientated clay coatings on grains in a *B* horizon with clay enrichment, and a *K* horizon. A *Km* horizon has developed with a single laminar horizon on its upper edge.
Mid-Pleistocene	The same with a distinct *Km* horizon with two or more laminar horizons (*K*21*m*) on top.

SOURCE: Gile and Hawley, 1969, p. 711

Arkley (1963) approached the study of soil age in a rather different way, as we have seen (section 2.3.2.*d.iii*). His calculations, which fitted quite well with the supposed age of the surfaces on which his soils were developed, showed that horizons of carbonate accumulation had minimum ages of between 4,000 and 15,000 years. This is not incompatible with the figures quoted by Gile and Hawley (1969).

Viewed theoretically, there are salts that will form soil horizons much more quickly than carbonates. We have referred in section 2.3.2.*d.iii* to Yaalon's model of horizon development, in which more mobile salts and ions are moved down to accumulate as soil horizons much more quickly than the less mobile. Chlorine and sulphate ions in

Yaalon's experimental columns moved down in appreciable quantities with fairly heavy water additions. In soils with material available for movement with predominantly downward movements of water and with limited but frequent wetting to about the same depth, one might expect a whole sequence of horizons to appear, some perhaps being dispersed before others arrived, whereas others would be superimposed. Sequences of horizon formation are due to other factors. For example, it appears that *B* horizons with clay accumulation may develop before *K* horizons in New Mexico soils and then be 'swamped' by the latter (section 2.3.2.*d.v*), and silica may accumulate in already formed *B* and *K* horizons.

Upward movement of salts is often much quicker than the processes we have so far discussed. Saline soils can develop with frightening speed. In areas of Pakistan, where the water-tables have been raised over a period of about 40 years, large areas were rapidly leaving production during the 1960s by salinization. A large part of the 40 years was taken up in the slow rise of the water-table and the actual salinization can take less than a season (Hunting Technical Services, 1961).

It appears that different rates of soil development are themselves related to the other factors of soil formation. Sites with large downward movements of water through the profile, as described on the upper parts of slopes in Texas by Goss and Allan (1968), where there is rapid flushing through the soil, will develop more quickly than sites where there is no such throughflow. Sites that have sandy parent materials which allow rapid through-flushing will also develop more quickly than less permeable soils. In the central Sudan many sandy soils of late-Pleistocene age have distinct clayey *B* horizons whereas clay soils have no such horizonation (Warren, 1970). Parent materials rich in mobilizable silica such as volcanic rocks will develop soils with silica enriched horizons quickly, whereas soils on granitic parent materials will only develop them slowly (Flach *et al.*, 1969).

(e) Soil classification in arid and semi-arid areas. Soil formation in arid and semi-arid areas, no less than in other parts of the world, is a complex process. Not only are there many possible combinations of different conditions amongst the basic soil-forming factors, but there are clear indications of secular changes in at least some of the factors, so that the soil-forming environment was probably very different at different periods in the history of almost every landsurface. Soil classification is thus difficult, and clearly the systematization of soil knowledge must follow from, rather than lead to, discoveries about soil-forming processes. Soil classifications are therefore very much creations of a particular

state of knowledge about soils, and are devised simply to hand on the large body of information to the student or the applied scientist.

The earlier classifications of desert soils were based on very inadequate knowledge. Many classifications almost completely ignored desert soils (e.g. the classification of African soils by D'Hoore, 1964, which classified desert soils as 'not-soil', and others which placed disproportionate emphasis on the saline and alkali soils of low, wet sites which are really rather unimportant). Many of the early classifications were 'zonal' in concept, and were inclined to group all desert soils under one heading; but more recent classifications, using as their basis the characteristics of the soil profile rather than of the soil and its environment, have led to the inclusion of desert soils in several soil classes. This is clearly preferable since, if there have been changes in climate in the recent past, there are likely to be many soils in deserts which have inherited their characteristics from prior conditions.

Two classification systems should be mentioned. The soils of arid areas have been included in a number of the primary groups of the U.S. Department of Agriculture 7th Approximation Soil Classification (1960). The system has been expanded for use in arid areas by Flach and Smith (1969). A new classification, which nonetheless uses soil terms and concepts that are more familiar than those in the 7th Approximation, is the FAO/UNESCO legend for the soil map of the world (e.g. Bramao, 1969).

(f) Palaeosols in deserts. Because of their interest in palaeoenvironmental reconstruction, many geomorphologists have paid more attention to ancient than to modern soils in deserts, and where soil scientists have worked closely with geomorphologists, they too have been attracted to the study of evidently ancient formations, because of their importance to the geometry of the landscape and therefore to the pattern of modern soils. There can be few soils that have not received the accolade of palaeoenvironmental speculation, as our discussion has shown, but fruitful speculation must feed on knowledge of processes acting under known conditions at the present day. The slowness of many of these processes and therefore the imperfection of the evidence, however, leads the investigator directly into a vicious circle of speculation. We have briefly touched upon some of the speculation in the preceding sections, so that we need here only briefly outline some of the generalities concerned with palaeosols.

We can divide the problem into two: there are weathering and soil formations in deserts on a large scale which appear to have only peripheral connection with modern pedogenesis; and there are a number of features in rather shallower profiles that can more easily be confused with contemporary soil formation.

Deep, horizonated weathering profiles are found on rocks in most of the more stable deserts. In Australia, the fresh rock is typically overlain by 17–33 m and even up to 83 m of altered material (Langford-Smith and Dury, 1965). Towards the top of this profile there is often found a 'duricrust' cemented with either silica, sesquioxides (particularly iron), carbonates or gypsum; the hardness of this horizon means that it stands out prominently on eroded slopes, and protects the softer material beneath from erosion. In an extensive summary of duricrusts, Langford-Smith and Drury (1965) noted that the duricrust itself may be from 1 to 13 m thick. Silica duricrusts are between 0·5 and 3 m thick in South Australia and between 1·5 and 9 m thick in central Australia (Mabbutt, 1967; Wopfner and Twidale, 1967). The duricrust is surmounted by a thin layer of material in which the modern soil has developed and is often underlain by a 'mottled' zone in which nodules of the cemented material are found in a softer, paler matrix, and by a 'pallid' zone which is pale and virtually without cementation. This may directly overlie the unweathered rock or may be underlain by variable depths of altered but less-leached material. These profiles appear to be very widespread in Australia (e.g. Mabbutt, 1967; Mulcahy, 1967; Wopfner and Twidale, 1967). They are virtually universal in the West African savanna and sahel zone (e.g. Dresch and Rougerie, 1960; Michel, 1969; Tricart, 1959; Urvoy, 1942), in the Sudan (e.g. Sandford, 1935; Delaney, 1954) and in the Kalahari (e.g. Flint, 1963). Deep-weathered profiles without duricrusts are also reported in the central Sahara (e.g. Rognon, 1961).

The great age of these formations is attested by several features. In semi-arid New South Wales they have sometimes suffered folding and are therefore probably of late Tertiary age (Langford-Smith and Dury, 1965). In other parts of central Australia they are overlain by more recent formations and the profiles often occur beneath ancient landsurfaces that have since been considerably dissected (e.g. Jessup, 1960*a*; Mabbutt, 1967). In Western Australia they can be seen to occur in a series of consecutive surfaces (e.g. Mulcahy, 1961). In the Ahaggar Mountains of the central Sahara they are overlain by more recent deposits (Rognon, 1967). In the southern Sahara, in the Sudan and in the Kalahari the ancient horizons are frequently covered by fixed sands (e.g. Goudie, 1970; Sandford, 1935; Warren, 1970). On the edges of the Ethiopian Massif they have been covered with lavas (Delaney, 1954).

The specific character of these duricrusts has been given palaeoclimatic significance by many workers. It seems to be generally accepted that lateritic duricrusts (in which iron is the most characteristic cementing agent) are indicative of a tropical savanna climate (e.g. Dresch, 1960; Flint, 1963; Tricart and Cailleux, 1964). The lateritic profiles indeed seem to be confined to the southern parts of the Sahara and to the northern and eastern parts of the Australian Desert (e.g. Dury, 1966*b*; Mabbutt, 1967; Michel, 1969; Urvoy, 1942) into which there may have been incursions of savanna climates in the past.

Silcretes seems to be associated in Australia with drier climates than laterites (e.g. Dury, 1966*b*; Mabbutt, 1967). Mabbutt and Dury imagined that silica became more mobile in drier climates, and so replaced iron as the chief translocated cementing agent, but it has been suggested that silica may actually have been mobilized in lateritic uplands and then moved considerable distances downslope into basins to form the silcrete there. The review by Langford-Smith and Dury (1965) reveals the widespread belief in Australia that the silcretes post-date the laterite and have replaced it in some areas. Mabbutt (1967) suggested this in central Australia, though elsewhere the evidence is far from clear. Langford-Smith and Dury (1965) quoted several authorities who associate silcretes with a lower zone in the profile or in a toposequence and laterite with a higher zone. It has also been observed that silcretes are associated with fairly siliceous rocks, and laterites with granites (e.g. Mabbutt, 1967).

Horizons cemented with carbonates have been very widely viewed as palaeofeatures (section 2.3.2.*d.v*). The widespread calcretes in the High Plains, particularly of Texas, are almost universally recognized to be of Tertiary age (Brown, 1956). Coque (1962), working in Tunisia, was of the opinion that the *encroûtement* type of *croûte calcaire* had been formed when the climate was wetter than it is now. Ball (1939), Capot-Rey (1947), Schoeller (1945), and Daveau (1966) saw calcareous tuffs in drier parts of north Africa as evidence of a wetter climate, and Flint (1963), in referring to similar features in southern Africa, was of the same view. The calcretes that have been replaced with silica in the Namib Desert must also be very old (e.g. Goudie, 1970). Gypsum crusts have been assumed to be ancient in central Australia mainly because of their topographic position (Wopfner and Twidale, 1967).

Both Tricart and Cailleux (1964) and Langford-Smith and Dury (1965) suggest that duricrusts change their character across the deserts from bauxitic in the wet tropics through lateritic and siliceous to calcareous in the poleward desert margins.

Whereas there can be little argument about the great age of some of these massive horizons, there is much more dispute about rather less distinct features in the soil. We have mentioned the arguments surrounding the palaeoenvironmental interpretation of red soils (section

2.3.2.*c*), clayey *B* horizons (section 2.3.2.*d.iv*), and *K* horizons (section 2.3.2.*d.v*).

The resolution of these problems depends upon which properties of the soil are chosen as indicators. In the first place, one cannot use as a palaeoclimatic indicator a property that is known to develop rapidly in the soil, or one that is ephemeral or liable to quick destruction. Most soils develop their major characteristics relatively quickly (e.g. Dickson and Crocker, 1957), and this is probably as true for desert as for more humid soils (section 2.3.2.*d.vii*). Thus a change in environment is likely to lead to a rapid change in some of the soil features and an eradication of some of the evidence.

Secondly, there are characteristics of the soil that will survive a climatic change in one direction but not another, and soil characteristics that will have the opposite property. For example a *K* horizon, or a reddened horizon is likely to survive well in a drier climate than the one in which it was formed, but less well in a wetter one. Organic matter will adjust to new conditions, whether they are wetter or drier.

Thirdly, one must distinguish between upward and downward translocation. If an horizon develops only by downward movements of material in the profile, it will not remount the profile. Thus a carbonate horizon well above the water-table, if moved down a profile by a wet period, will not move up again in a dry one, although it may grow upwards by further translocation. Since there are likely to have been several climatic shifts in the recent geological past, the depth to carbonate horizon may have little climatic significance. The fact that carbonate horizons occur deeper in the profile in wetter areas today, probably means only that relative positions in the climatic gradient have been preserved throughout the Quaternary.

Fourthly, one must always distinguish between micro-climates and macro-climates. This confusion has been most obvious in the case of red coloration, where the refusal to accept that red beds could form in deserts was a result of ignorance about the micro-climate within some desert soils and particularly within sandy soils.

Fifthly, the examination of the time element in desert soil formation needs to be much more careful. It may well be that many of the translocational and transformational processes in deserts are the same in their basic pattern as those in more humid soils, but take place much more slowly. The varying rates of desert varnish formation and of red bed formation have been mentioned. On very ancient stable surfaces in deserts there may be time to develop soil features that are formed much more quickly in more humid areas. The very long period postulated for the development of solodized-solonetz soils in New South Wales is a case in point (Hallsworth and Waring, 1964). Accession of salt to soils in wetter parts of the same State has produced alkaline soils in very much shorter time (Collis-George and Evans, 1964). Over long periods of time the depth to the top of an illuvial horizon may progressively increase. The depth will probably be related to the maximum inflows of water and, as the time period increases, larger inflows with longer and longer return periods will be involved, so that ancient horizons will be deeper for this reason as well (Arkley, 1963).

Sixthly, and perhaps most appropriately in a discussion of the geomorphological importance of soils, is the complication of soil and landscape development. New surfaces are constantly appearing in deserts either by erosion or by deposition. The complexity of the pattern that results from this in alluvial areas of New Mexico has been studied by Gile and Hawley (1966 and 1969). In the Sudan, younger soils that have not been leached and are therefore relatively alkaline have been found on a series of relatively young fixed dunes. On more ancient ones there is found a deeper acid soil with signs of an argillic *B* horizon. Whether this argillic horizon is the result of a wetter climate in the past, or whether it is the result of processes that are active today but which have acted over long periods, is a difficult problem to resolve (Warren, 1970). In Israel the interdependence of soils and landscapes has been stressed by Dan and Yaalon (1968). They term these inter-related forms 'pedomorphic-surfaces'. It is to these inter-relationships we now turn.

(g) Geomorphic importance of desert soil formation. The soil mantle can be viewed as part of the morphogenetic system of deserts from two vantage points. The first and more general, in which the soil is seen as a direct contributor to the system of desert surface degradation, has been examined in the first section of this book. The second, in which the processes of soil profile formation are seen as interacting with this system, is the subject of this short discussion.

The ways in which the soil mantle in general can interact with the denudation processes of the landscape have been examined by Dan and Yaalon (1968) and by Beckett (1968) in a review and extension of the *Aufbereitung* concept of Walther Penck. Beckett's ideas can be considered in the context of the desert landscape.

If the early stages of soil development are not hampered by surface erosion, a weathered mantle can develop and deepen to a point where further deepening is very slow, and the deepening then continues at more or less the same rate as the slow surface erosion. In these conditions, a soil profile will develop.

The principal processes of soil profile development are the translocation of transformed materials from one horizon to another. In the present context, the importance of this is that the permeability and the mechanical strength of the various horizons is altered by these migrations. In general, the *A* horizon is weakened since it is more highly weathered and since it tends, however slowly, to lose binding and flocculating agents such as the carbonates, iron and clays. The case of soil in which there is upward migration of salts is rather different; here translocated salts either lead to a loose, puffy surface easily disturbed by wind erosion, or to a firm, moist surface (if deliquescent salts are concerned) which will strongly resist soil blowing. Water erosion is not very active on the low angle slopes of saline soils.

In most cases the upper horizons of the soil are mechanically weakened, and made more permeable. These two processes have different effects on mechanical resistance. In arid and semi-arid areas where rainfall may be intense, the results of the development of the *A* horizon may lead to faster erosion by slopewash and gullying. Permeability is of less importance to wind erosion, but the effects of weathering should lead to increased vulnerability of the surface to wind erosion.

Weathering is accompanied, especially where it is more pronounced on the wetter fringes of deserts, by the accumulation of organic matter in the surface horizon: plant roots tend to bind the soil, and algae and bacteria tend to form resistant crusts. These two processes may have opposing effects. Plant-root binding would lead to a more resistant surface, and plants themselves cut down windspeeds near the ground, whereas bacterial crusts, although binding the soil, would lead to greater surface runoff, and therefore to greater erosion by wash and by gullying.

The later stages of *A*-horizon development are more difficult to speculate about. Soils might be more acid and less cohesive in the later stages of ecological succession, but the biological community would then be more adapted to the environment and better able to resist disruptive events. Speculation about changes in resistance of desert surfaces with increase of soil age are very tenuous; the question is, of course, highly academic, since very few desert surfaces remain undisturbed for long enough for clear answers about the later stages of time sequences and biological successions to be resolved. An acceptable hypothesis might be that *A* horizons are weakened to the extent that they become highly unstable and would be eroded if serious outside disturbance such as vegetation removal were introduced. We must balance against this the evidence of the creation of highly stable pavement surfaces (section 2.4).

There is little doubt that most illuvial horizons are considerably strengthened during soil profile development. Material removed from the *A* horizon accumulates in and cements the illuvial horizons. Indeed many *B* and *K* horizons are readily recognized by their induration, and by the fact that they resist erosion when they have been exposed at the surface by wind or water erosion. The decreased permeability of illuvial horizons, often reduced to very low figures (e.g. Flach *et al.*, 1969; Gile, 1966), is usually an unimportant contribution to their erosion-resistance since, although runoff is high from an exposed illuvial horizon, mechanical resistance to erosion is very high in compensation. Several authorities have proposed a sequence of *A* horizon stripping and *B* or *K* horizon exposure at the surface to explain the occurrence of clay-, carbonate- or silica-rich horizons at the surface in deserts (e.g. Durand, 1959 in north Africa; Gile, 1967 in New Mexico; Goudie, 1970 in southern Africa; Jessup, 1961 in Australia).

One interesting effect, perhaps unexpected in deserts, is that soil development in some instances leads to quite marked throughflow of water in the soil. Where the contrast in permeability between the *A* horizon and the *B* horizon is especially marked, water penetrating the *A* will have to flow laterally downslope through the soil over the surface of the *B*. There is strong circumstantial evidence for this process in solodized-solonetz soils, both in Australia, where Hallsworth and Waring (1964) observed a very wet layer over the *B* after rain, and in Tchad, where Bocquier (1968) observed that the top of the *B* was sharpened by throughflow. The laminar character of the *K21m* horizon in calcareous soils has been attributed to such lateral movement over *K* horizons (e.g. Flach *et al.*, 1969).

The probable, and in fact frequently found, result of soil development in deserts is, therefore, first to weaken the soil surface, supplying in this way large amounts of detritus to the aeolian and fluvial erosional systems, and then to provide a hardened carapace which preserves, more or less intact, ancient desert surfaces.

2.4 Particle Concentration: Stone Pavements

2.4.1 PAVEMENT FORM AND STRUCTURE

Stone pavements are armoured surfaces composed of angular or rounded fragments, usually only one or two stones thick, set on or in matrices of finer material comprising varying mixtures of sand, silt or clay (Plates 2.1 and 2.12). They occur on weathered debris mantles, alluvial deposits and soils in many different environments ranging from periglacial (e.g. Tedrow, 1966) and mountain areas to recently dug gardens. But they are extremely common in hot deserts where they adopt a variety of forms and have acquired many local names such as *gibber plains* or *stony mantles* in Australia, *hammada* and *reg* in north Africa and the Middle East, *saï* in central Asia, and *desert pavement* in North America.

Pavements are commonest in areas where particle concentration at the surface is uninhibited by vegetation, and in areas where coarse and fine particles coexist in a deposit. In deserts, they are particularly common where relatively mixed and *generally* unsorted alluvial deposits occur, such as in the areas of smooth, gently sloping surfaces between channels on the upper part of alluvial fans, and on the debris aprons adjacent to mountain fronts. Pavements occur in areas ranging in size from a few square metres to several hundred hectares: the size of a pavement area is a reflection of the nature and extent of sediment types, and the relation of the surface to local and regional particle-sorting processes.

The stone mosaic is composed of coarse particles which are either on, or lie embedded in, the finer matrix. Even in the most developed pavements there are gaps between particles: the density and spacing of particles is extremely variable, reflecting the original composition of the deposit, its degree of superficial disintegration, and the extent to which particle concentration has progressed. In California, Cooke recorded densities up to 1·36 particles per cm^2 and noted that up to 17 per cent of the groundsurface at sample pavement sites was not covered by coarse debris in areas of well-developed pavement; particle density increased and spacing of particles decreased with increase of slope (Cooke, 1970*b*). In Australia, silcrete gibber cover ranges from 'complete' to 50 per cent (Dury, 1968). Gaps between coarse particles are occupied by finer material; if this contains a proportion of clay or salt, as it often does, the fine material is usually indurated to form a thin, strong carapace; if there are no binding agents, the material is loose.

The coarse particles are either *primary* or *secondary*. Primary particles are similar in almost all respects to the coarse particles found in the underlying materials; these are particles which have always been at the surface or have subsequently appeared on the surface; they may include particles of caliche or silcrete. Secondary particles are derived from primary particles by processes of disintegration. Split boulders and granular disintegrates are common on pavements and the weathering products are usually angular, often lying freely on the surface. They are generally produced by the weathering processes described in section 2.2.2. Artifacts are a form of secondary particle common in pavement areas which were formerly suitable human habitats or sources of stone for implement manufacture (Ascher and Ascher, 1965; Cooke, 1970*b*;

Davis and Winslow, 1965). Superficially weathered fragments are often moved downslope from their parent particles.

A coarse particle may be veneered with a desert varnish (section 2.3.2.*d.i*); the patina is a dull black rind on the exposed portion of fragments, on the buried portion it is usually an orange-brown stain, and at ground-level it is a pronounced black band (Engel and Sharp, 1958).

An important distinction may be drawn between pavements that are underlain by soils, and those that are not. The former are the more common (Cooke, 1970*b*; Engel and Sharp, 1958; Mabbutt, 1965), but in such circumstances it is not necessarily correct to conclude that

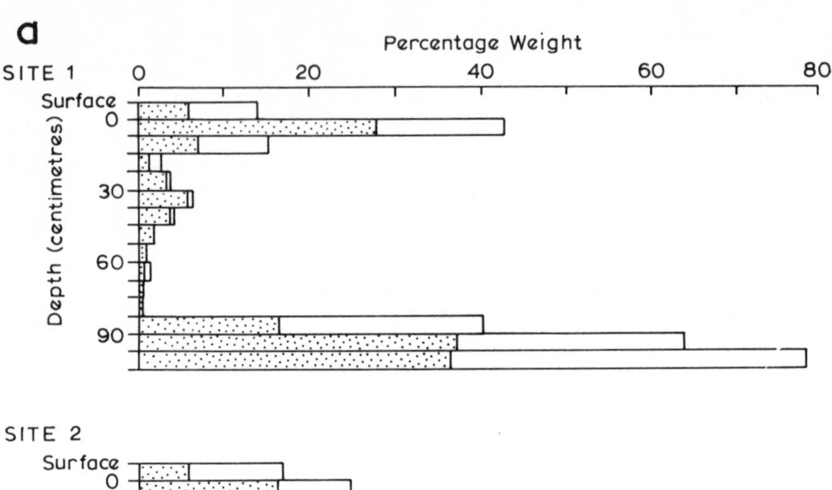

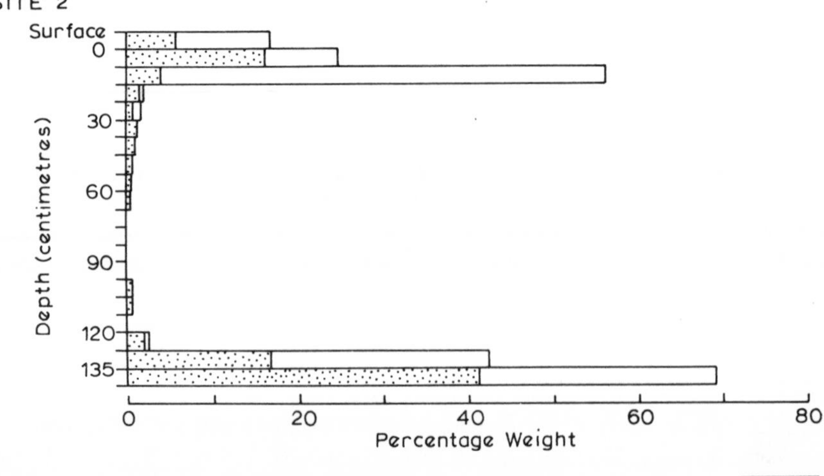

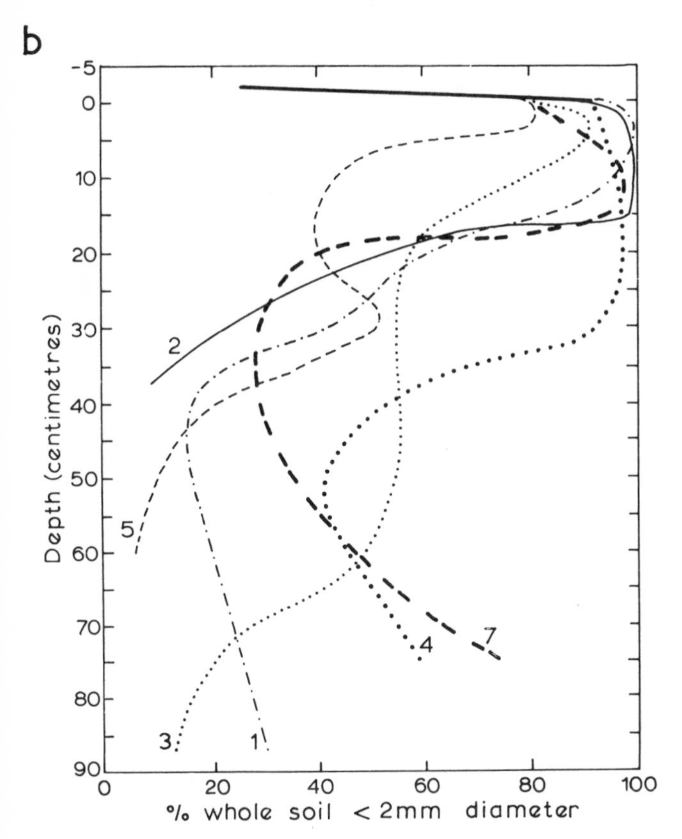

2.19 Contrasted examples of stone distribution in pavement soils: (*a*) is from central Australia (after Mabbutt, 1965); (*b*) shows six samples from Nevada (after Springer, 1958).

the pavement results from development over a long period of time on a relatively stable groundsurface by processes acting within the soil profile. A fairly common feature of many soils beneath pavements is the absence of coarse material in the upper part of the profile (Fig. 2.19). It must be emphasized, however, that although this feature is common, coarse-particle distribution can vary from being uniform throughout the profile, to the extreme situation shown in Fig. 2.19. Stone pavements on material without soil development may be produced on relatively unstable groundsurfaces by superficial processes. The distinction may therefore mark a genetic difference between, for instance, *autochthonous* and *allochthonous* pavements.

2.4.2 PROCESSES OF PARTICLE CONCENTRATION

Three groups of particle-concentration processes may be deductively recognized: *(a)* deflation of fine material by wind, *(b)* removal of fines by water at the surface and *(c)* processes causing upward migration of coarse particles to the surface.

(a) Deflation. Until recently, stone pavements in hot deserts have usually been explained by the deflation of fine material from the surface, which leaves a residue of coarse particles: the concentration of coarse particles is seen as a function of their distribution in the original sediment and the extent of deflation. It is assumed that the accumulation of coarse particles at the surface represents a layer of soil similar in depth to a layer of uneroded soil with a similar quantity of coarse particles. Blake (1858 and 1904), Free (1911), Gilbert (1875), Loew (1876), Moulden (1905) and Walther (1924), and more recently Brüggen (1951), Chao (1962), Clements (1957), Durand (1953), Symmons and Hemming (1968) and many others have identified deflation as the principal process of coarse-particle concentration.

Simple and persuasive though the deflation hypothesis is, conclusive evidence of deflation is rare. The proportions of sand, silt and clay in superficial sediments compared with the proportions for the sedimentary material as a whole may give an indication of the extent of deflation. For instance, Lustig (1965) showed that in certain areas of Deep Springs Valley, California, silt : clay ratios are unusually high, and he suggested that these areas are related to the paths followed by winds which have removed much of the clay material. Evidence of wind abrasion on pebbles in pavement areas may suggest the probability of local deflation, but such evidence is scarce.

There is no doubt, of course, that loose, fine material *can* be removed by the wind, as the extended and precise experimental work of Chepil (Chepil and Woodruff, 1963) demonstrated (section 4.2). Indeed, it is possible that in circumstances where deflation is the only process responsible for the superficial concentration of coarse particles, the distribution of coarse particles in a stable pavement surface (Fig. 2.20) may be expressed in terms of Chepil's (1950) 'critical surface-roughness constant'* (defined as the ratio of height of non-erodible surface projections to distance between projections which will barely prevent movement of erodible fractions by the wind). In some areas, fine material beneath the pavement may be of aeolian origin (e.g. the *parna* deposits in Australia).

* Subsequently called by Chepil the 'critical surface barrier ratio'.

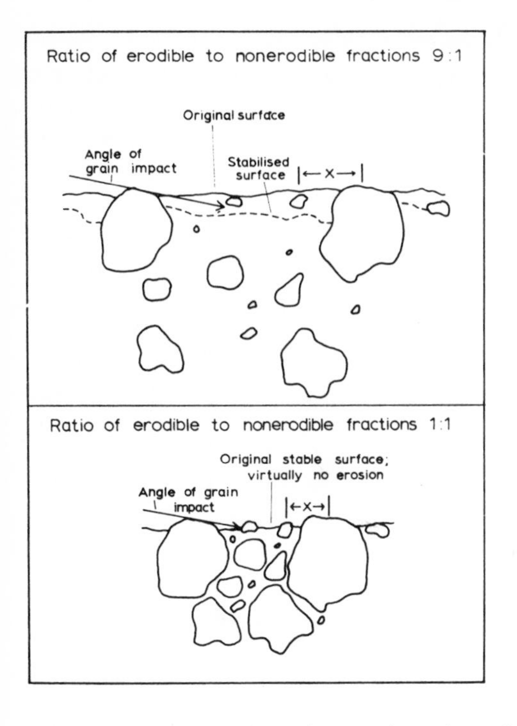

2.20 Diagrammatic representation of amounts of erosion with two different proportions of erodible to non-erodible fractions (cross-sections through the maximum diameter of the non-erodible fractions). (After Chepil, 1950.)

Arguments can be advanced against the ubiquitous application of the deflation hypothesis. Although desert winds are probably strong enough to move most loose fine material, in semi-arid areas scrubby vegetation may reduce winds to below the requisite threshold velocities, and surface vegetable litter may protect loose fines. Even more important is the fact that *loose* fine material is *unusual* in most deserts. Fine material is abundant, but it is usually formed at the surface into a thin carapace or crust (the *Staubhaut* of Mortensen, 1927), which is strong, dense, relatively impermeable and protects the underlying fines from erosion, especially by the wind. The crust is up to about 3 mm thick and is composed of a 'washing-in layer' and a thin upper coating of clay particles orientated parallel to the surface (McIntyre, 1958*a*; Tackett and Pearson, 1965). Beneath the crust a vesicular layer is often developed (section 2.3.2.*d.ii*). Crust formation may also be promoted by algae and lichens (section 2.3.2.*b*). Wet soil aggregates are first broken down by raindrop impact, and fines are washed into surface pores, reducing their volume; after aggregate destruction, raindrop impact causes compaction of the surface, and produces the thin coating which is extremely impermeable (McIntyre, 1958*b*). The compaction is a function of the size and terminal velocity of raindrops, and hence it might be expected to be extensive in desert areas experiencing intense rainfall (Dregne, 1969). The crust has been observed in many deserts (e.g. Mabbutt, 1965; Mortensen, 1927; Sharon, 1962) and in pavement areas it commonly shows little sign of having been deflated. Deflation will only occur if the carapace is broken by, for example, prior wind abrasion or vehicular traffic. In their experimental verification of the deflation process, Symmons and Hemming (1968) first prepared their sites by deliberately destroying the superficial carapace.

(b) Water sorting. Experimental observations show beyond doubt that some stone pavements are composed, in part, of coarse particles remaining after finer materials have been dislodged by raindrop erosion and removed by running water. Sharon (1962) demonstrated that surface runoff is the dominant process of pavement formation at sites in Israel. At cleared pavement sites in California, surface fines and some buried coarse particles were removed by surface runoff and collected in sediment traps downslope; after only four years, stone pavements were becoming re-established at these sites (Plate 2.13). Experimental work on stony soils by Lowdermilk and Sundling (1950) revealed that the amount of fine-sediment removal under conditions of uniform rainfall intensity increased directly with gradient; furthermore, they showed that as the experiments proceeded, the amount of material removed diminished and a stone pavement emerged, first on steeper slopes and later on gentler slopes, until the test sites were ultimately covered with a stable pavement surface.

The frequency of runoff, and the extent of sediment dislodgement and removal, is largely a matter of speculation in most desert areas. But two points deserve emphasis: *(i)* the effect of raindrops on particle detachment may be considerable in arid and semi-arid areas where rainfall intensity and raindrop momentum are occasionally high (Rose, 1960 and 1961; Williams, 1969) and there is little vegetation to inhibit direct access of raindrops to the ground; *(ii)* runoff necessary for the removal of dislodged particles may be promoted in pavement areas by the relative impermeability of the surface crusts. These two features are not, however, unrelated, for although raindrops cause detachment, they also promote crust development which, as a rain continues, reduces soil-splash (e.g. McIntyre, 1958*b*). Reduced permeability related to crusts may be offset partly by increased permeability due to cracking in clay materials.

The occasional saturation of pavements may cause fine material at the surface—and especially in stone-free upper soil horizons—to flow slightly. For example, if coarse particles are removed from a saturated pavement, the depressions which are left tend to fill with mud, adjacent particles jiggle slightly, and the surface settles down to a new equilibrium (e.g. Denny, 1967). This settling phenomenon probably explains the overall smoothness of many pavement surfaces.

(c) Upward migration of coarse particles. The concentration of coarse particles at the surface and at depth, and the relative scarcity of coarse particles in the upper soil profile (Cooke, 1970*b*; Mabbutt, 1965; Springer, 1958; Fig. 2.19) prompts the suggestion that stones may have moved upwards through the soil to the surface. Two main mechanisms might be involved in such movements: cycles of freezing and thawing, and cycles of wetting and drying.

Laboratory experiments have shown that alternate freezing and thawing of saturated mixed sediments causes coarse particles to migrate upwards. There is, however, some disagreement on the precise mechanism. Corte (1963) demonstrated that the movement of particles may depend on the amount of water between the ice-water interface and the particle, the rate of freezing, the distribution of particles by size, and the orientation of the freeze-thaw plane to the surface. Inglis (1965) described a simple mechanism as follows (Fig. 2.21): in a slurry frozen from above down to line A, the adhesion of the ice to the top of the large sphere will be strong enough to support its weight; subsequent freezing

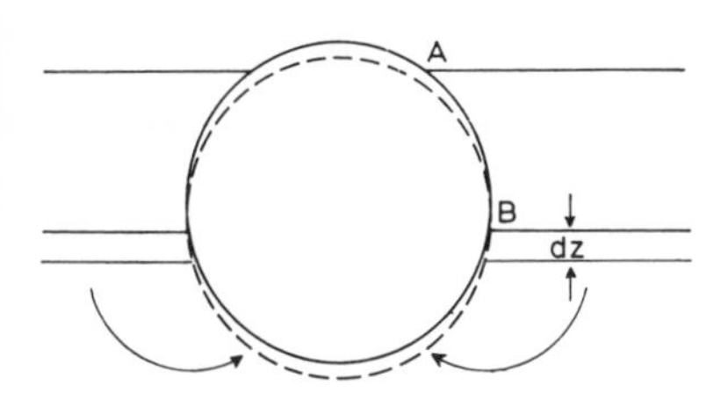

2.21 Vertical movement of a spherical stone in a partially frozen material (after Inglis, 1965). For explanation, see text. Copyright 1965 by the American Association for the Advancement of Science.

of a layer of thickness z causes its expansion by the amount δz (where δ is the coefficient of expansion of water and sediment) which lifts the frozen slurry and the sphere upwards; because of the rigidity of the sphere, a cavity would be left beneath it except for the fact that unfrozen slurry will move in to fill the cavity. With thawing from above down to level B, the sphere will be supported by frozen slurry beneath it; as the thawing surface advances a distance z, contraction permits the whole mass of material above it to fall by the amount δz. But the sphere, still unmoved, protrudes into this descending material. The net effect of both movements of the freezing plane is for the large sphere to move upwards through the slurry.

Although sorting by freeze-thaw action may seem improbable in hot deserts today, it cannot be entirely ruled out, especially in high-altitude deserts such as those in Bolivia and central Asia. It may have been more effective in some places during cooler, moister periods of the Quaternary (e.g. Lustig, 1966; Rognon, 1967*b*). In addition, it is possible that in very saline localities, the role of water, ice and a freezing plane in bringing coarse particles to the surface is replaced by water, salt and a desiccation plane.

A much more effective and widespread migration mechanism in deserts is thought to be related to wetting and drying of the surface soil. Laboratory experiments by Springer (1958) on pavement soils from Nevada showed that the distance between the top of the soil and the tops of some particles placed in it decreased by up to 2·2 cm after 22 wetting and drying cycles. Jessup (1960), using stony tableland soils from Australia, demonstrated maximum upward movement of coarse particles after 22 wetting/drying cycles of one centimetre. And Cooke induced four pebbles to emerge at the surface from a depth of 1·2 cm in a pavement soil sample from the Mojave Desert, California after only four wetting/drying cycles. The precise nature of this process is uncertain, but Springer (1958) logically suggested that when the soil is wetted, it expands and a coarse particle is lifted slightly; as the soil shrinks on drying, cracks are produced around the particle and within the soil; because of its large size the coarse particle cannot move down into the cracks, but finer particles can. The net effect is an upward displacement of the coarse particle. An important pre-requisite for the operation of this process is the presence of expanding clay minerals in the fine material, and the process is probably most effective where such minerals are most abundant. If upward migration alone concentrated coarse particles at the surface, the density of surface particles would be a function of the density of particles within the wetting/drying zone, and the length of time the process had been at work.

(d) Evolution of pavements. Most stone pavements appear superficially to be similar. Yet it is clear from this account that the form can be produced by several particle concentration mechanisms and can be embroidered by various superficial weathering processes. It is unlikely that all processes operate at one site or that pavement at one site is produced by only one process; most pavements result from the activity of several processes. Furthermore, the duration, intensity and effectiveness of the processes vary and each process varies from place to place according principally to the nature of the topography (e.g. whether the surface is active or relict), the sediment characteristics, and climatic conditions.

Whatever the processes and their relative importance may be, it seems probable that once a superficial mosaic of coarse particles is established, it protects the underlying material from further erosion by wind or water. In places it might be possible to speak of such established mosaics as being 'windstable' or 'waterstable'.

The age of pavement initiation may vary greatly from place to place. Springer (1958) studied pavements which were possibly of pre-Wisconsin age; Sharon (1962) created new pavements in a few years. If the stone mosaic is removed, then the surface would probably become unstable, and particle concentration processes would work to reconstitute the stable surface. The success of pavement renewal will depend largely on the availability of coarse particles: for wind erosion and upward migration processes, the local supply is limited to that immediately below the surface; runoff alone can bring new coarse particles on to the pavements from other areas.

2.5 Volume Changes: Patterned Ground Phenomena

Patterned ground phenomena—sorted or non-sorted varieties of circles, nets, polygons, steps and stripes—have been widely reported in polar, subpolar and alpine areas, where they have generally been explained in terms of frost action (Washburn, 1956). Similar features are, however, found in other climatic environments (e.g. Bremner, 1965; Costin, 1955; Verger, 1964) and they are quite common in some arid and semi-arid lands. Although frost action cannot be completely disregarded at present or in the past in some arid and semi-arid lands, ground patterns in such areas appear to be generally attributable to two different, and often related, groups of processes, both of which involve changes at or near the groundsurface: wetting and drying, and solution and recrystallization of salts. The effects of the first group of processes are most apparent in fine sediments or soils (e.g. in the lower plain areas on Fig. 2.2), especially where they contain a high proportion of swelling clays such as montmorillonite; the effects of the processes associated with salts are manifest in zones of salt concentration such as around the margins of ephemeral water bodies, along drainage channels, and within the capillary zone of water movement in the soil.

2.5.1 WETTING AND DRYING PHENOMENA

(a) Gilgai. Gilgai, an Australian aboriginal word, is a term applied to certain small-scale surface undulations. Most gilgai appear to be related to the differential expansion and contraction of the soil with periodic wetting and drying. Although gilgai are usually associated with fine-

grained alluvial soils, they also occur in stone pavement areas where particle sorting may occur (e.g. Hallsworth and Beckman, 1959; Hallsworth, Robertson and Gibbons, 1955; Ollier, 1966). Gilgai have been reported from a few arid and several semi-arid areas including Coober Pedy (Ollier, 1966) and New South Wales (Hallsworth *et al.*, 1955) in Australia, the central Sahara (Meckelein, 1959), the Middle East (Harris, 1959; White and Law, 1969), on tropical blackearths in Kenya and elsewhere in east and central Africa (Stephen *et al.*, 1956), South Dakota (White and Bonestall, 1960) and elsewhere (Verger, 1964).

Elements of gilgai topography often include puffs or mounds, depressions ('crab-holes' or 'melon holes') or channels, and the shelf areas between them (Fig. 2.22). Not all elements are always present in any one area. Verger (1964) constructed a simple classification of puffs and depressions according to their shape, and their grouping in plan (Fig. 2.22). This classification includes most of the forms recognized by Hallsworth *et al.*, (1955), Hallsworth and Beckmann (1969), Harris (1968) and others. Round gilgai (a, $\propto$) and some network gilgai (c, γ) occur on flat ground; lattice (e.g. b1, γ1), wavy (e.g. c1, γ2) and some network gilgai occur on gently sloping ground and their puffs or depressions are generally orientated parallel to slope contours.

The dimensions of the different elements are extremely variable. The vertical interval between puff, crest or depression floor and shelf may vary from a few centimetres to approximately three metres; the diameter of non-linear positive or negative features may be up to about 50 m; linear features may be many metres long and as much as 12 m wide.

Explanations of gilgai are numerous and range from the bizarre and ridiculous to the probable. As we have found with other forms, it is improbable that all gilgai result from the same process, but many areas of gilgai display diagnostic characteristics. A feature common to most gilgai areas is the presence of soils that expand and contract differentially on wetting and drying. Beyond the recognition of this simple fact, the analyses of Hallsworth and Beckmann, (1968) and Hallsworth *et al.* (1955) provide several generalizations which are probably of wider relevance.

(i) The vertical interval (Fig. 2.22) in well-developed gilgai soils is often significantly, directly and principally, related to the swelling capacity of the clay (which provides the force for soil movement), and to the sodium saturation of the exchange complex (the sodium ions adsorbed to the clay complex produce larger, more resistant clods, and the larger the clods moved by swelling, the greater the amplitude of the undulations).

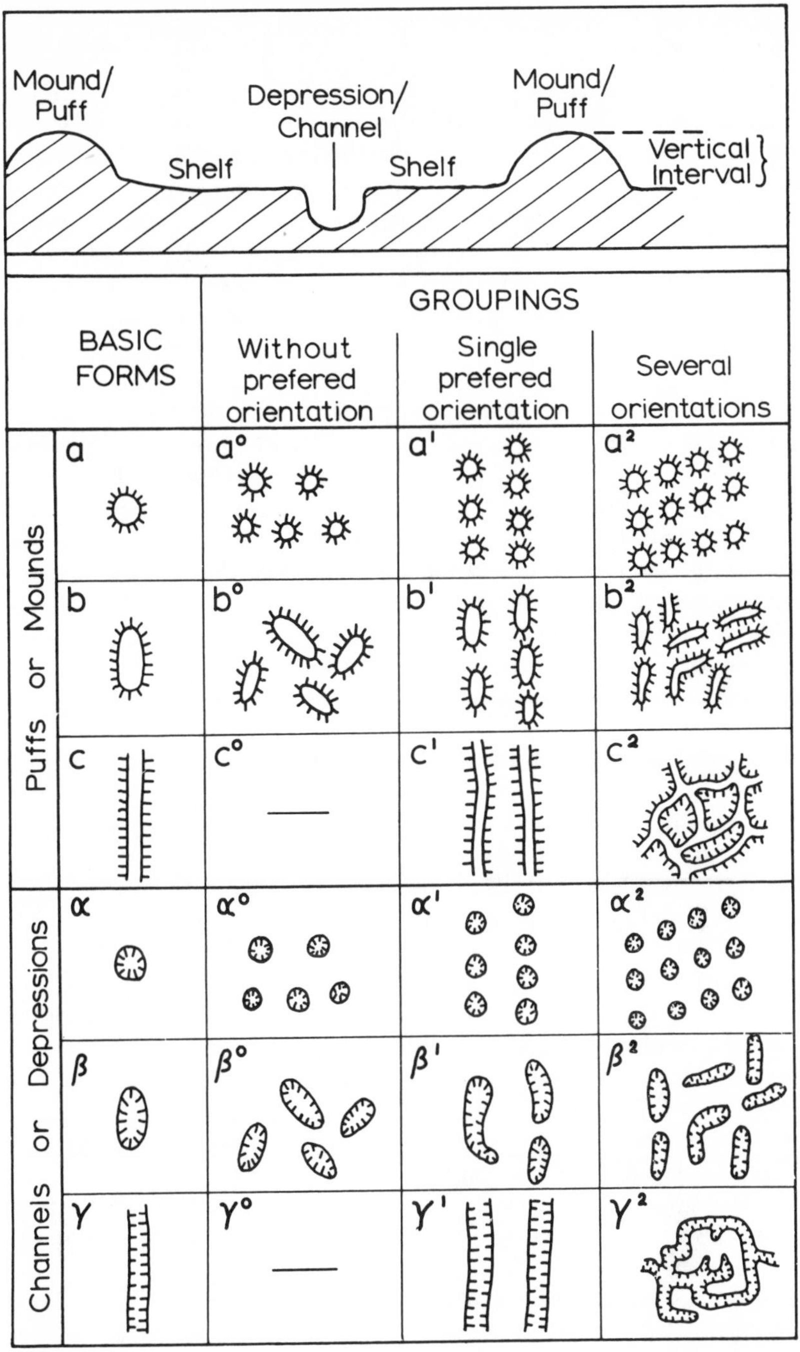

Mound/ Puff

Depression/ Channel

Mound/ Puff

Shelf

Shelf

Vertical Interval

	BASIC FORMS	GROUPINGS: Without prefered orientation	Single prefered orientation	Several orientations
Puffs or Mounds	a	a°	a¹	a²
	b	b°	b¹	b²
	c	c°	c¹	c²
Channels or Depressions	α	α°	α¹	α²
	β	β°	β¹	β²
	γ	γ°	γ¹	γ²

2.22 Terminology and classification of gilgai topography (after Harris, 1968; Verger, 1964 *et al.*).

(ii) Gilgai usually occur in soils where clay content, montmorillonite content and the proportion of sodium on the exchange complex all increase in depth, and where subsoil swelling in consequence is significantly greater than surface swelling (which is itself often over 10 per cent). (Not all gilgai soils include montmorillonite as a major clay mineral, however.) The type of gilgai that forms is related to the depth of the layer of maximum expansion and to the variations in expansion between different layers. The surface layer should be sufficiently coherent to allow subsoil cracks to extend to the surface and thin enough to allow the drying front to extend into the swelling horizon.

(iii) In addition to soil properties, gilgai formation depends on an external set of variables, of which the climatic conditions (controlling the nature of wetting and drying) and the soil moisture conditions are probably the most important. For instance, if the swelling layers never dry out, movement of the wetting front will be minimal and gilgai would be unlikely to form. Gilgai formation is most likely in areas with a marked alternation of wet and dry seasons.

(iv) A key consideration in explaining the patterns of gilgai development is the *differential* nature of the process. Hallsworth *et al.* (1955) offered the following explanation:

> As the soil dries out, the topsoil dries first, shrinking and cracking as drying proceeds. This cracking aids the drying out of deeper layers, which, as drying continues, also shrink, ultimately to a considerably greater extent than the topsoil . . . When re-wetting takes place . . . the whole soil swells again. If the drying out and subsequent shrinkage had been uniform, and had decreased uniformly from surface to subsoil, re-wetting and consequent swelling would merely have restored the soil level to what it was before. In a deeply cracking soil, however, uniformity of drying and shrinkage is unlikely, since the moisture content will show a horizontal gradient from the face of each crack into the interior of the block, as well as the general vertical gradient from surface to subsoil. Re-wetting will also be uneven. Light showers will merely wet the surface, whilst heavy rains . . . will wet and swell the subsoil at the base of the cracks long before the rest of the subsoil at the same level has had any change in moisture content. Uneven re-wetting is consequently likely to be the usual occurrence on such soils.
>
> The cracks themselves also contribute indirectly to the uneven distribution of pressures on re-wetting. During the dry period pieces of surface soil fall down the cracks, and the fact that the subsoil shrinks more than the topsoil, giving cracks that are wider below than

they are at the surface, makes it easy for large pieces of the surface soil to break off and to slip down into the crack below. Subsequent re-wetting and re-hydration would lead to excess pressure causing the fracturing of blocks of soil from that lying below. Repetition of this process would cause the blocks to move upwards towards the surface, since subsequent cracking and fracturing would start to occur again about the position of the earlier cracks, which would have become lines of weakness. (Hallsworth *et al.*, 1955, pp. 25–26)

On the basis of this argument and the similar one of Verger (1964; Fig. 2.23) it is probable that the production of isolated pressure zones and the subsequent development of gilgai may be closely related to the prior formation of cracks and crack patterns.

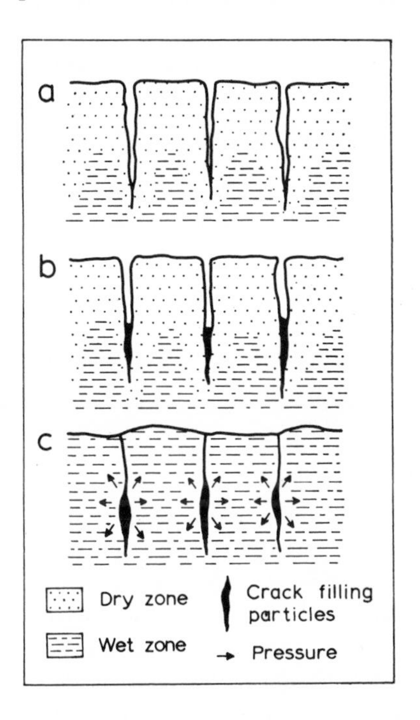

2.23 The formation of gilgai by wetting and drying (after Verger, 1964).

Patterns of interlocking gullies in enclosed drainage depressions on the Mesopotamian plain of Iraq and on the Nile flood basins in the Sudan appear similar in certain respects to network gilgai described in Australia, and a similar origin has been attributed to some of them (Harris, 1959). But White and Law (1969) identified significant differences between the two areas of gully patterns (*tabra* channels), and they pointed out that in places they lacked some of the features noted by

Hallsworth *et al.* (1955) in Australia, such as high expanding-clay content and increase of exchangeable sodium with depth. They suggested that in Iraq and the Sudan the patterns have formed entirely by the local intensification of compaction caused by water standing for long periods in pre-existing surface depressions, rather than by heaving of soil in the areas between channels. The pre-existing depressions were rills which have become inactive, abandoned irrigation ditches, or features associated with desiccation cracks. As with other explanations, periodic wetting and drying out is essential to the effectiveness of the process. This alternative explanation reminds us yet again that similar phenomena in different areas are not necessarily produced by the same processes.

Ollier (1966) recognized two types of gilgai in gibber plains of the central Australian Stony Desert both of which involve a degree of particle sorting: *circular gilgai* and *stepped gilgai*. Circular gilgai consist of bare clay topsoil shelves surrounded by slightly raised rims of pebbles, are up to nine metres wide, and occur on flat surfaces; stepped gilgai occur on slopes, run parallel to slope contours, are up to three metres wide, and are composed of bare clay topsoil shelves with pebble accumulations on their upslope and downslope sides. Both features are associated with soils containing considerable quantities of swelling clays. Fig. 2.24 summarizes the evolution of these features: if a patch of soil is wetted, it expands, and a small dome is formed; pebbles on the surface of the dome move towards its edges and settle there; when the soil dries, the clay contracts, and the clay shelf is slightly lower than the pebble rims; when dry, wind may remove some of the clay particles from the shelf. On flat surfaces (Fig. 2.24*a*), pebbles accumulate around the dome; on sloping surfaces (Fig. 2.24*b*), stones move downslope, and even the step itself may move *en masse* (with a gypsite layer at the base of the soil profile perhaps acting as a lubricating layer).

It is not clear, on present evidence, whether these or other gilgai forms are being continuously formed and destroyed, whether they are being continuously modified over a long period of time, whether they soon reach a condition of equilibrium, or whether they are fossil forms. It is known, however, that in some areas where gilgai have been eliminated by cultivation, they will re-form in several years, so that the processes are apparently active at present (Hallsworth *et al.*, 1955).

Features similar to stepped gilgai are the small steps which interrupt generally smooth but sloping pavement surfaces. They are characterized by risers of pebbles, and treads of finer material which is often capped by a thin veneer of stone pavement. In some areas, such as in Panamint Valley, California, the steps appear to be produced by the accumulation

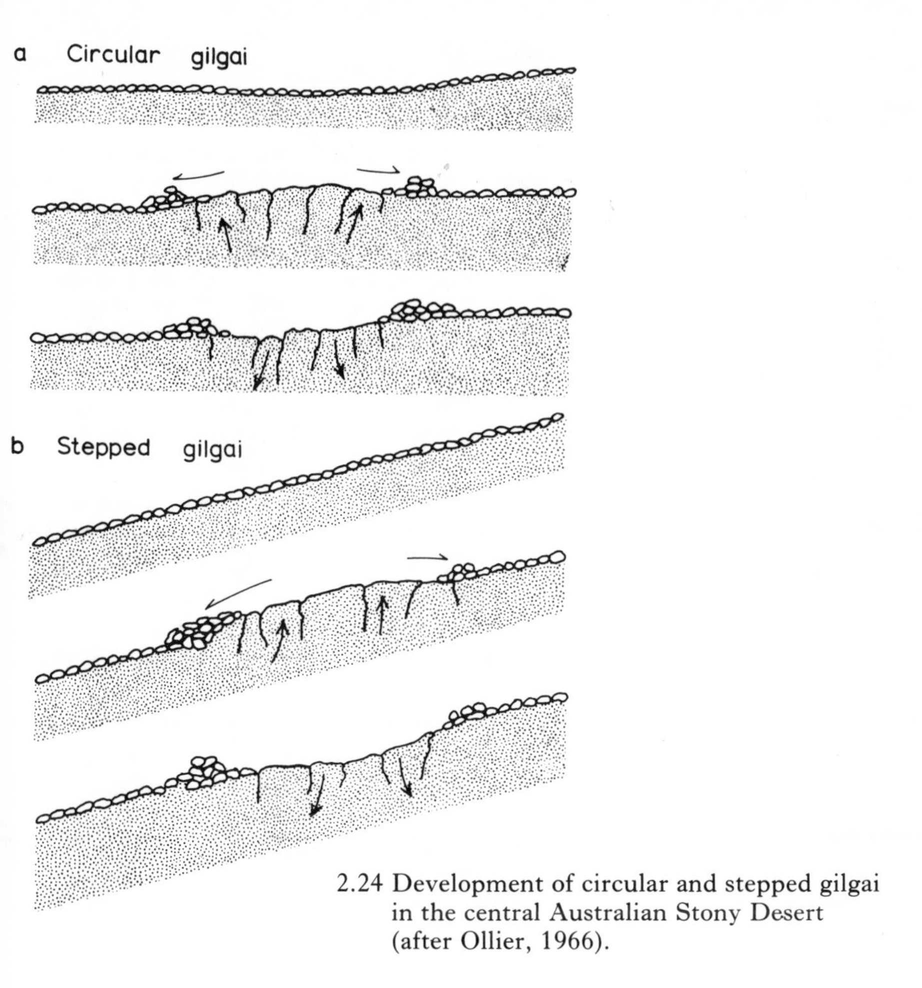

2.24 Development of circular and stepped gilgai in the central Australian Stony Desert (after Ollier, 1966).

of fine material upslope of large boulders that extend through the soil profile; elsewhere, in Death Valley, California, for example (Denny, 1965 and 1967) downslope creep of fine, relatively stone-free material beneath the pavement when it is saturated may lead to the development of steps; and Hunt and Washburn (1960) indicated that the treads of some steps had up to 10 times as much salt as stable ground around the steps, which might indicate that the salts may play a role in developing the features.

(b) Desiccation cracks. As a saturated fine-grained, cohesive sediment dries out, it may pass through liquid, plastic and brittle-solid phases,

and its volume is reduced until its *shrinkage limit* is reached. Volume reduction up to the shrinkage limit caused by evaporation of water from the sediment is often accompanied by sufficient tensional stress for rupture to occur and cracks to be formed. In general, the morphology of rupture patterns depends mainly on the intrinsic conditions of the material—such as moisture content, structure and degree of packing—and on extrinsic conditions of the environment—temperature, humidity, rate of desiccation, etc. (Corte and Higashi, 1964). There has been surprisingly little experimental study of desiccation cracks, and our comments here rest largely on the work of Corte and Higashi (1964), Lachenbruch (1962) and the early work of Kindle (1917).

The microtopography of a desiccated sediment is composed of patterns of *cracks*, and of polygonal *blocks* (*flakes* or *cells*). Cracks are usually fairly straight or smoothly curved in plan; their lengths, depths, widths, numbers and patterns vary greatly (e.g. Chico, 1963). The plan shape of blocks is determined, of course, by the crack pattern; the profile of blocks may be flat, convex, concave, or irregular (plates 2.14 and 2.15). Many of the morphometric properties of cracks and blocks may be significantly related. Corte and Higashi (1964), for instance, in a series of experiments involving stone-free tills of different initial degrees of compaction, demonstrated relationships between the mean size of blocks and the total length of cracks and thickness of the material. The mean size of the blocks is proportional to a power function of the thickness of the material depending on the material at the base of the experimental trough and dry density of the material; and the total length of cracks is also proportional to a power function of this thickness depending on the same factors.

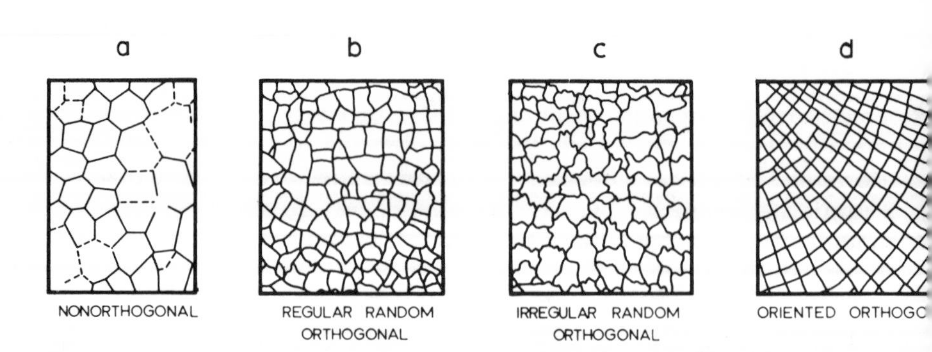

2.25 Common systems of desiccation cracks (after Lachenbruch, 1962, and Neal, 1965*b*).

Lachenbruch (1962) recognized two common systems of cracks: an orthogonal system, in which cracks meet at right angles, and a non-orthogonal system which is characterized by tri-radial intersections commonly forming obtuse angles of about 120 degrees (Fig. 2.25); in a single area, both systems may occur. Crack systems of different dimensions may also be found together. On the surface of Panamint Playa, California, for example, there coexist networks of small mudcracks and 'giant' desiccation polygons (Cooke, 1965; Neal, 1965*b*).

Some generalizations concerning the nature and origin of desiccation cracks and blocks follow.

(i) Desiccation patterns are developed according to the principal of least work in which polygons are compelled to have a maximum area (Hewes, 1948). The area of polygons probably varies directly with the total work done and inversely with the square of the tension (Hewes, 1948):

$$\frac{W}{A} = \frac{T^2}{E} \qquad . \quad . \quad . \quad (2.3)$$

where W = total internal work, E = modulus of elasticity, T = tension, and A = area.

(ii) The spacing of cracks may increase with the rate of desiccation and the proportion of clay present in the material (Kindle, 1917). The type of clay present (Chico, 1963) and the cohesive properties of the material may also influence the extent of rupture. For instance, montmorillonite-rich sediment is likely to contract more than a sediment with a comparable proportion of kaolinite; in some sediments, such as sands, there may be little contraction and no rupture.

(iii) According to Lachenbruch (1962), *orthogonal systems* are probably characteristic of inhomogeneous or plastic media in which stress builds up gradually, with cracks forming first at loci of low strength or high-stress concentration. As cracks do not form simultaneously, a new crack tends to join a pre-existing one orthogonally. Orthogonal systems are of two types (Fig. 2.25): those showing preferred orientation ('oriented orthogonal'), and those without preferred orientation ('random orthogonal'). *Nonorthogonal* systems probably form in very homogeneous, relatively non-plastic media which are dried uniformly. Lachenbruch suggested that under these circumstances, 'cracks propogate laterally until they reach a limiting velocity . . . and then branch at obtuse angles. The branches then accelerate to the limiting velocity and bifurcate again' (Lachenbruch, 1962, p. 59). All elements in this network are thus generated virtually simultaneously. Corte and Higashi (1964) independently reached a similar conclusion: cracks in orthogonal

systems develop sequentially, whereas those in nonorthogonal systems form virtually instantaneously.

Geometrically regular crack patterns, as in a hexagonal system, are unusual in reality because uniform conditions of material and its desiccation are rare. It is probably more realistic to assume a more or less random distribution of high-stress points or weak points from which complex, irregular polygonal patterns develop, as Smalley (1966) has suggested for basalt flows.

(iv) Observations of interfacial fracture markings by Corte and Higashi (1964) suggested that cracking begins a little below the surface and propagates either to the surface or to the bottom of the sample trough with non-uniform speed.

(*v*) The occurrence of stones in and on fine-grained sediments affects the nature of cracking in several ways (Corte and Higashi, 1964). Firstly, when they are appropriate distances apart, surface stones act as starting points for cracks. Secondly, shape and porosity of stones significantly affect cracking patterns. For instance, Corte and Higashi (1964) showed that porous cubes developed cracks radiating from their corners, whereas impervious and semi-porous cubes often became surrounded by semicircular cracks. A third interesting phenomenon noted by Corte and Higashi is 'habituation'—the repetition of crack patterns when wetting and drying are repeated. Habituation occurred when there were layers of shale on the surface but not when there was a surface cover of gravel. A possible explanation of this difference is that under the influence of wind and rain the shale pieces moved only a little into the cracks and were left tilted on the depressed surface of the old cracks, whereas the gravel moved further down into the cracks. The mixture of clay and gravel is probably more cohesive than the clay, and the next generation of cracks therefore developed in new, weaker positions; the clay beneath the shale responded to the next desiccation in the same way as before. Finally, Corte and Higashi demonstrated that wind and water spray on unvegetated, cracked surfaces can lead to particle concentration in the cracks to produce stone nets etc.: some stone patterns in deserts may result from this process.

(vi) The surface of *blocks* between cracks may in profile be concave, convex, flat or irregular. The concave surface is attributed to the more rapid drying and shrinking of the surface layer than the material immediately beneath (Longwell, 1928). In its extreme form, the concave upper layer may break away from the underlying sediment and produce a *mud curl* (Plate 2.15); occasionally the cleavage line may represent a boundary between unlike sediments. Optimum conditions for mud curl development appear to be the rapid drying of a very thin mud layer.

Kindle (1917) demonstrated that convex block profiles could be formed when a considerable proportion of salt is present in the fine-grained matrix. When such a mixture dries out, a contraction-crack network develops, but there is also slight expansion which produces up-arching and which is sometimes accompanied by salt efflorescence on a 'puffy' surface. Field observations suggest that these blocks are often moister than flat or concave blocks nearby: one possible explanation may lie in the hygroscopic nature of some salts.

A possible explanation of flat blocks is suggested by the previous comments on other profiles: they may be formed when relatively thick salt-free material is slowly dried.

For more detailed theoretical consideration of the geometry and mechanics of cracking, the reader is referred to the work of Corte and Higashi (1964).

Giant desiccation fissures. Playas make splendid airfields and race-tracks; but the occasional formation of giant desiccation fissures in them poses considerable problems for vehicular movement (Plate 2.16). Thus, considerable effort has been expended in studying this phenomenon in recent years (Neal, 1965*a* and 1965*b*; Neal, Langer and Kerr, 1969).

Giant desiccation features have been recorded in detail in playas of former Quaternary lakes in the western United States that are usually characterized by little ground-water discharge from the capillary fringe, rigorous desiccation, and hard, dry compact crusts composed largely of clays experiencing considerable volume changes on drying. The fissures may be up to a metre wide, are sometimes over a metre deep, and may be up to several hundred metres long. They may occur in isolation, or they may form irregular random orthogonal, striped, or other patterns. Fissures may be continuous, discontinuous or merely a series of aligned holes. Cavities beneath the surface between holes suggest that the fissures may be initiated beneath the surface.

There can be little doubt that giant desiccation fissures, like other cracks, result from the shrinkage of playa material, but their great size suggests that high-magnitude forces are responsible for them. It seems probable that the large stress required to produce the fissures builds up gradually, perhaps over several years, and is concentrated beneath the surface in the slowly changing capillary fringe. Intense evaporation extending deep into the playa material may play an important part in lowering the capillary fringe and promoting fissure formation (Fig. 2.26). In some areas of the United States, the water-table may have fallen in recent years as a result of man's exploitation of underground water resources (Neal and Motts, 1967). As most fissured playas in the western United States are located in tectonically unstable areas, it is

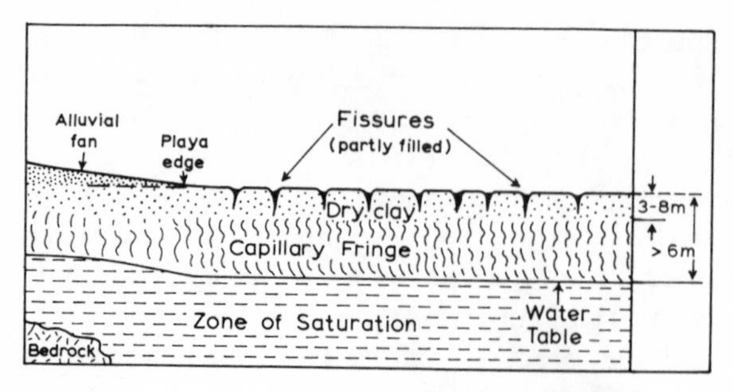

2.26 Hypothetical cross-section of a giant-fissure environment showing the relative positions of the water-table, capillary fringe, and fissures (after Neal, 1965*b*).

possible that earthquakes may trigger the release of tensile stress (Neal, 1965*b*).

Another feature described by Neal (1965*b*; Fig. 2.27) is the 'ring fissure' which is produced around a phreatophytic plant that lowers the water-table and capillary fringe as it grows, dries out surface material, and causes shrinkage and cracking.

Once they are formed, cracks are usually modified fairly rapidly. A crack may suffer wind abrasion, and gradual in-filling by wind-blown material and by debris washed into the crack when the surface is flooded or runoff occurs. Cracks may be zones of slightly greater moisture, and plants may grow in them. Eventually, cracks may be completely eliminated and their only trace may be lines of bushes or slight surface discolorations.

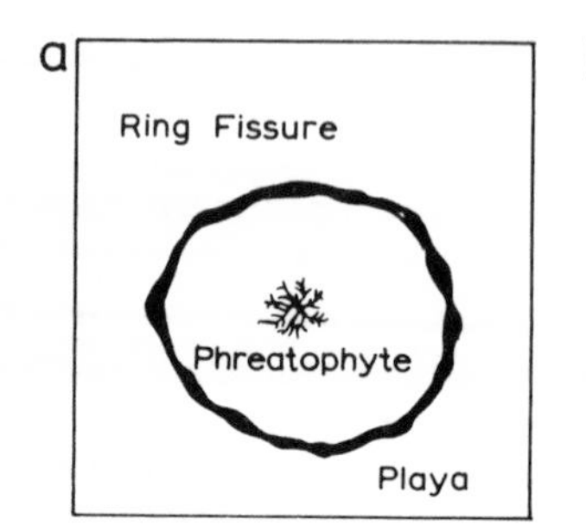

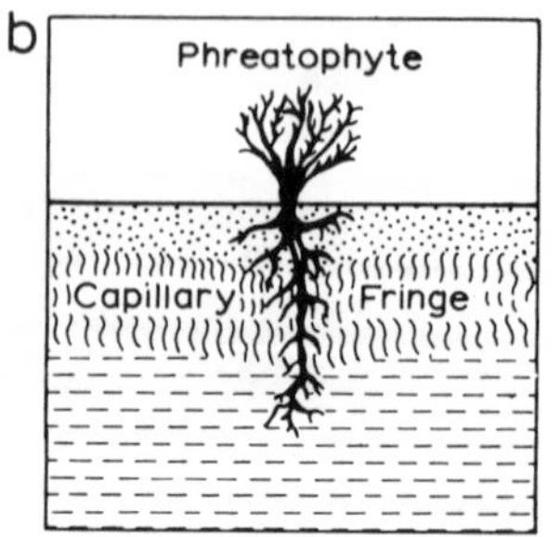

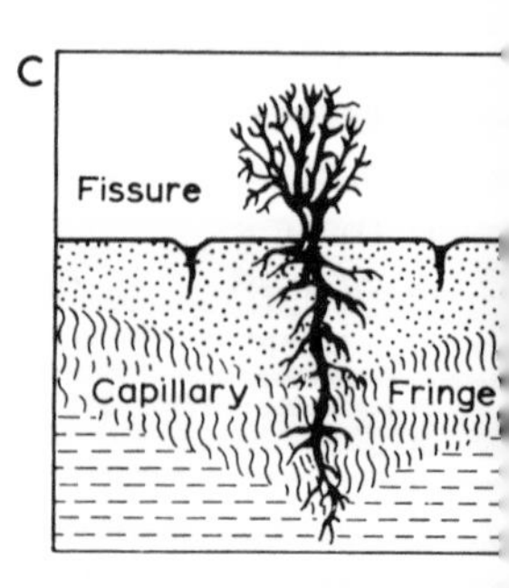

2.27 Ring fissures. (*a*) Plan view; (*b*) before and (*c*) after a sequence leading to fissure formation showing the lowering of water levels with increased growth of the phreatophyte and associated drying of superficial sediment (after Neal, 1965*b*).

2.5.2 PIPING AND SUBSIDENCE PHENOMENA

Piping is a term used to describe subterranean channels developed by water moving through incoherent and insoluble clastic rocks. The phenomenon is a fairly common and perhaps characteristic feature of certain dry lands. It has been recorded in many areas, such as Arizona (Parker, 1963), north-east Victoria, Australia (Downes, 1946), and New Zealand (Gibbs, 1945).

Pipes are commonly formed adjacent to arroyo walls in Arizona (Plate 2.17) and elsewhere as a result of rapid water movement down steep hydraulic gradients, through a permeable, easily dispersed clay, silt, or other sediment. Initially, water seeping through the arroyo bank carries with it dispersed and disaggregated clay and silt particles, and a small hole develops. The hole is enlarged by this process, but as it becomes larger it acquires more water. The increased flow leads to more rapid development of the pipe by corrasion and wall caving. Eventually, the roof of the pipe collapses, creating a sink hole down which surface-water may be funnelled augmenting the flow and the erosion of the pipe still further (Plate 2.18). Ultimately the sink holes merge and a ragged gully pattern is created (e.g. Buckham and Cockfield, 1950). Parker (1963) identified four basic criteria for this type of piping: sufficient water to saturate part of the material above base-level; hydraulic head to permit subterranean water movement; permeable and erodible soil above local base-level; and an outlet for flow.

Pipes also occur in materials which may not appear, at first sight, to be very permeable, such as clays, silts, loess and volcanic ash. All these materials may contain swelling clays, usually montmorillonite, but often illite or bentonite (Parker, 1963). Swelling clays have two important qualities in this connexion: when they are dried they shrink and create cracks, thus rendering impermeable materials permeable; and when wet, they become highly dispersed, non-cohesive and, therefore, easily removed in suspension by water, even by slow-moving water. For both of these effects to be realized, wetting and drying is essential, and this is most likely in arid and semi-arid areas.

Terrestrial subsidence in arid and semi-arid lands may also be related to irrigation. The finest expressions of this phenomenon are in the relatively recently irrigated Central Valley of California, where records are extensive and precise (Bull, 1964; Lofgren and Klausing, 1969). On alluvial fans in western Fresno County, on the west side of the Central Valley, Bull (1964) described two important features—subsidence of land, and the formation of cracks between the subsiding and the stable ground. About 212 km^2 are affected, and the subsidence is up to 3·3 m in places. In this area, subsidence results mainly from compaction of deposits by an overburden load as the clay bond supporting the voids is weakened by water passing through the deposits for the first time: in short water-deficient deposits are compacted by wetting. Tests showed that maximum compaction occurs where clay content (mainly montmorillonite) is about 12 per cent.

An alternative cause of terrestrial subsidence related to irrigation has been analyzed in the south-eastern Central Valley by Lofgren and Klausing (1969). They showed that intensive pumping of water for irrigation and consequent lowering of ground-water levels led to subsidence in an area of over 2,000 km^2 (Fig. 2.28). The subsidence here was due to the compaction of water-yielding deposits as fluid pressures declined and intergranular effective stresses increased. Maximum subsidence by 1964 was 3·6 m, and maximum decline of ground-water levels was 61 m. Rates of subsidence varied greatly and directly with rates of seasonal pumping; The ratio of subsidence to head decline ranged from $0{\cdot}5 \times 10^{-2}$ m of subsidence per metre of head decline to $5{\cdot}0 \times 10^{-2}$ m of subsidence per metre of head decline. Compaction of sediments accompanying water-table decline can lead ultimately to the formation of surface fissures (Robinson and Peterson, 1962).

2.5.3 SALT PHENOMENA

Literature on the role of salts in forming patterned ground in deserts is remarkably thin, although it can be traced from Huntington's (1907) observations in the Lop Desert to the thorough investigation by Hunt and Washburn (1966) in Death Valley, California. Perhaps one reason why the role of salts has not been considered extensively is that the patterned ground is commonly associated with wetting and drying, as well as with salts, and the former offers a more obvious and better understood explanatory mechanism.

Salts occur at desert surfaces through the surface evaporation either of runoff, or of saline solutions raised to the surface by capillarity. In the former case, evaporites are produced in a regular crystallization sequence that, for example, begins with carbonate precipitation, and is followed by the formation of sulphates and chlorides. This sequence may be evident in both vertical stratification and annular rings around an evaporation pan. Such separation of salts of different physical and chemical properties means that patterned ground features associated with them may show a similar areal segregation.

An important example of salt segregation is the well-known occurrence of nitrate and iodate deposits on the flanks of the coastal mountains adjacent to valley floors and evaporation pans in the central valley of the Atacama Desert (Fig. 2.29). Of the many attempts to explain this

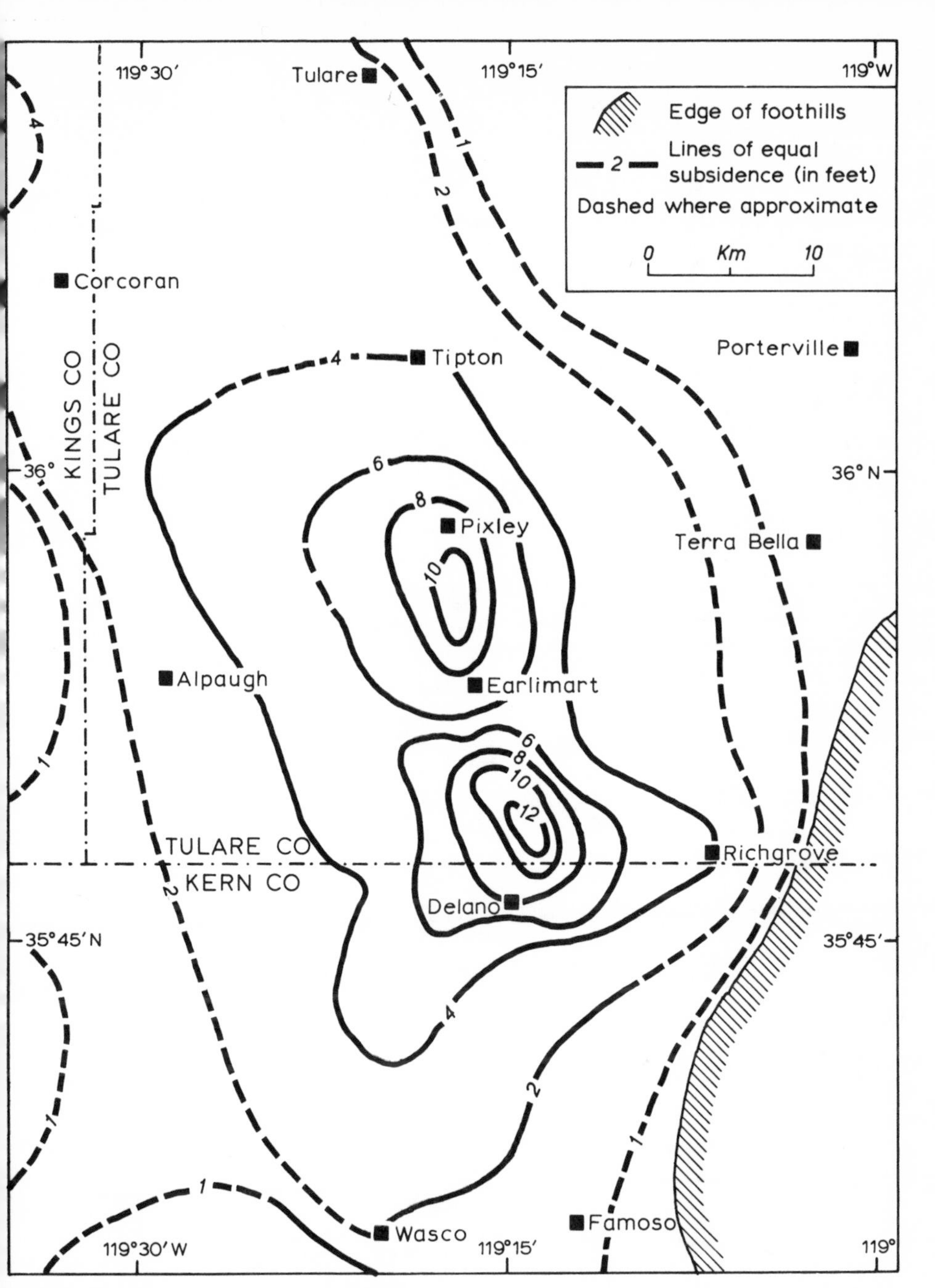

2.28 Land subsidence in the Tulare-Wasco area, California, as a result of lowering of ground-water levels, between 1926 and 1962. Compiled from comparison of topographic maps (1926–54), and levelling of the United States Coast and Geodetic Survey (1954–62) (after Lofgren and Klausing, 1969).

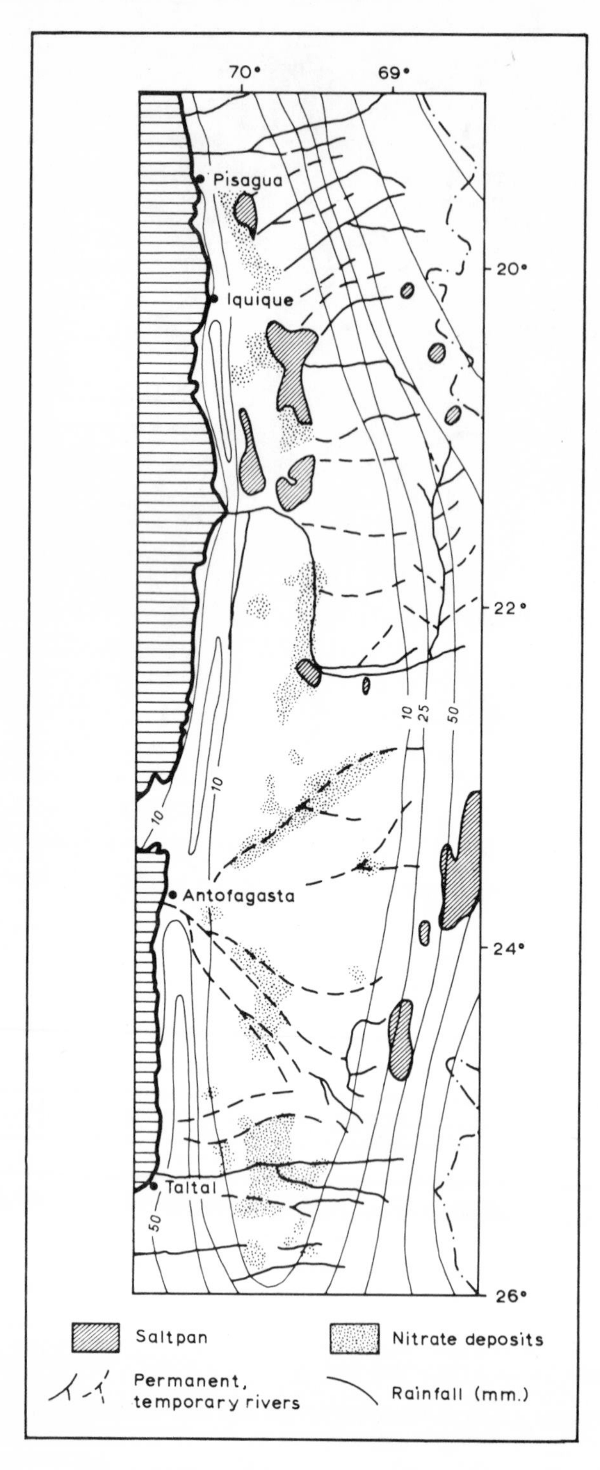

2.29 Location of nitrate deposits in the Atacama Desert, Chile (after Mueller, 1960).

phenomenon, the most satisfactory hypothesis is that of Mueller (1960 and 1968). He demonstrated that, as saline waters from the Andes evaporate in the floor of the central valley, residual solutions containing (amongst other things) nitrates and iodates move away from waterlogged reception areas *upslope* through the soil of the coastal mountains' lower slopes by capillary migration, and eventually they evaporate to complete dryness. The relative mobility of anions under given circumstances influences the distribution of salts. SO_4^{--} ions, for example, migrate slowly and concentrate near the waterlogged zones; Cl^- ions migrate quickly and travel furthest from the waterlogged zone; NO_3^{--} ions appear to move at intermediate rates as the nitrate deposits occur in an intermediate position. Intense aridity in these areas prevents washing of the salts back into the lower ground. The question of salt migration has been further reviewed in section 2.3.2.*d.iii*.

Patterned ground phenomena associated with salts include sorted and nonsorted nets, polygons, steps and stripes.

Hörner (1936) described salt polygons in several areas of central Asia. The body of the polygons probably consisted of porous gypsum, and occurred beneath a stone pavement; cracks between the polygons were filled with fine sands. Cracking occurred when the salts were dried. Filling of cracks with sands prevented lateral expansion of the polygons on wetting, so that they expanded upwards, and their surfaces rose slightly.

In Death Valley many patterned ground features are associated with the various zones of the salt pan and with gravel fans marginal to them. The following description is abstracted from the detailed account by Hunt and Washburn (1966).

On the flood plain of the salt pan the patterned ground is ephemeral and reflects the hydrological regime: in places where water collects, polygons develop by contraction of a thick salt crust, and complex nets of ridges and blisters may be superimposed on these as a result of evaporation from ground-water which creates new salts in the cracks, and between the crust and the underlying muds; where flooding is infrequent, salts accumulate in polygonal cracks and produce salt ramparts around mud polygons.

In the chloride zone beyond the limit of flooding, which is characterized by varieties of rock salt, a salt crust is developed on damp mud: smooth surfaces of silt saturated with salt have polygonal patterns of desiccation cracks, and solution pits are common at crack junctions. Nets and polygons have developed in the gypsum deposits of the sulphate zone. For instance, polygons in dry, massive gypsum consist of primary cracks that extend through a thin surface layer and the underlying anhydrite to the crumbly, porous gypsum beneath, and of secondary cracks confined to the surface layer. Sorted and nonsorted nets occur in the carbonate zone towards the edge of the saltpan. Debris has been washed into cracks to produce the sorted nets. Sorted polygons occur where there is a layer of rock salt beneath the surface (Fig. 2.30): the salt is polygonally cracked, and the cracks are reflected in shallow surface troughs in the overlying silt in which coarse debris accumulates. On older gravel fans there are sorted steps, sorted polygons and stone stripes.

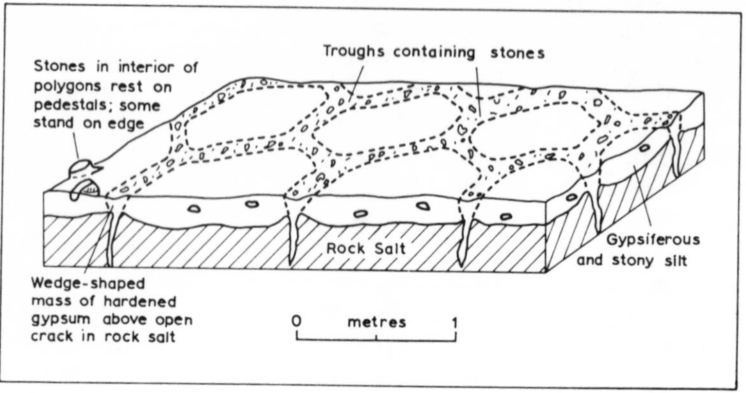

2.30 Sorted stone polygons related to ridges and cracks in an underlying layer of rock salt, Death Valley (after Hunt and Washburn, 1960).

Hunt and Washburn's conclusions are of general significance. They may be summarized briefly as follows. The patterned ground features—sorted and nonsorted steps, stripes, polygons and nets—occur in an area characterized by salt crystallization, clearly suggesting a genetic relationship. Although, the precise processes are not well understood, their effects are very similar to those of ice in high latitudes. Salts may affect the form of patterned ground in several ways: the geometry of cracks is influenced by the nature and amount of the salts found in the muds; salt accumulating in cracks tends to widen and accentuate them; and salt crusts, by retarding evaporation, favour upward salt migration in the capillary fringe, and this produces heaving. Volume reduction of salts, and hence cracking, result mainly from either drying or cooling. Volume expansion, and hence some heaving features, occur as the result of crystallization, thermal expansion or hydration. The areal patterns of patterned ground phenomena associated with salts depend firstly on the nature and vertical and horizontal disposition of salts and, secondly, on important climatic and hydrologic variables such as the frequency of wetting and drying, the depth of the capillary fringe, the relative importance of surface and capillary water, and the nature of temperature changes.

Krinsley (1970) described some patterned ground phenomena associated with salt crusts on the Great Kavir, Iran. Here, ground-water is the main source of salt and it reaches the surface through desiccation cracks in the crusts. In areas near the centre of the basin, where polygonal cracks persist between inundations, salt accumulates more rapidly than it is removed, and fresh brine introduced along the desiccation cracks is less saline than the indigenous brine and thus evaporates more easily. Fresh brine may bubble out along the cracks during the day, and its rapid evaporation may result in salt blisters or 'blossoms' along the cracks. The salt cells or plates between cracks grow by salt crystallization on their borders, and the growth may be most rapid on their windward sides. Windward edges may eventually overlap the leeward edges of adjacent plates, giving the illusion of overthrusting. In addition, black saline muds may be extruded through the cracks—perhaps as the result of thermal expansion or of pressure from the overlying crust. The muds may quickly solidify to form dykes along the cracks. The uneven escape of muds may be responsible for tilting salt plates and for the creation of an extremely irregular surface.

2.5.4 OTHER PATTERNS

In addition to the patterned ground phenomena attributed to wetting and drying or the action of salts, several other patterns have been described from various deserts. Many of these are barely perceptible topographically, but are emphasized through vegetation patterns associated with them.

Of several distinctive patterns recognized by MacFadyen (1950) in Somalia, it is the 'vegetation arcs' and the 'water lanes' which have attracted most attention. Vegetation arcs are concentrations of dense grass and scattered trees, with an average positive relief of about 0·4 m, that are spaced at an average interval of some 158 m with their chords *normal to drainage directions* on broad alluvial plains sloping at between 1:166 and 1:460. The ratio of bare ground to vegetation in the areas of vegetation arcs is 3 : 1 or 4 : 1. The 'water lanes' run normal to the contours MacFadyen tentatively attributed the arcs to the deposition of loose organic material in the form of 'strandlines'. A more detailed and attractive hypothesis was suggested by Hemming (1965). In his view, vegetation arcs are fundamentally remnants of a more extensive vegetation cover which has been reduced by overgrazing, cutting and burning.

This activity created bare patches where permeability was reduced and where, because of grazing pressures, stone pavement development and high clay content, recolonization by vegetation is difficult. Runoff in the form of sheetflow characterizes the patches, and it nourishes vegetation on the downslope sides of patches. Thus less xerophytic vegetation, such as the relatively unpalatable *Andropogon*, becomes established. Arcs only occur where ground slope permits sheetflow but is insufficient to generate channelled flow. The vegetation arcs migrate upslope by colonization along their upper edges, at a rate which may be approximately 15–30 cm p.a.

'Water lanes' separated by 'vegetation stripes' which run *normal to the contours* on parts of the Somaliland Plateau have been studied by Boaler and Hodge (1962). Some of the stripes of vegetation along the water lanes are of riparian origin, associated with small stream courses; they are similar to features found in many deserts (e.g. in the Papago Reservation, southern Arizona). The multiple, parallel tracks of vegetation are different. They occur parallel to the greatest slopes (1:190–1:350) on alluvial outwash plains, in both *Acacia* bush areas and grassland areas; they are straight or gently curved, and continuous for several kilometres. They are 70–180 m wide and separated by lanes approximately 45–80 m wide. Contrasts between vegetation in the stripes and the lanes is attributed to soil differences. In the vegetation stripes, the soils are heavier, topsoil is thinner, salts increase with depth more rapidly than in the intervening areas, and a micro-relief of potholes and mounds occurs. Soil differences are probably related to variations in the nature of the superficial debris. Surface sorting of debris, it is suggested, may be accomplished by wind, water, or wind and water together.

MacFadyen's hypothesis of 'strandline' deposition was slightly modified by Worrall (1959) in his more detailed examination of similar patterns of grass in the Butana area of the Sudan: he suggested that rhythmic sedimentation might occur on a large scale where a gentle sheetflow of water develops an attenuated waveform of great length and small amplitude, and plant deposition concentrates beneath wave crests. The origin of grass patterns in the Sudan is not explained completely by this hypothesis, however, and an alternative hypothesis is that soil cracking occurs parallel with the grass bands and 'may be due to gravitational pull downslope resulting in alternate compression and expansion of soil in rhythmic form without the surface being corrugated. The expanded parts would permit greater depth of water penetration and so more favourable conditions of plant growth' (Worrall, 1959, p. 52). In terms of this hypothesis, the features may be regarded as incipient gilgai. The grass bands apparently migrate upslope annually.

Vegetation patterns have also been recognized in semi-arid West Africa where arcs similar to those recognized by MacFadyen are known graphically as *brousse tigrée* (Clos-Arceduc, 1956).

In the wanderrie country of Western Australia, Mabbutt (1963) analyzed the patterns of sandy wanderrie banks on plains tributary to major drainage lines, which are separated by flats of alluvium with fairly high silt and clay content. The banks are transverse, longitudinal or oblique to drainage axes, and they may be up to approximately three kilometres long, 400 m wide, and over a metre high; spacing varies between 100 and 250 m; the ratio of bank area to flat area varies from less than one to four. The features are ascribed to the combined action of wind and sheetflow in sorting materials in the alluvial environment. Transverse patterns were developed parallel to postulated dominant north-west and south-east winds; the longitudinal pattern occurs where drainage is cross-wind; the oblique pattern is intermediate. Where drainage ways are transverse to the dominant winds they have acted as sources of aeolian material, leading to the formation of longitudinal banks in their lee. Between such banks and on windward slopes, 'stepped profiles were formed by declining sheetflow transport under increasing interference by wind. Steeper slope zones of accumulation of wind-sorted sands alternated with flatter alluvial slope sectors. With increasing aeolian activity, the steeper zones become loci of transverse band growth' (Mabbutt, 1963, p. 540).

Arcuate ripples, up to 175 m long, up to 18 m high, and arranged roughly *en echelon*, cover large areas of desert in Utah, Arizona and New Mexico (Ives, 1946). The ripple crests support sagebrush which accentuates the patterns, and troughs between ripples are plated with caliche. The ripples are in fact small, transverse clay dunes produced by dominant winds blowing, in these areas, from the north-west.

PART 3

The Fluvial Landscape in Deserts

3.1 Desert Drainage Systems

3.1.1 NATURE OF DRAINAGE SYSTEMS

Since the outstanding early contributions of G. K. Gilbert, R. E. Horton and others, there has been an accelerating interest in the geomorphological study of drainage systems and their principal components, namely slopes and channels. In the past two decades progress has been particularly rapid, thanks notably to research workers in the United States Geological Survey. Much of the recent work has been reviewed and is readily accessible (e.g. Leopold, Wolman and Miller, 1964; Carson and Kirkby, 1972). Our purpose here is not to repeat this information, but to select and summarize some of that which is pertinent to drainage systems in arid and semi-arid lands.

A basic tenet of most studies in fluvial geomorphology is that the drainage basin is an open system characterized by input, throughput and output of energy and materials; and by self-regulation amongst the component variables. Some of the variables are independent, others are dependent, and the relations amongst them vary according to the time span that is considered. Table 3.1, derived from Schumm and Lichty (1965), suggests the status of drainage basin variables at three different time spans: 'cyclic' time (the time encompassing an erosion cycle), 'graded' time (the time when grade and a condition of dynamic equilibrium exists) and 'steady' time (a fraction of graded time). The variables on the table are listed in approximately increasing degrees of dependence. In the following paragraphs, we look briefly at one or two of the few studies that have attempted to describe precisely some of the characteristics of these drainage basin variables and the relationships between them.

The drainage basin and its constituent parts can be analyzed in terms of linear, areal, and volumetric attributes. In recent years a few authors have studied some of these morphometric properties in dry lands and related them to other environmental variables. The following examples illustrate the benefits of this approach.

In a multivariate analysis of 22 drainage basins in the south-western

TABLE 3.1 *Status of Drainage Basin Variables During Timespans of Decreasing Duration*

Drainage basin variables	*Status of variables during designated timespans*		
	Cyclic	*Graded*	*Steady*
1 Time	Independent	Irrelevant	Irrelevant
2 Initial relief	,,	,,	,,
3 Geology (lithology, structure)	,,	Independent	Independent
4 Climate	,,	,,	,,
5 Vegetation	Dependent	,,	,,
6 Relief (or volume of system above base-level)	,,	,,	,,
7 Hydrology (runoff and sediment yield per unit area within system)	,,	,,	,,
8 Drainage network morphology	,,	Dependent	,,
9 Hillslope morphology	,,	,,	,,
10 Hydrology (discharge of water and sediment from system)	,,	,,	Dependent

SOURCE: Schumm and Lichty (1965).

United States, Melton (1957) convincingly demonstrated that drainage density (the number of streams per unit area) is high on rocks with low infiltration capacities (e.g. shale and schist), and low on rocks with high infiltration capacities (e.g. granite, clastic, and acid-volcanic rocks); and that it also varies inversely with Thornthwaite's precipitation-effectiveness (*P–E*) index, and directly with per cent bare area and runoff frequency-intensity. The mechanisms controlling these relationships principally concern runoff: high volumes of runoff need high stream densities to remove them. In addition, he showed that steep valleyside slopes are associated with high infiltration capacity, *P–E* index and relief ratio, and with low wet soil strength and runoff intensity-frequency. Melton's study still stands as one of the few that have thoroughly examined the precise relationships between morphometric properties of drainage basins and the general causative factors of climate, mantle characteristics, vegetation density and bedrock lithology.

Schumm and Hadley (1961) also investigated several aspects of drainage-basin morphometry in semi-arid areas of the United States. From a study of 59 small (0·25 km^2 to 43 km^2) drainage basins they showed that mean relief ratio (relief of drainage basin/basin length, measured in a straight line approximately parallel to the major drainage channel) correlated closely with mean annual sediment yield (measured from accumulations in stock-water reservoirs). In other words, steeper slopes were being eroded more quickly. Similar results were obtained by Hadley and Schumm (1961) in eastern Wyoming and by Brice (1966) in Nebraska. Data from the Cheyenne River basin in Wyoming also indicate that sediment yield per unit area decreases with increasing size of drainage basin, probably because of water losses by infiltration through channel beds, and larger bottomlands and lower gradients in larger basins. These studies clearly indicate that sediment yield from drainage basins in semi-arid areas is in part related to morphometric properties of the basins.

Sediment-yield data are a useful indication of the rates of removal of material from drainage basins, but much more information is required for the preparation of a sediment budget. A detailed attempt to construct a sediment budget in a semi-arid drainage basin was made by Leopold, Emmett and Myrick (1966). They monitored erosion, transportation and deposition rates near Santa Fé, New Mexico, over a period of seven years. Amongst other things, they measured changes in channel dimension, changes in scour and fill in channel beds (using scour chains), the extent of slope erosion (using nails inserted into the ground through washers); soil creep (by means of carefully located iron pipes); the rate of headcut erosion; and the rate of sediment accumulation in reservoirs. Valuable results were obtained from each set of measurements. For example, the scour-chain evidence showed convincingly that the period 1958–1964 was in general a time of net channel aggradation. More importantly, it is possible to use data from the individual studies to build up a picture of the sediment budget in the area. Extrapolation of experimental-plot data to the whole contributing area was difficult, and in any case the observations were not comprehensive. But, by making realistic estimates, Leopold, Emmett and Myrick compiled the sediment budget shown in Table 3.2.

TABLE 3.2 *Sediment Budget for Areas near Santa Fé, New Mexico*

	Estimated average rates (metric tons p. km²/p.a.)	*per cent*
Total erosion	5452·9	100
Surface erosion	5335·2	97·8
Gully erosion	78·5	1·4
Mass movement	38·4	·7
Total deposition	1205·5	22
Deposition in channels	564·9	10
Trapped in reservoir	640·6	12

SOURCE: Leopold, Emmett and Myrick, 1966.

Clearly sediment production in this area is dominated by surface erosion (slopewash), and only a small proportion of the material is being trapped in channels and reservoirs. Other studies that have pursued these themes in deserts are often primarily concerned with specific features of drainage basins such as slopes, channels and pediments, and they are therefore more appropriately considered in the following sections.

3.1.2 SLOPES

(a) Introduction. Slopes are the dominant feature of most landscapes and they have naturally attracted an enormous amount of attention from geomorphologists. But until the post-war years little of the observation of desert slopes was precise, and because of the absence of measurements, much of it was also inaccurate. Melton (1965*b*) has shown, for example, that many of Bryan's (1923*a*) estimates of slope angles in Arizona were gross overestimates. The following short discussion of desert slopes is based therefore almost wholly on the measurements and observations in a few recent studies. Other comments on slope studies are presented in sections 3.3 and 3.4.

A useful definition of a slope derives from the work of Horton (1945): a slope is the portion of a drainage basin in which there is overland flow; that is to say, an area where water is not concentrated into channels. This definition is not altogether unambiguous since it is not easy to distinguish between a rill (usually taken as an ephemeral form) and a channel (taken as a perennial feature), and equally the distinction between a rill and a concentration of flow is not always clear. However these difficulties are alleviated if we adopt the scale approach (Schumm and Lichty, 1965): a slope is a feature which should be studied at the graded scale or, in the terms used in Part 1 of this book, a slope is a unit to be studied at the local scale.

Slopes can be classified according to the nature of the controlling process. On steep slopes, debris is removed by the effect of gravity as soon as it is produced: these are bare-rock or *gravity-controlled slopes*. On gentler slopes the debris produced by weathering passes downslope by mass movement: these are *debris-controlled slopes,* with a thin coarse debris mantle. On still gentler slopes the debris is finer and is moved by slope wash: these are *wash-controlled slopes*. On many desert hillsides the slope profile from the drainage divide to the stream is composed of slope units from each of these types. The upper slope may be gravity-controlled, the middle slopes will be debris-controlled and the lower slopes will be wash-controlled. Even more commonly, complex alternating lithologies give rise to recurring sequences of the slope types (e.g. Schumm and Chorley, 1966). Distinct breaks of slope may separate these elements. The actual values of the slope angles involved will depend on a number of factors.

(b) Gravity-controlled or bare-rock slopes. In parts of deserts in which there are outcrops of resistant rocks and in which there is high available relief, bare-rock slopes are common. The nature of 'free-face' slope units—which occur near the tops of composite hillsides or where there has been deep incision by channels—depends very much on the character of the local bedrock. Relatively unresistant rocks tend to form rounded upper slopes; more resistant ones form more rectilinear faces (Schumm and Chorley, 1966). Rounding, when it occurs, is due to weathering. Jointing is another control on the type of face.

A common type of free face is composed of vertical columns or plates which break away from the slope and add debris to the lower hillside (Plate 3.1). The kinds of process that may control the modification of such slopes have been described by Schumm and Chorley (1964). Threatening Rock in New Mexico fell in 1941 after a measured slow movement away from the cliff. The measurements of rock movement and weather information suggest that as moisture accumulated from precipitation, rock movement increased. It was particularly fast in the moister winter months when frost action and the wetting of underlying shale by snow melt were active. The process involved here is basal sapping of a harder cap-rock and it is a common one (Schumm and Chorley, 1966).

The rock fragments or blocks dislodged from the cliffs may accumulate at the base of the gravity slope as a scree or talus slope on which debris-control operates (section 3.1.2.*c*). Theoretically one can envisage the scree building up in front of the 'free-face' and ultimately almost burying it and protecting a gently sloping rock-cut slope as it does so (Bakker and Le Heux, 1952). This rock-cut slope may later be exhumed. Profiles containing free-faces and a debris-controlled slope have commonly been described in deserts (Plate 1.12).

However, in parts of many deserts, gravity-controlled slopes meet much gentler slopes along boundaries where there is very little debris accumulation. On the Colorado Plateau, for instance, only volcanic rocks have pronounced debris slopes whereas the more common sandstones have very few. It is suggested that the sandstone shatters when it falls from the cliffs and its porosity helps rapid weathering and disintegration thereafter. The disintegration is considerably promoted by seasonal freeze-thaw processes. A measure of whether talus will accumulate is the talus-weathering ratio. If the rate of production of talus from a cliff exceeds its weathering, talus accumulates, but if the situation is reversed it is removed (Schumm and Chorley, 1966).

Another characteristic gravity-controlled slope is that influenced by unloading phenomena (see also section 2.2.1). These have been described in some detail from several arid areas, and notably from Australia.

Unloading is a well-known geomorphological process. It is observed that cracks develop in a rock parallel to the surface and open up as the overlying load is removed although it may be that many large-scale curved crack systems are due as much to compressional stress as to unloading (Twidale, 1964). Blocks are formed between other joints and these cracks, and as the blocks are removed, smooth curved rock faces are exposed at the surface (Plate 2.4). The character of these slopes can be seen to depend largely on the character of the joint planes. This type of phenomenon is found in a wide variety of rock types and in many different climates. It is very obvious in many desert rocks, particularly in granites (Plates 2.4 and 2.6) but has also been observed in sandstones (Bradley, 1963). The characteristic rounded mountains resulting from these processes are known as domed *inselbergs* or *bornhardts* (Plate 3.2). They are perhaps more typical of savanna than of desert landscapes and it has been suggested that in deserts they are palaeoforms, although this view is not universally accepted (e.g. Ollier and Tuddenham, 1962).

The free-face and the unloading structures are often considerably modified in detail by small-scale tafoni and weathering phenomena, as described by Ollier and Tuddenham (1962). On Ayers rock in central Australia, numerous caves and pockets are found, especially where the rock-surface intersects a bedding plane in the rock. These appeared to have originated by pitting and spalling of the rock where moisture was more abundant in an overhang or along a joint. These processes are described in greater detail in section 2.2.1. The angle between the domed inselberg and the plain on which it stands is sometimes very marked and this has been attributed to weathering there and subsequent stripping (see section 3.4).

(c) Debris-covered slopes. The number of measurements of debris-covered hillside slopes in deserts is rather small in relation to the number of speculations about their genesis. In view of this disparity we shall ignore much of the literature and concentrate on two recent studies of hillsides in the south-western United States.

In a study in Arizona it was found that the angles of the straight portions of desert debris-covered hillslopes vary from 12 degrees to 37·5 degrees on granitic rocks and from 17·5 degrees to 34 degrees on volcanic rocks (Melton, 1965*b*). In both cases the mean is about 26 degrees and there is a very distinct change in the character of the slope at angles just above this (at 28·5 degrees): steeper slopes are unstable (a block of a size common on the slope rolls down it), while gentler slopes are stable. Above this angle the slopes become increasingly unstable until an angle of between 34 degrees and 38 degrees is reached; this last figure is a limiting angle for debris-covered slopes, since slopes steeper than this have no debris cover. Carson (1971) found a similar distribution of slope angles in Wyoming with a modal group between 25 and 28 degrees.

These observations alone contradict earlier studies of desert hillslopes, some of which were actually made in the same area as Melton's study (e.g. Bryan, 1923*a*). They make it clear that an explanation involving the angle of repose of the debris alone is untenable, since most slopes are below this angle. Two kinds of explanation have been offered. Melton (1965*b*) suggested that the steeper angle (34–37 degrees) was indeed the angle of *static* friction of the debris (i.e. the angle of repose), but that the limiting angle for stable slopes (about 26 degrees) was the angle of *sliding* friction of the debris. He found that experimentally produced values for these angles were very similar to the angles that he measured in the field. However Melton's experimental results only concerned the movement of single cobbles over a rocky surface, and, as Carson (1971) pointed out, this is not representative of the field situation where the blocks interlock in a mantle. Carson maintained that the real relationships should be sought in the angle of internal friction of the debris mantle. But there is a contradiction here: the angles of internal friction for gravel-soil mixtures are known to lie in the 42–45 degree range, which is considerably above the angles found in the field. Carson explained this anomaly by invoking positive pore-water pressures in the mantle, and showed that if one postulates saturated mantles, the stability angle is indeed close to 26 degrees. He maintained that although storms which could produce such saturation are rare in Wyoming (return periods in the order of 1,000 years), they may still be important in controlling slope angles. It is here, of course, that palaeoclimatic interpretations may begin to be invoked.

There is a considerable spread of slope angles about the modal slope value. One might suppose that the size of the debris would have an effect on slope angle, but Melton (1965*b*) found that the relationship between angle and boulder-size was very weak in Arizona. He suggested that the association (if any) would be due to weathering rather than frictional considerations: on steeper slopes smaller blocks roll away more easily, and finer material is washed away more quickly to leave only the coarser blocks. Carson (1971) related the angle of slopes to the state of weathering of the debris mantle. The steepest slopes (near the angle of repose of the coarse weathered blocks) have the least weathered mantles; the modal group of angles are in mantles that are partly weathered, and the low-angle slopes are in very weathered material. Locally steeper angles are often maintained near to stream channels, where these undercut the slopes (Melton, 1965*b*).

The processes taking place on debris-covered slopes are probably dominated by mass movement and by movement of individual blocks at the angles between 37 degrees and 28 degrees. Many blocks roll down the slope to the foot of the steep, straight unit. They cannot move beyond this point because of the gentler wash-controlled slopes beyond. They are therefore broken up by weathering. As the slope angles decline, wash becomes more and more important. Large blocks detached from the rock by weathering are disintegrated *in situ* and the resulting material is removed by the wash. Melton (1965*b*) found that there were linear block-fields on some slopes running normal to the contours; he attributed these to local washing away of the finer debris to leave behind only the larger blocks.

(d) Wash-controlled slopes. In fine-grained and incohesive materials, such as weathered mantles, and at the foot of debris-controlled slopes in more resistant materials, there are slopes on which wash processes predominate. For example, Schumm and Chorley (1966) described how shales on the Colorado Plateau form wash- (or creep-) controlled slopes, whereas porous, stronger sandstones give gravity-controlled slopes.

Some of the processes that occur on the wash-slopes below debris-mantled slopes will be described in sections 3.3 and 3.4; this section is concerned only with a brief description of the wash process and of the kinds of slope that develop in loose fine-grained materials where wash is the most important process. Slopewash of course is particularly characteristic of arid areas where lack of vegetation results in rapid runoff (e.g. Schumm, 1964).

We can base our discussion on the work of Emmett (1970), who examined the effects of wash by assuming that it obeyed similar laws to those obeyed by waterflow in channels (section 3.1.3).

In his field experiments Emmett found that when rain fell on a slope, the upper part of the slope transmitted runoff as a thin sheet in which the flow was laminar. Lower down the slope the flow still had properties of laminar flow, but rain falling on to the surface considerably added to the efficacy of the water as an eroding agent, and here the flow could be described as 'disturbed'. Near the foot of the slope the flow became truly turbulent.

This flow can be characterized by the same equations as characterize the flow of water in channels, namely:

$$d \propto Q^f \qquad \ldots (3.1)$$
$$v \propto Q^m \qquad \ldots (3.2)$$
$$s \propto Q^z \qquad \ldots (3.3)$$

where d is depth of flow, v is velocity of flow, s is slope, Q is discharge and f, m and z are exponents. (The equation $w \propto Q^b$ in which w is width and b is an exponent does not apply to slopes, since width can be considered constant; increase in discharge is therefore assumed to be absorbed by the other variables.) Emmett then made use of the probabilistic concepts of Leopold and Langbein (1962) to produce a solution to these equations.

The variances of d, v and s are defined respectively as f^2, m^2 and z^2, the variance of shear may be taken as $(f+z)^2$ and the variance of friction as $(f+z-2m)^2$. The most probable way in which an increase in discharge can be accommodated is for the sum of these variances to be minimized, or

$$f^2+m^2+(f-m)^2+(f+z-2m)^2 \rightarrow 0 \qquad \ldots (3.4)$$

Horton's theoretical value for f tallied closely with Emmett's experimental results: for turbulent flow the value is 0·60 and for laminar flow it is 0·33. For constant widths (as on slopes) $m = 1-f$ (since $m+f+b$ should $= 1$, and here $b = 0$, see section 3.1.3). For turbulent flow, therefore, $m = 0{\cdot}40$ and for laminar flow $m = 0{\cdot}67$. With these values the value of z which gives the minimum in equation 3.4 is $z = -0{\cdot}20$ for turbulent flow and $z = +0{\cdot}33$ for laminar flow. Thus, returning to equation 3.4, it can be seen that for turbulent flow the slope angle must decrease downslope (the slope must be concave), while for laminar flow the slope angle must increase downslope (the slope must be convex). Emmett suggested that for disturbed flow the value of z would be intermediate between that for turbulent and laminar flow, or about $z = 0{\cdot}0$, so that where there was disturbed flow one would find straight slopes with no change in angle downslope. This, of course, is a description of common slope profiles found in nature.

Emmett's observations show that overland flow can be an effective agent in erosion, and confirm the results of Leopold, Emmett and Myrick (1966) quoted in section 3.1.1. It is evident that laminar flow is an effective eroding agent in contradiction to the assertions of King (1953). In particular, disturbed flow is effective in erosion: the general pattern in such flow is for there to be distinct anastomosing concentrations of flow separated by zones of very thin sheetflow (Fig. 3.1).

Emmett's observations of wash processes on slopes in Wyoming suggest that sheetflow without rills is the dominant process on slopes there. These processes are in dynamic equilibrium with the rest of the erosional system and with the prevailing conditions of vegetation (*Artemesia* scrub) and rainfall. Rills do not develop in this part of Wyoming, and Emmett suggests that they only occur when there is a change in the controlling conditions; and even then the system might adjust as a whole to eliminate the need for the high erosion rates associated with rills. However, where vegetation is inhibited rills may well be the dominant form (Plates 3.2 and 3.3).

Wash processes on slopes sort the debris produced by rock weathering. On a sandstone talus slope and pediment in arid Australia it was found that wash on the talus slope was only capable of moving particles smaller than 5 mm. These finer particles were moved *en masse* across the pediment without much sorting, but when the sediment entered small rills the material finer than 100 mm was washed out (Walker, 1964).

We return in section 3.4 to further discussion of the efficacy of overland flow in forming pediments.

(e) Other hillslope processes in arid areas. Mass movement of the type found on slopes in humid climates is not generally thought to be important in arid areas, although we have referred to the possibility of its occurrence rarely as an explanation of the 26 degree angle on debris-covered slopes. However, some observations were made by Schumm (1964) of mass movements in a soil mantle in Colorado and on mont-

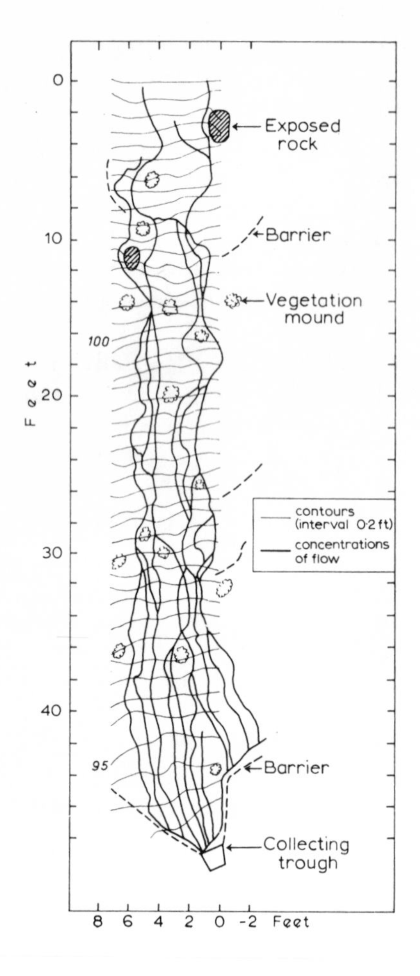

3.1 Pattern of concentrated flow in overland water movement (after Emmett, 1970).

morillonitic clay slopes in South Dakota (1956*a*). The process is probably confined to slopes on soft rocks in semi-arid (and cool) rather than truly arid (hot) areas. Schumm (1964) noted that creep erosion was important in Colorado on the upper convex and straight portions of slopes. The lower concavity did not erode because of the transport of material across it, although bedrock was near the surface. Schumm found that slopewash was more important in summer and creep was more important in winter when there was infiltration and freeze-thaw activity. Rates of creep were almost as high as those in high-latitude environments.

Larger-scale mass movements are not unknown in arid and semi-arid areas. It is surprising, for example, to find obvious slumping features on the scarp of the Tadmaït Plateau in Algeria, in one of the very arid parts of the Sahara. In the United States many slumps have been thought to be palaeofeatures (see Schumm and Chorley, 1966) and this would be one explanation of the Tadmaït slump. However modern slumping has been recorded in the arid United States (Reiche, 1937). These slumps probably occur only after rainstorms with long return periods.

Kirkby and Chorley (1967) have pointed to the possible development of slopes in temperate climates by throughflow, and we have already noted that throughflow has been thought to occur in arid areas (section 2.3.2.*g*). In eastern Australia and in Tchad where throughflow may occur in solodized-solonetz soils, the slopes involved are usually long pediment or alluvial fan slopes whose angles are controlled by other processes (Bocquier, 1968; Hallsworth and Waring, 1964), but in Western Australia, throughflow has been observed on steeper slopes (Bettenay and Hingston, 1964) and here we might expect that it would have effects similar to those described by Kirkby and Chorley. Ruxton (1958) made more specifically geomorphological studies of steep debris-mantled slopes in the Sudan. He found a debris slope beneath a free-face in which there was evidence of throughflow. The throughflow apparently removed finer particles from within the coarse matrix, and deposited them downslope where it came to the surface. Ruxton found red coarse soils on the debris slope, but black clayey soils on the lower pediments. This process of finer particle removal is rather similar to that postulated for the block-fields in Arizona by Melton (1956*b*; section 3.1.2.*c*).

We might close this discussion by noting that for many of the more important slope processes described in deserts it is long return-period events which seem to have been most important. This is the case for the processes on debris-controlled slopes (Carson, 1971), on gravity-controlled slopes (Schumm and Chorley, 1966) and for slumping (Reiche, 1937). As the return periods of events are increased we enter more and more into the field of the palaeoclimatic speculation that has cluttered most discussions of desert slope processes (e.g. Daveau, 1964), but the return-period concept does seem to release geomorphologists from the *necessity* of postulating different climates to avoid their difficulties.

3.1.3 CHANNELS

The nature of rivers and channels, like the drainage basin of which they form a part, is also determined by a group of variables whose relationships may vary through space and time. The variables relevant to the rivers and channels and their relationships in different timespans, as hypothesized by Schumm and Lichty (1965) are summarized in Table

3.3. The three time periods considered are 'geologic' (approximately the Quaternary), 'modern' (approximately the last 1000 years) and 'present' (defined as one year or less).

TABLE 3.3 *The Status of River Variables During Time Spans of Decreasing Duration*

River variables	*Status of variables during designated time spans*		
	Geologic	*Modern*	*Present*
1 Time	Independent	Irrelevant	Irrelevant
2 Geology (lithology and structure)	,,	Independent	Independent
3 Climate	,,	,,	,,
4 Vegetation	Dependent	,,	,,
5 Relief	,,	,,	,,
6 Palaeohydrology (long-term discharge of water and sediment)	,,	,,	,,
7 Valley dimensions	,,	,,	,,
8 Mean discharge of water and sediment	Indeterminate	,,	,,
9 Channel morphology	,,	Dependent	,,
10 Observed discharge of water and sediment	,,	Indeterminate	Dependent
11 Observed flow characteristics	,,	,,	,,

SOURCE: Schumm and Lichty, 1965.

Although much of the literature on fluvial landforms and processes in deserts is implicitly concerned with these variables, there are few studies which illustrate precisely their nature and relationships in the present or in the past. In this section we shall look first at the nature of flow in ephemeral channels, then at the contemporary relations amongst discharge, channel morphology and the drainage net, and finally at two examples of channel changes during different timespans.

(a) Nature of flow in ephemeral-stream channels. The varied nature of flow in ephemeral-stream channels reflects the full range of climatic and drainage basin characteristics, notably type and distribution of rainfall, infiltration capacity of surface material, antecedent moisture conditions and local topography. In deserts where precipitation occurs as short, intense showers, the flash flood is a typical result (Plates 3.5, 3.6). The hydrographs of such floods usually rise steeply, show a brief period of peak flow, and decline relatively slowly. The steep rise often has one or more vertical sections, each denoting the passage of a bore. The flood-rise hydrographs shown in Fig. 3.2 were constructed by Schick (1970) from a five-year study in the Nahal Yael research watershed in the desert of southern Israel, and they indicate some of the variety displayed by rising flash floods. Schick also suggested that in these typical desert streams with their considerable alluvial storage, high peak discharges are associated with short durations of peak flow and relatively long recession periods; and smaller peaks each have correspondingly shorter total flow duration but relatively longer duration of peak flow.

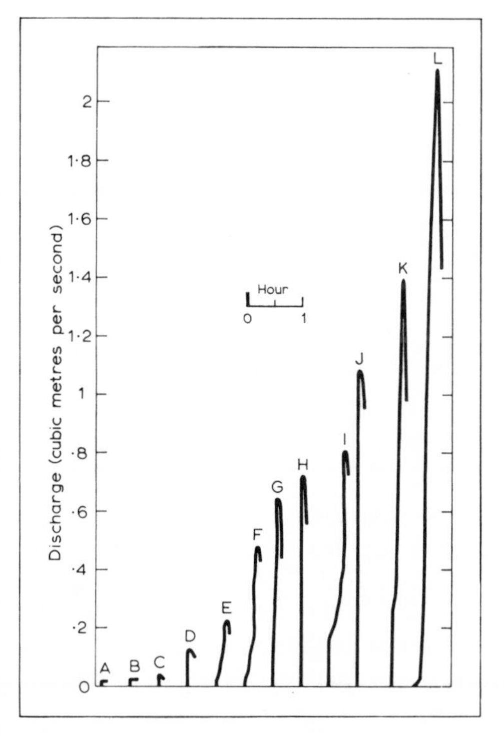

3.2 Hydrograph rises for various stations and watersheds within the Nahal Yael research watershed, southern Israel (after Schick, 1970).

Such short-lived floods may effectively modify their channels, and especially their channel beds. They certainly carry debris both in suspension and by traction. Suspended-sediment concentration characteristically increases downstream as the flow comes into contact with progressively more alluvium, and as infiltration increases (see below; Leopold and Miller, 1956; Schick, 1970). A clear example of flow reduction resulting from infiltration is shown in Fig. 3.3. The hydrographs refer to measurements of discharge at two flumes 10·9 km apart, along a stretch of Walnut Gulch, near Tombstone, Arizona, after a storm in 1964 (Renard and Keppel, 1966). The loss of water between the two flumes is some 57 per cent (or 142,542 m^3/km). Bedload movement may considerably exceed suspended-sediment movement (Schick, 1970). Leopold and Miller (1956) monitored the movement of large bedload particles in New Mexico and it was noticed that even flows no deeper than half the diameter of cobbles rolled such particles along channel beds. There is no doubt that flash floods effectively modify channel beds, but the extent of their influence on channel banks is open to some doubt. This doubt arises because firstly, ephemeral flows only occasionally impinge on channel banks and secondly, in fine alluvium, bank caving after flood recession may be more effective than erosion by floodwater itself (Leopold and Miller, 1956).

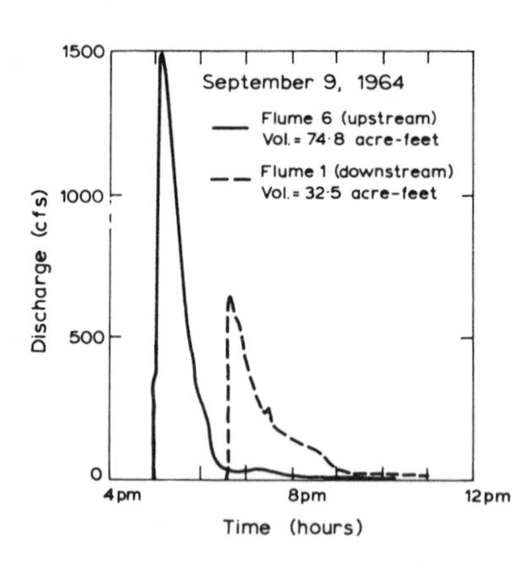

3.3 Transmission losses for a flood in Walnut Gulch, Arizona, represented by the hydrographs for two flumes 10·9 km apart in the ephemeral-stream channel (after Renard and Keppel, 1966).

(b) Inter-relationships among channel properties and the drainage net. Perhaps the most important single study of the drainage net in dry areas is that by Leopold and Miller (1956). Their analysis of ephemeral-stream channels in the semi-arid area of Santa Fé, New Mexico, yielded conclusions of wide significance and provided a basis for continuing investigation of the relations amongst properties of drainage nets in deserts.

Extending earlier work by Leopold and Maddock (1953), Leopold and Miller first demonstrated the relations between water discharge and channel depth, channel width, flow velocity and sediment discharge, at particular stations and in a downstream direction. These relations are summarized in the general equations:

$$w = aQ^b \quad \ldots (3.5)$$
$$d = cQ^f \quad \ldots (3.6)$$
$$v = kQ^m \quad \ldots (3.7)$$
$$L = pQ^j \quad \ldots (3.8)$$

where w = mean channel width, d = mean channel depth, v = mean flow velocity, Q = bankfull discharge, L = suspended-sediment load, a, c, k and p are constants, and b, f, m and j are exponents. (As $Q = wdv$, by definition, $Q = aQ^b \times cQ^f \times kQ^m$, and $b+f+m = 1$). Of particular interest are variations in the values of the exponents at particular stations and downstream, and comparisons between these values and those from areas with perennial streams.

Table 3.4 shows the values given by Leopold and Miller (1956) and Bruce (1966).

TABLE 3.4 *Values for Exponents b, f, m and j*

	Average downstream relations				*Average at-a-station relations*			
	Ephemeral streams		*Perennial streams*		*Ephemeral streams*		*Perennial streams*	
	New Mexico[1]	*Nebraska*[2]	*Average*[3]	*Nebraska*[2]	*New Mexico*[1]	*Nebraska*[2]	*Average*[3]	*Nebraska*[2]
b	0·5	0·03	0·5	0·69	0·26[4]	0·35	0·26	0·24
f	0·3	0·48	0·4	0·12	0·33[4]	0·43	0·40	0·56
m	0·2	0·45	0·1	0·19	0·32[4]	0·22	0·34	0·20
j	1·3		0·8		1·30[4]		1·5–2·0	

[1] SOURCE: Leopold and Miller, 1956. (The authors express reservations about the small number and accuracy of their observations, but the data probably give a reasonable indication of reality.)
[2] SOURCE: Brice, 1966. (Data from the Medicine Creek basin. At-a-station data may be considered reliable; downstream data only indicate the correct order of magnitude.)
[3] SOURCE: Leopold and Maddock, 1953.
[4] Unadjusted median values.

An important contrast between ephemeral-stream channels in a semi-arid area and 'average river channels' arising from these data is that in the case of the former, suspended-sediment concentration increases downstream more rapidly than discharge (e.g. $j = 1{\cdot}3$) whereas in the case of the latter the reverse is usually true (e.g. $j = 0{\cdot}8$). It seems probable that in ephemeral-stream channels the increased sediment concentration downstream may arise largely from infiltration into the channel bed. It is also possible that the higher downstream rate of velocity increase observed in the ephemeral-stream channels may be a response to the increase in sediment concentration. Schick (1970) has also recorded increasing sediment concentration with distance downstream in ephemeral-stream channels in southern Israel.

The second phase of Leopold and Miller's enquiry was an analysis of the drainage nets. They demonstrated that stream order, as defined by Horton (1945), was proportional to the logarithm of stream length, stream number, drainage area and channel slope.

i.e. $$O \propto \log N \qquad \ldots (3.9)$$
$$O \propto \log l \qquad \ldots (3.10)$$
$$O \propto \log s \qquad \ldots (3.11)$$
$$O \propto \log A_d \qquad \ldots (3.12)$$

where O = stream order, N = stream number, l = stream length, s = channel slope, A_d = drainage area.

Such relationships have also been established, using similar ordering systems, in other arid and semi-arid areas (e.g. Brice, 1966) and in other climatic situations. The precise relationships, as expressed on semi-logarithmic graphs by the slope of regression lines and the intercept values on the Y axis when the X axis value = 1, may vary from area to area.

Leopold and Miller used these relationships to demonstrate the connections between channel net and hydraulic-geometric properties. From equations 3. 5–8, it follows that:

$$l \propto A_d^{\,k} \qquad \ldots (3.13)$$
$$s \propto A_d^{\,k} \qquad \ldots (3.14)$$
$$s \propto l^k \qquad \ldots (3.15)$$

where k is an exponent having a particular value in each equation. Furthermore, as:

$$Q \propto A_d^{\,k} \qquad \ldots (3.16)$$

then:

$$O \propto \log Q \qquad \ldots (3.17).$$

And, as w, d, v and L are related to Q (equations 3.1–4) it follows that these variables are also related to stream order in the form:

$$O \propto \log w \qquad \ldots (3.18).$$

Similarly, because:

$$L \propto Q^j$$

and:

$$Q \propto A_d^{\,k}$$

then:

$$L \propto A_d^{\,jk} \qquad \ldots (3.19).$$

Thus it can be shown that all the variables discussed are interrelated in the form of power or exponential functions. Such relationships strongly suggest a mutual adjustment amongst the variables which is a response to general environmental conditions. In short, there appears to be a tendency towards the creation of a quasi-equilibrium condition in the system.

(c) Channel changes. The recognition of mutual adjustment between variables in response to environmental conditions is of great value in the study of channel changes. In principle, a change of any one variable is likely to disturb the established equilibrium and to cause an appropriate adjustment in the channel system. In what is patently a complex multivariate situation, similar adjustments—in terms of visible landforms—may result from changes in different variables or combination of variables within the system. The changes may be considered at various time scales. Here we shall examine two superficially contrasted but fundamentally similar examples. The first concerns channel metamorphosis along the Murrumbidgee River on the Riverine Plain in Australia as the result of climatic change during the Quaternary. The second relates to channel initiation during the last hundred years in valley floors of the south-western United States.

(i) Channel change on the Riverine Plain, Australia. The form and metamorphosis of an established, stable channel in alluvium should be controlled by the quantity of water and the type and quantity of sediment moved through it. In a series of articles, Schumm (1960, 1961, 1963*b*, 1968 and 1969) has described empirical relationships between channel characteristics and water and sediment discharge for 36 stable alluvial channels in semi-arid and sub-humid regions of the Great Plains (U.S.A.) and the Riverine Plain (Australia). He used empirical data to develop equations which express simply the general relations between channel characteristics and water and sediment discharge and, more particularly, the changes likely to occur to the former with changes of the latter. These derived equations are presented below, without the detailed data and argument upon which they are founded, as a basis for discussion.

Firstly, water discharge (Qw) is directly related to channel width (w), depth (d), and meander wavelength (L) and is inversely related to channel gradient (S). Thus:

$$Qw \simeq \frac{w, d, L}{S} \qquad \ldots (3.20).$$

In addition, at a constant discharge of water,

$$Qs \simeq \frac{w, L, S}{d, P} \qquad \ldots (3.21)$$

where Qs is bedload sediment discharge, and P is sinuosity (ratio of channel length to valley length).

Four further equations describe the effect of changing, separately, water and sediment discharge. A plus or minus exponent is used to indicate how the various aspects of channel morphology will change.

$$Qw^{+} \simeq \frac{w^{+}\ d^{+}\ L^{+}}{S^{-}} \qquad \ldots (3.22)$$

$$Qw^{-} \simeq \frac{w^{-}\ d^{-}\ L^{-}}{S^{+}} \qquad \ldots (3.23)$$

$$Qs^{+} \simeq \frac{w^{+}\ L^{+}\ S^{+}}{d^{-}\ P^{-}} \qquad \ldots (3.24)$$

$$Qs^{-} \simeq \frac{w^{-}\ L^{-}\ S^{-}}{d^{+}\ P^{+}} \qquad \ldots (3.25)$$

In practice, of course, environmental changes often lead to contemporaneous changes in both water and sediment discharge. In the following equations, which describe such changes, Qt represents the *proportion* of bedload in the total sediment load, F is the width-depth ratio, and the plus and minus signs indicate an increase or decrease in a variable.

$$Qw^{+}\ Qt^{+} \simeq \frac{w^{+}\ L^{+}\ F^{+}}{P^{-}}\ S^{\pm}\ d^{\pm} \qquad \ldots (3.26)$$

$$Qw^{-}\ Qt^{-} \simeq \frac{w^{-}\ L^{-}\ F^{-}}{P^{+}}\ S^{\pm}\ d^{\pm} \qquad \ldots (3.27)$$

$$Qw^{+}\ Qt^{-} \simeq \frac{d^{+}\ P^{+}}{S^{-}\ F^{-}}\ w^{\pm}\ L^{\pm} \qquad \ldots (3.28)$$

$$Qw^{-}\ Qt^{+} \simeq \frac{d^{-}\ P^{-}}{S^{+}\ F^{+}}\ w^{\pm}\ L^{\pm} \qquad \ldots (3.29)$$

At present, it remains to be shown how dependable these equations are; Schumm himself has employed them in examining the Murrumbidgee River and its palaeochannels on the semi-arid Riverine Plain of New South Wales.

Long-term channel changes are ultimately associated with climatic changes, except where tectonic activity is significant. Often, however, evidence of former channels and channel changes in the Quaternary has been eliminated, and the nature of former channels and their changes is a matter of speculation. But the Murrumbidgee River system provides a notable exception: here there is preserved evidence of successive channel systems, and the problem is to describe differences between them and to seek an explanation of the differences. Certainly the explanation lies ultimately in climatic changes, but the nature of these changes has been a matter of controversy (e.g. Butler, 1960; Langford-Smith, 1962).

Although the local and regional sources of water and sediment in the basin of the Murrumbidgee River have remained largely unchanged, there is evidence on the Riverine Plain of three distinct channel systems (Schumm, 1968). The modern river has a low width-depth ratio, a low gradient and a moderately high sinuosity, and it carries only a small, fine-grained (suspended) load. An older channel, that of the 'ancestral river', has a similar width-depth ratio, gradient and sinuosity, but it is larger, and has a higher meander wavelength; its sediment load is also fine-grained (suspended) and small. A still older channel system, that of the 'prior streams', contains relatively wide and shallow channels which have high gradients and width-depth ratios, low sinuosity, and relatively high sand (bed)loads.

The prior-stream channels were probably produced during a drier period than at present in which there were high flood peaks (estimated from calculations based on measurements of channel cross-sections, slope, and mean velocity) but less total runoff that at present. Evidence for a drier period includes the correspondence between soil salinity and the distribution of prior-stream channels, weathering characteristics of palaeosols, aeolian deposits and hydrological information (Schumm, 1968, p. 48). Transformation of the prior-stream channels into those of the ancestral river could have been accomplished by an increase of water discharge and a decrease in the proportion of bedload, as described in equation (3.28). Such changes may have accompanied a change of climate to more humid conditions on the plain and to higher precipitation in the headwaters of the Murrumbidgee, as occurred during the Australian 'Little Ice Age' 3000 years ago. The change from prior channel to ancestral channel as deduced from equation (3.28) requires a significant reduction of channel gradient, and Schumm (1968) showed that this was achieved largely through lengthening of the channel by an increase in sinuosity.

To create the present channel of the Murrumbidgee a further change was necessary: as the present channel is similar in most respects to that of the ancestral river except that it is smaller and has a lower meander wavelength, a reduction of water discharge is required (see equation 3.23). This deduction leads to the conclusion that the present climate is 'drier' than that prevailing at the time of the ancestral river.

(ii) Recent channel trenching in the south-western United States. The alluvial stratigraphy of many valleys in the south-western United States reveals abundant evidence of alternating phases of erosion and deposition since the last glaciation (e.g. Haynes, 1968; Fig. 1.7), and the changes in fluvial processes can confidently be associated with changes of climate. Documentary sources indicate that many alluviated valley floors which were unchannelled in the early part of the nineteenth century have become entrenched with vertical-walled channels called *arroyos* (Plate 3.8) since the beginning of white settlement. Much attention has been given to the causes of this recent entrenchment and it is clear that, unlike earlier entrenchments, the phenomenon may have resulted from different causes in different areas (Cooke and Reeves, in preparation).

The erosion of valley floors to produce arroyos may have been accomplished by increase in flow velocity, increase in the erodibility of valley-floor materials, or a combination of both. Increase in flow velocity could have resulted from increase of discharge or slope, reduction of surface roughness, or increase in depth of flow. Increased erodibility of valley-floor materials might have arisen through the reduction of riparian vegetation. Depth of flow might have been increased by concentrating a given discharge through the construction of embankments or the cutting of irrigation canals. A good example is Greene's Canal on the Santa Cruz River, Arizona, which collected and concentrated floodwater to such an extent that the original canal has now been transformed into an arroyo. Surface roughness could have been reduced by removal of vegetation—especially the grasses—on the floodplain by cattle or by road traffic. An illustration of this change occurs in the Altar Valley, Arizona, where an arroyo has been formed in part along the line of an early road. Increase of slope could locally have been effected by deposition arising from increased sediment concentration as discharge declined downstream: slope could have been increased to the point where gullying was initiated (e.g. in Wyoming and New Mexico: Schumm and Hadley, 1957), and along a single valley there could have been several gullies which ultimately became integrated into a single arroyo. Such slope changes would have been essentially random, and need not have been determined by specific environmental changes in the basin.

But the most commonly advocated cause of arroyo formation is that of increased discharge, achieved either through a change of climate or through vegetational changes from other causes leading to an increase of surface runoff and valley-floor discharge. Several different secular climatic changes have been proposed. Some have argued that, during the critical period of trenching towards the end of the nineteenth century, climate may have become drier (e.g. Antevs, 1952; Bryan, 1925*b*); others have suggested that it may have become wetter (e.g. in Arizona: Huntington, 1914); and there are those who argue for a period of relatively low frequency of small rains and high frequency of large, heavy rains, especially in summer (e.g. in New Mexico: Leopold, 1951). Each of these changes, of course, would have been accompanied by vegetation changes. Vegetation removal and alteration could also have been achieved by grazing of cattle and sheep, and perhaps by fire and deforestation, and such changes would probably lead also to greater surface runoff and increased discharge along the valley floors (e.g. in Colorado: Duce, 1918); and a further increase might result from reduction of infiltration capacity by animal trampling (e.g. Leopold, 1921).

In any given area there are likely to be several changes, some perhaps independent of one another, others consequent upon a single change. Melton (1965*a*), for example, argued that in some Arizonan *cienegas* (marshy, grass-covered valley-floor areas) reduction of grass cover in the valley led (*a*) to deposition of sediment on the cienegas and to increased inclination of their *transverse* slopes (slopes normal to line of flow) and thus to concentration of flow; and (*b*) to reduction of critical erosion velocities of superficial sediments and reduction of hydraulic roughness. The net effect of these changes would have been to increase both flow velocities and the erodibility of the cienega deposits, so that entrenchment occurred.

This very brief review and sampling of the extensive 'arroyo' literature serves to emphasize the variety and possible complexity of arroyo origins. Some explanations—climatic change for instance—are regional in character; others—such as concentration of flow in irrigation channels—are of only local significance. For each hypothesis there have often been both advocates and adversaries. It appears that much of the controversy has arisen because of the complexity of the problem, because there has been inadequate recognition of the fact that different changes in different areas can lead to similar results, and because precisely documented local explanations are rather scarce.

3.2 Mountains and Plains: Introduction

Our basic desert profile (Fig. 3.4a) comprises a *mountain* and a *piedmont plain*. Although such a profile characterizes most deserts, it may vary greatly in detail from place to place (Plates 2.9 and 2.10). For instance, in

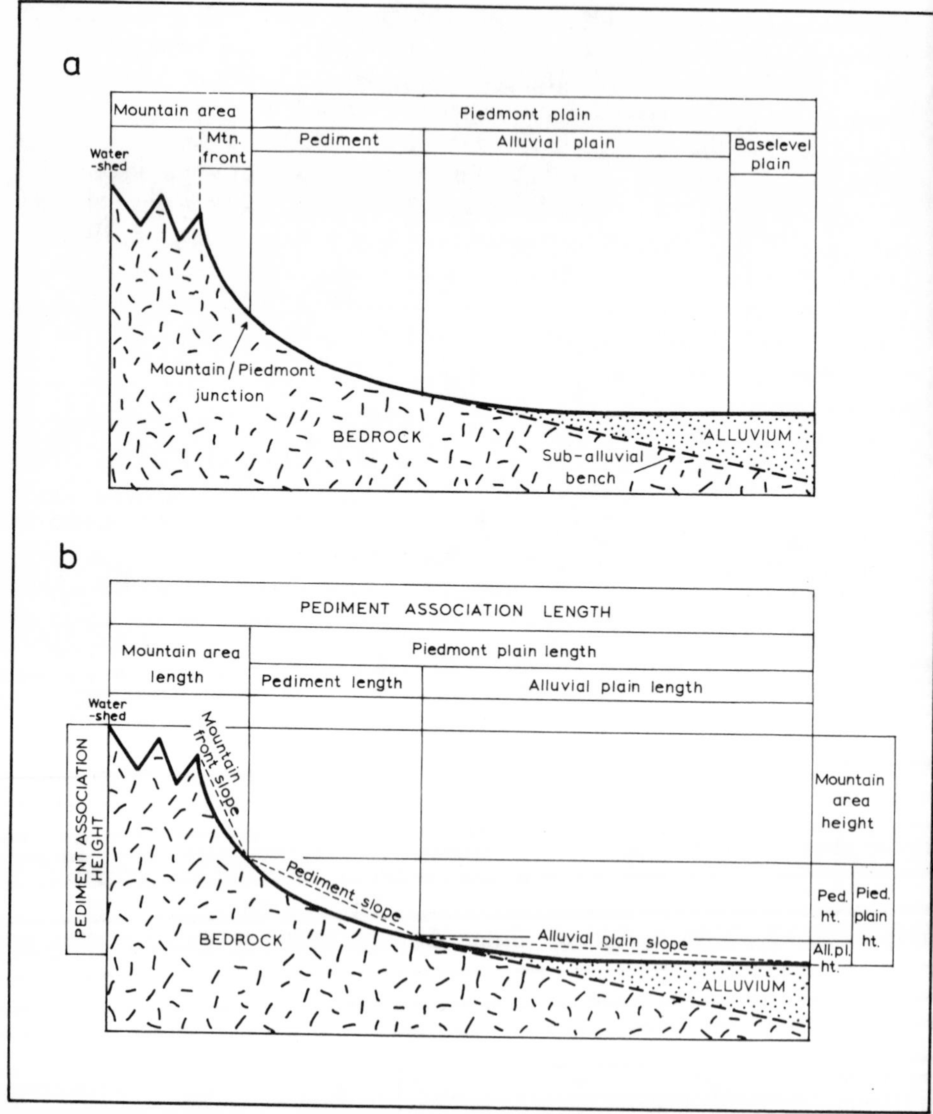

3.4 Components and morphometric properties of the characteristic desert profile.

basin-range country the ratio of mountain area to plain area may approach one, whereas in the open country of the central Sahara or South Australia it may be very much smaller. Furthermore, the piedmont plain may be composed of a variety of different landforms. A basic contrast is between the bedrock *pediment* and *alluvial plains*. The latter may include *base-level plains* (such as a playa), *bajadas* (alluvial zones often of considerable stratigraphic and topographic complexity), or *alluvial fans*. The origins of, and relationships among, these various landforms are major concerns of the desert geomorphologist.

Unfortunately there has been relatively little morphometric analysis of the mountain-piedmont plain profile and its constituent parts, and this is certainly a field for profitable enquiry in the future. A significant exception is the study by Lustig (1969*a*) in which he measured 11 topographic parameters of mountain ranges in the Basin-Range Province of the United States, determined their areal variations by trend-surface analysis, and identified significant geomorphological consequences of these distributions. We shall refer again to this study, and we hope that similar studies may be possible in other areas.

In the following pages we propose to build our discussion of the nature and origin of mountain-piedmont plain profiles in deserts around the examination of three distinct subsystems within desert-drainage systems: alluvial fan systems, pediment systems, and playa systems. The discussion will not be comprehensive; neither will it be impartial. But it will impinge upon most aspects of landforms in desert drainage systems and their evolution, and it will emphasize those studies which have used precise measurements of form and process. Of the features we fail to consider in detail, the alluvial plain (bajada) without distinct alluvial fans is the most significant: this omission is deliberate, for the feature has received relatively little attention in recent years.

3.3 Alluvial Fan Systems

One type of desert drainage system that has attracted considerable attention includes an alluvial fan, the mountain source area to which it is tributary and the playa area tributary to it: the alluvial fan system.

Most of the recent precise and important work on alluvial fans and their associated landforms has been accomplished in various parts of the western United States, notably by Anstey (1965), Beaty (1963 and 1970), Blissenbach (1954), Bull (1962, 1963, 1964*a* and *b* and 1968), Denny (1965 and 1967), Drewes (1963), Hooke (1967 and 1968), Lustig (1965) and Melton (1965*a*), and this commentary draws heavily upon their work. Emphasis on these features is justified, not only because they are common in many desert areas, but also because their study throws light on many morphometric relationships and on geomorphological processes in desert drainage systems.

3.3.1 DEFINITION AND OCCURRENCE OF ALLUVIAL FANS

Alluvial fans are deposits with surfaces that are segments of cones radiating downslope from points which are usually where streams leave mountains, but which may be some distance within the mountain valleys (e.g. Murata, 1966), or may lie within the piedmont plain (Plates 3.11 and 3.12).

The fan shape is determined by many features, of which the two most important are a point of drainage exit from mountains, and divergent paths of separate flows (Beaty, 1963). As Bull remarked, fans are fundamentally formed by deposition of alluvial material beyond the limits of mountain valleys as the result of 'changes in the hydraulic geometry of flow after the stream leaves the confines of the trunk stream channel' (Bull, 1968, p. 102). Stream discharge is by definition equal to the product of the mean depth, width and velocity of flow. Thus when a stream reaches the end of a channel on a fan, it spreads out. The increase in width of flow is accompanied by decrease in depth and velocity, which causes deposition of sediment. In addition, discharge often decreases as the stream travels over permeable deposits, thus increasing sediment concentration and leading to further deposition (Bull, 1968). These comments are true for alluvial fans generally, but as the eyewitness accounts recorded by Beaty (1963) show, individual flows may not widen greatly or as rapidly as the fan itself. Thus decrease of depth and velocity and overall decrease of discharge combine to focus deposition near to the limits of channels leading from mountains.

Alluvial fans are by no means confined to hot deserts. They occur in cold arid areas such as northern Canada (Leggett, Brown and Johnston, 1966), and also occasionally in humid areas. But in humid areas of perennial drainage, streamflow tends to remove the potential fan debris through the drainage system.

Although they are praticularly common in the western United States and have been most studied there, fans are common features in many deserts, especially basin-range deserts. For example, they occur in the Atacama Desert, in deserts of Pakistan and neighbouring areas, and in Iran.

Several features of deserts have been invoked to explain the occurrence of alluvial fans. Firstly, lack of vegetation means that positions of drainage channels are relatively unfixed. Secondly, according to Beaty (1963), ideal conditions for fan formation—typical of many desert areas

—are long periods when debris can accumulate in the mountains, and occasional heavy thunderstorms which lead to rapid and massive evacuation of material. But Melton's (1965*a*) evidence from Arizona—that blocky fragments are not being produced rapidly from bedrock at present and large quantities of coarse debris are not therefore available for fan formation—suggests that 'aridity' alone is not sufficient for the formation of alluvial fans.

Other factors are perhaps more important. Alluvial fans seem to occur mainly where the ratio of depositional area to mountain area is small: that is, where large highlands border small lowlands. This is true, for instance, in the northern part of the Basin-Range province in California, where the mountain-basin relationship largely arises from tectonic movements and where fans are common; but in the Mojave Desert to the south, the ratio is large, mainly because of extensive erosional modification of tectonic relief, and fans are rare (Hooke, 1968; Lustig, 1969*a*). The juxtaposition of mountain and lowland may result, in general, either from tectonic movements or from fluvial exploitation of boundaries between rocks with strongly contrasted lithologies.

In addition, dissected mountain areas with channel systems focusing on the boundary between mountain and plain are an important prerequisite for fan formation. Fan formation and fan morphometry will also be partly determined by the location, number and spacing of drainage-channel ends on the mountain-plain boundary and by the plan-shape of the boundary.

3.3.2 SOME MORPHOMETRIC CHARACTERISTICS OF ALLUVIAL FANS

In their simplest form, alluvial fans are discrete features which coalesce along the base of a mountain front: such fans are 'unsegmented'. More commonly, an alluvial fan is composed of several segments that have resulted from erosional and depositional changes over a period of time.

This important distinction has given rise to some confusion in the earlier literature on fans, and makes precise generalization of fan dimensions difficult. Anstey (1965) measured fans in the western United States and Pakistan with radii of up to 9·6 kms, and overall gradients up to about 10 degrees (Fig. 3.5) and these figures seem to be generally in accord with those of other workers. Further precise generalizations especially in terms of fan area, are not possible at present.

In detail, the morphology of alluvial fans is composed of three basic elements: channels, abandoned areas of former fan-surfaces, and depositional surfaces downslope of channels. Of these features, the channels play an important role. There are often two distinct groups. The first includes distributaries from the main channel at the apex of the fan.

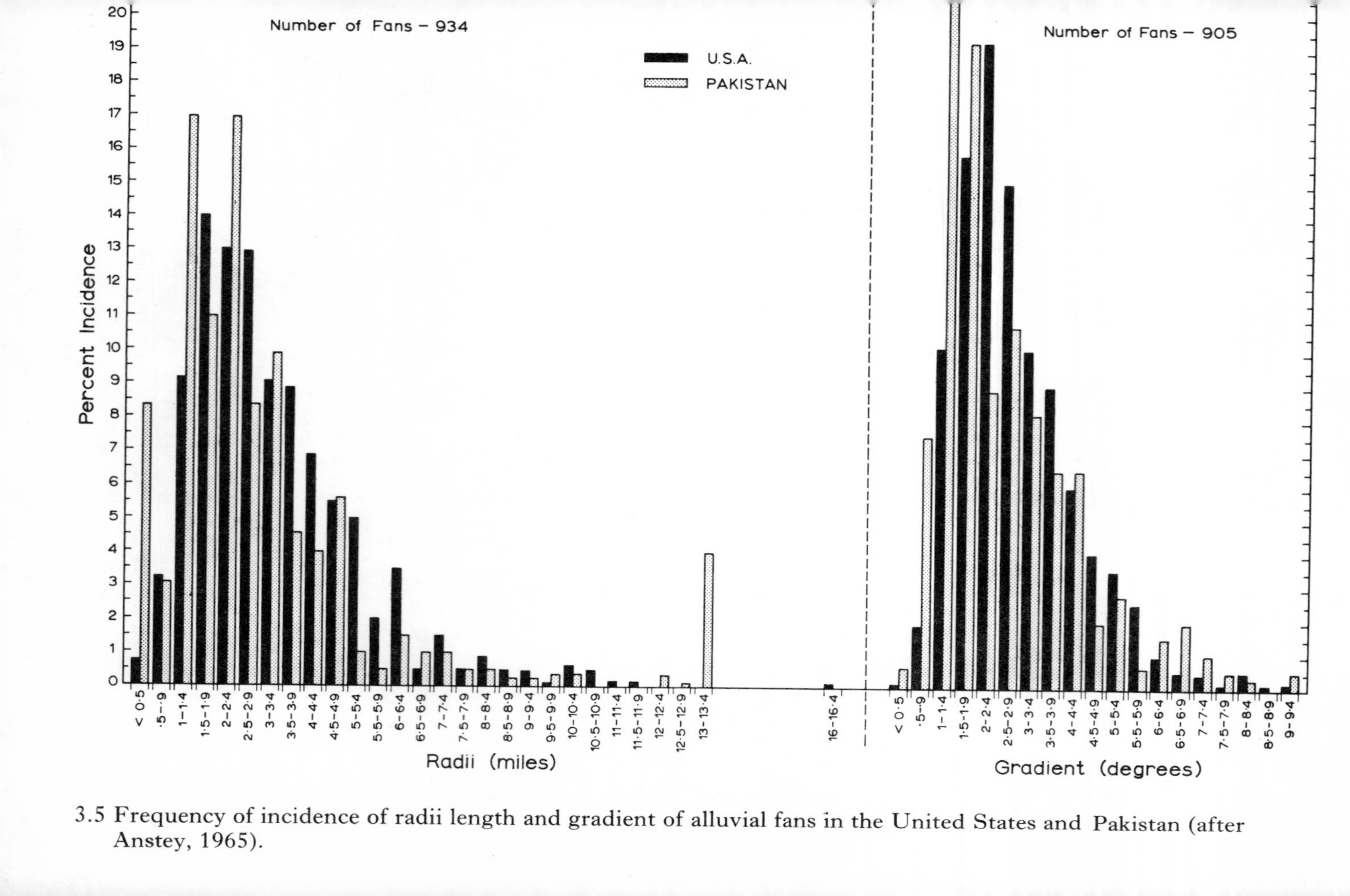

3.5 Frequency of incidence of radii length and gradient of alluvial fans in the United States and Pakistan (after Anstey, 1965).

These may form a generally radial, braided pattern, in which channels can be shown to be of widely different ages. The second group consists of channels, often forming a dendritic pattern, that rise on the fan itself and result from local rather than regional runoff. Most fans, even unsegmented ones, are entrenched near their original apex at the mountain front.

An inherent feature of fan development is the continuously changing pattern of channels and loci of deposition. These shifts may occur within a single flood, or with successive floods, and arise largely from the progressive filling and overflow of channels often as the result of blockage by boulders or mudflows. Over a long period of time these changes ensure the maintenance of fan form by distributing material widely over the surface. Channel piracy is a common feature of alluvial fans and arises in large measure because of the juxtaposition of numerous channels of differing bedloads and gradients at different levels. Mountain-derived channels, for example, may have coarser load and steeper gradients than gullies confined to the fans which receive only local water and sediment (Denny, 1967).

(a) Morphometry of unsegmented fans. Three morphometric properties of unsegmented fans which have been considered are shape, area and slope.

(i) Shape. As their name implies, alluvial fans are characteristically fan-shaped in plan. Several authors have attempted to describe fan shape precisely, using simple equations (Bull, 1968; Lustig, 1969*b*; Murata, 1966; Troeh, 1965). For instance, the surface of a fan may be compared to that of a segment of a cone characterized by concavity of the longitudinal profile and convexity of the transverse profile, and an appropriate equation, suggested by Troeh and adopted by Bull and Lustig, is:

$$Z = P+SR+LR^2 \qquad \ldots (3.30)$$

where Z = the elevation of any point on the fan, P = elevation at the apex of the fan, S = is the slope of the fan at P, R = the radial distance from P to Z and L = half the rate of change of slope along a radial line.

Application of this equation by Bull (1968) to a single fan demonstrated that it provided a reasonable approximation of the fan form. In the future, as Lustig (1969) observed, it should be possible to find a general solution of the equation on a regional basis and thus provide an index for regional comparison.

(ii) Area. Bull (1964*b*), Hooke (1968) and others have demonstrated

that fan area A_f is related to drainage basin area A_d by the general equation:

$$A_f = cA_d^n \qquad \ldots (3.31)$$

where the constant c is the area of fan with a drainage area of one square kilometre, and the exponent n is the slope of the regression line (Fig. 3.6). Anstey's (1965) data do not show a similar relationship between fan area and drainage basin area and the reason for this contrast appears to lie in differences in methods of delimitation.

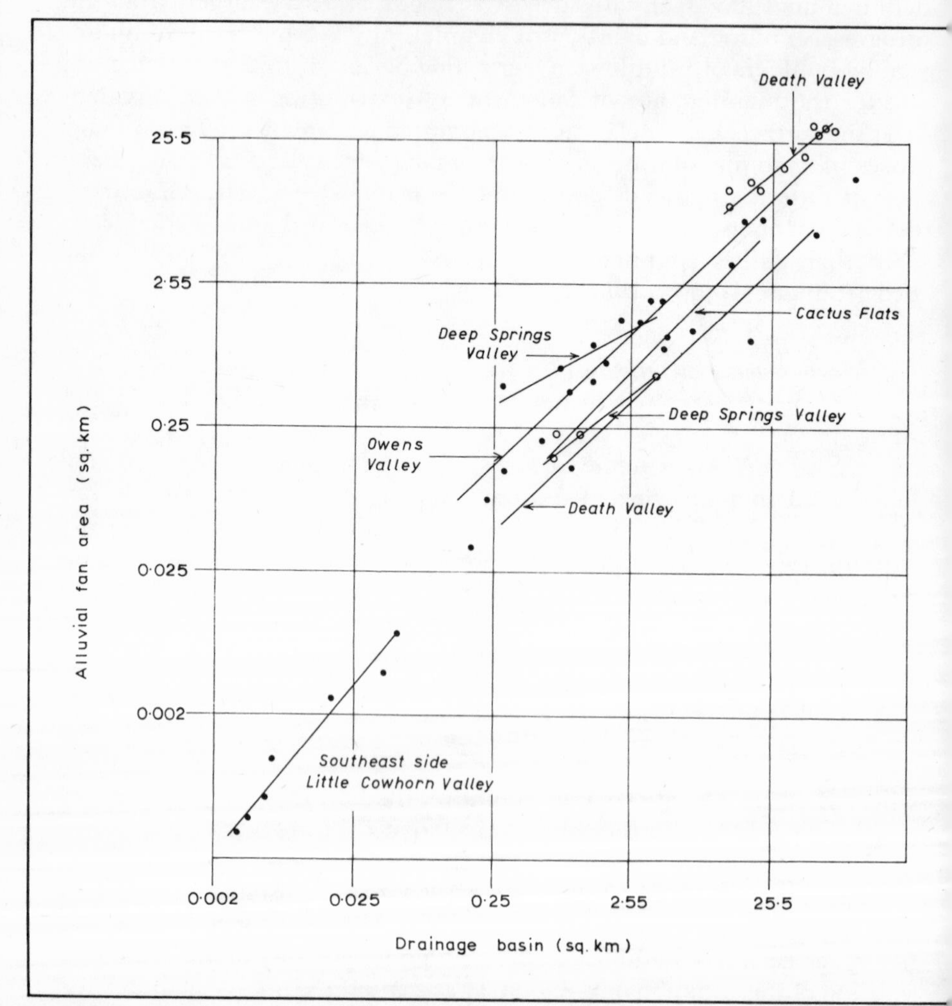

3.6 Relation between alluvial fan area and drainage basin area in different regions of eastern California (after Hooke, 1968).

Hooke argued that the general equation results from a tendency towards a steady state area among coalescing fans in similar environments; that they are, in short, parts of space-sharing systems. His argument rested on laboratory and field observations which suggest that *relatively uniform* deposition may occur over a fan-surface. Thus, if a fan is too small for the volume of debris being supplied to it, for instance, it will increase in thickness (and area) faster than adjacent fans until a steady state is established (or re-established). Because the volume of debris being supplied increases with drainage area, the larger drainage areas are generally associated with larger fans. This hypothesis therefore requires that the area of each fan is proportional to the volume of material supplied to it per unit time.

In addition, Hooke (1968) argued that in an enclosed basin containing an aggrading playa surrounded by alluvial fans, the rate of playa deposition beyond the fan should equal the rate of fan deposition. For instance if the playa is too large for the amount of debris being supplied, it would increase in thickness more slowly than the fans, which would therefore encroach on it, reduce its area and thus increase its rate of thickening. Because of the ephemeral nature of change within the system, a steady state may not be achieved at any one time, and the tendency towards it may continue for a protracted period. Over longer time periods, fans in enclosed basins will continue to grow, and they will progressively bury the mountain front.

Five important factors affect c, other than area: *(i)* the ratio of depositional area to erosional (drainage basin) area, with which c is directly and positively correlated; *(ii)* sediment yield, as determined by rock types and relief in the drainage basin. For instance, fans derived from easily erodible rocks are larger than those derived from resistant rocks; *(iii)* tectonic tilting may affect c by altering basin area and relief as in Death Valley, California, where fans associated with the dip-slope of upfaulted blocks tend to be larger than those adjacent to fault-scarp mountain fronts; other variables of significance are *(iv)* climate, on a regional scale and *(v)* the amount of space available for fan deposition, on a local scale.

The value of n, the regression coefficient, is usually less than one, implying that large drainage basins yield proportionally less sediment on to fans than smaller ones. Hooke suggests several reasons for this: *(i)* large basins may be less frequently covered by a single storm; *(ii)* more sediment may be stored on valleyside slopes or in the channels of large basins and *(iii)* valleyside slopes tend to be lower in larger basins and may therefore yield less sediment.

The type of steady-state relationship envisaged by Hooke contrasts with that of Denny (1965 and 1967) who suggested that rate of fan deposition equalled rate of fan erosion. There is no doubt that both erosion and deposition occur on fans. To take a simple example, new deposition on to fans is associated with the main channel in the mountains; but downslope there may be quite separate channel systems signifying erosion confined to the fan itself.

Denny's hypothesis has not been established—indeed it would be almost impossible to demonstrate empirically—and it seems rather unlikely, unless material eroded from the fan is *removed entirely from the system*. For if eroded material merely accumulated on the adjacent playa, the playa would influence the form of the fan. For example, if the playa encroached on the fan, fan deposition rate would increase, but erosional processes on the fan would operate over a smaller area and erosion rate would decrease until deposition rate was positive on fans and equalled that on the playa (Hooke, 1968).

(iii) Slope. Fan slope, which rarely exceeds 10 degrees, is determined in much the same way as longitudinal stream profiles by a suite of variables, of which debris-size, water discharge and the type of depositional process are the most important. The profile of an alluvial fan probably tends towards the attainment of a steady-state condition.

The longitudinal profile of an unsegmented alluvial fan is generally a smooth exponential curve: there is not, as it sometimes asserted, a sharp break of slope on the profile at the junction between mountain and plain; it is an hydraulic profile determined by variables in the drainage system.

Bull's investigations in the Central Valley of California suggested that the longitudinal profile of fans with source areas having high rates of sediment production are steeper than those from areas with low rates of sediment production (Bull, 1968). Fan slope is generally inversely proportional to fan area, drainage-basin area and discharge. The latter is especially important because large discharges can transport debris on a lower slope than small discharges as they have higher flow velocities and higher bed-shear stresses (Hooke, 1968). One consequence of this situation is that increase of discharge across a single fan may cause erosion or even trenching near the fan head, and deposition towards the fan toe, thus decreasing slope overall; smaller discharges will tend to deposit higher up the slope: actual fan slope will reflect a balance between these tendencies. Melton (1965*a*) showed that for a sample of 15 alluvial fans in southern Arizona, upper-fan gradient (S) is positively correlated with basin relative relief ($H/\sqrt{A}$)—a dimensionless measure of basin ruggedness defined as the ratio between vertical relief above fan apex (H)

1.1 *Vent de sable* near Adrar Bous, Niger. Photo: J. Rogers.

1.2 Dust-devil, Kansas Settlement, Willcox Playa, Arizona.

1.3 A stretch of intermittent flow along the allogenic Mojave River, near Barstow, California, 1967. Flow is from the left, across the road.

1.4 View north across the Huasco Valley, Atacama Desert, Chile, showing fluvial terraces cut by the allogenic, exoreic Huasco River (cf. Fig. 1.4).

VOLCANIC LANDFORMS

1.5 Geysers and sinter in a region of ignimbrite deposition, El Tatio, northeastern Atacama Desert, Chile.

1.6 Composite volcano and blocky lava field, Volcan Licancabur, eastern Atacama Desert, Chile.

FAULT FEATURES

1.7 Recent fault scarp developed across an alluvial fan, Panamint Valley, California. One drainage line has been maintained across the uplifted zone.

1.8 Salt lake created by impounded drainage at the base of a fault scarp in the San Andreas rift zone, Carrizo Plain, California.

PALAEOFORMS AND CLIMATIC CHANGE

1.9 Relict landform: overflow channel between Silver Lake and Searles Basin in the Mojave River system, Mojave Desert, California.

1.10 Evidence of climatic change: rock drawings of savanna-land animals – rhinocerous and giraffe – in sandstone of the Tibesti Mountains, Tchad, now a desert area. Photo: C. Vita-Finzi.

CONTRASTS IN DESERT SCENERY: Chile, the United States, and the Sahara.

1.11 Incised meanders of the San Juan River, the 'Goosenecks', in the plateau country of Colorado, U.S.

1.12 Monument Valley, Utah – alternating 'free-face' and 'constant' slopes in horizontally-bedded sediments of different lithologies.

1.13 'Badlands' developed in marls, clays and ignimbrites along a tributary of the Rió San Bartolo (Puna Slope), Atacama Desert, Chile.

1.14 Rotational slump masses developed in ignimbrite overlying marls along the Rio Grande, northeastern Atacama Desert (Puna Slope), Chile.

1.15 The Atakor Massif, Algeria.

2.1 Stone pavement (reg) surface and ventifact near Ouargla, Algeria.

2.2 Boulder of agglomerate set in a stone pavement, which has been split along several sub-parallel planes, southern Atacama Desert, Chile.

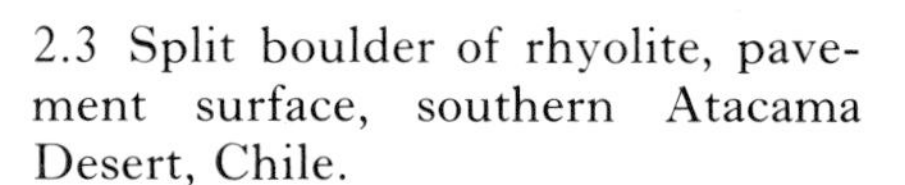

2.3 Split boulder of rhyolite, pavement surface, southern Atacama Desert, Chile.

2.4 Massive exfoliation on a granite mountain front, Granite Mountains, Mojave Desert, California.

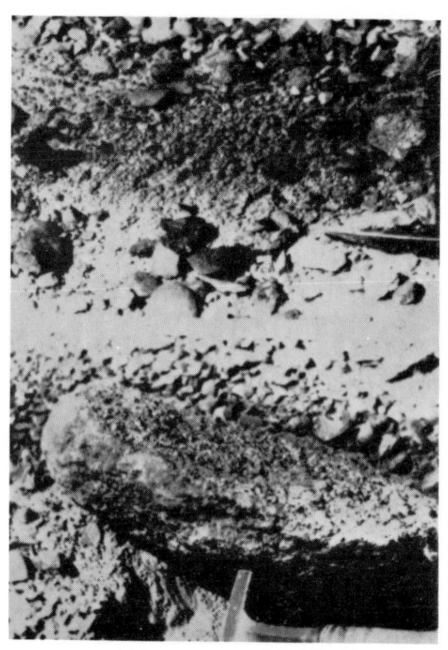

2.5 Granitic boulder disintegrated down to ground level, pavement surface, southern Atacama Desert, Chile. Upper illustration – the boulder *in situ*; lower illustration, the boulder removed.

2.6 Weathering cavern in agglomerate, near Copiapó, southern Atacama Desert, Chile.

2.7 Tafoni in granite near Ein Ekker, Algeria.

2.8 Weathering cavern in quartz monzonite of a pediment channel, Lucerne Valley, Mohave Desert, California.

2.9 Weathering pits with flared rims in sandstone, Casa Grande National Monument, Arizona.

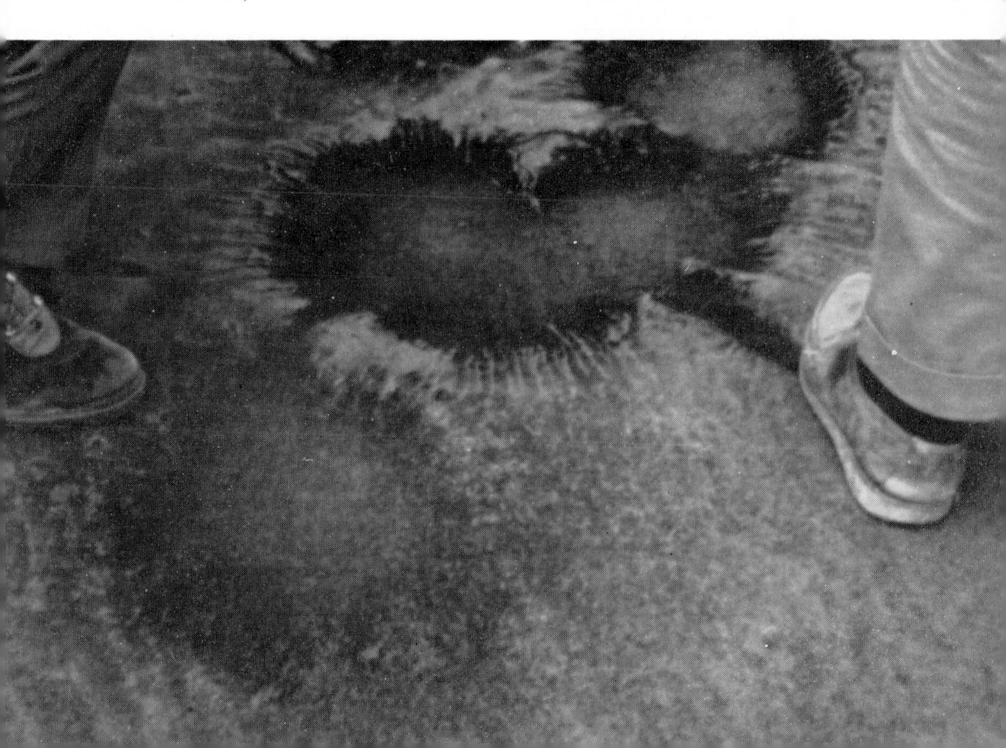

2.10 Caliche (K) horizon (over one metre thick) beneath the 'airport' surface, Las Cruces, New Mexico.

2.11 Stone pavement overlying a thick K horizon north of Tacna, southern Peru.

2.12 Stone pavement at the 607-m level of pluvial lake Coyote, Mojave River drainage basin, Mojave Desert, California.

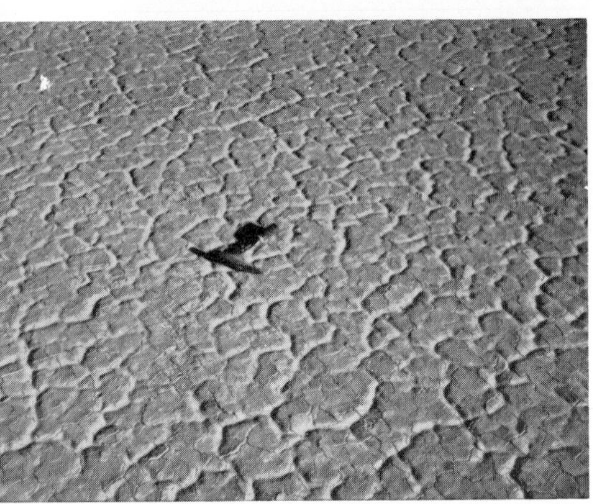

2.13 *Top:* cleared pavement site, near Hodge, Mojave Desert, California, after four years. The pavement has been partially reconstituted, and fine material has accumulated in the trough, largely as the result of surface runoff.

2.14 *Above*: convex mud blocks, and a nonorthogonal crack pattern with a younger nonorthogonal crack pattern superimposed upon it. Panamint Playa, California.

2.15 *Above*: mud curls on an unnamed playa west of Chuquicamata, Atacama Desert, Chile.

2.16 Giant desiccation fissure, Panamint Playa, California.

2.17 *Left*: pipe outlet in the base of an arroyo wall, San Pedro Valley, southern Arizona. Material washed from the pipe forms a small fan on the arroyo floor.

2.18 Plugged depressions and open pipes in montmorillonitic clays, Canyon de Chelly, Arizona.

2.19 A domed inselberg in the extremely arid part of the Sahara at Ein Ekker, Algeria. The rock is covered with a well-developed desert varnish, indicating that modern erosion is slow.

2.20 Stone pavement over a silty soil with few pebbles on the Tadmaït Plateau, Algeria. The surface pebbles are coated with desert varnish, whereas sub-surface pebbles are unvarnished.

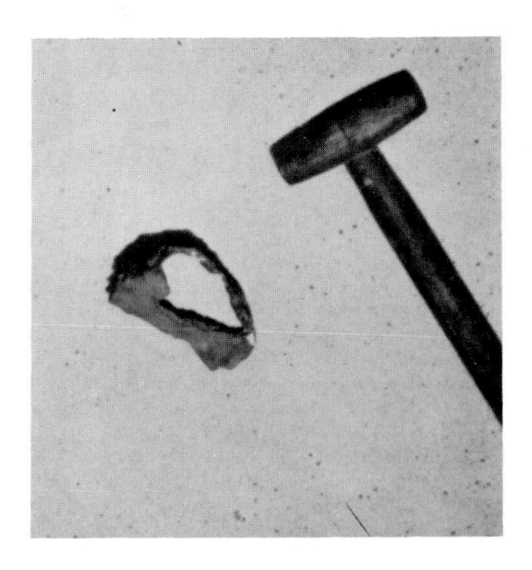

2.21 A varnished, case-hardened boulder whose interior has become softer than the patina and has been removed following fracture of the boulder. From the Kufra area, Libya.

2.22 Boulders disturbed by road making in the Atakor Massif, Algeria. Note the dark colour of the varnish on the parts of boulders formerly above ground level, and the paler (reddish) colour of the patina in the formerly buried portions.

2.23 A disintegrating varnish cover on a quartz monzonite mountain front, Mojave Desert, California.

3.1 The Navajo Twins in Bluff Sandstone, southwestern United States: an example of free-face disintegration by boulder fall.

3.2 Domed inselberg south of Tamanrasset, Algeria. Most of the surface is covered with thick desert varnish, but tafoni are being actively formed on the edges of joint blocks.

3.3 Rill erosion on hillside slopes of vegetation-free shales near Quetta, Pakistan.

3.4 Rills with levées on alluvial hill slope, near Copiapó, southern Atacama Desert, Chile.

3.5 Typical ephemeral stream channel on the margins of the Tucson Mountains, southern Arizona.

3.6 Arroyo of the Santa Cruz River near Tucson, Arizona: this feature has been developed since 1890.

MOUNTAINS AND PLAINS

3.7 An enclosed drainage basin, Tyler Valley, in the Mojave Desert, California. The lowest part of the system is marked by a playa.

3.8 Closed drainage system created by a plug of volcanic ash, central Atacama Desert, Chile. The basin is composed of marginal hills, fringing pediments, alluvial plains and a central playa.

ALLUVIAL FANS

3.9 Alluvial fan, dissected at its lower margins, in the Domeyko Hills, southern Atacama Desert, Chile.

3.10 Alluvial fans, western side of Panamint Valley, California.

PROCESSES AND PEDIMENTS

3.11 Successive mudflows along an alluvial fan channel, Pampa del Tamarugal, Atacama Desert, Chile.

3.12 *Left*: The beginning of a sheetflood on an alluvial surface after a hailstorm, Boqueron Chañar, southern Atacama Desert, Chile.

3.13 Pediment developed on quartz monzonite, Lucerne Valley, Mojave Desert, California.

3.15 *Left:* dissected pediment surface, in an extremely arid area. Jebel el Bisnra, Libya.

3.14 *Above:* granite pediment dome (skyline profile), near Boron, western Mojave Desert, California. The dome is not entirely an erosional feature, for it is associated with uplift along a fault line.

3.16 Dissected alluvial plains and pediments, Pampa del Tamarugal, Atacama Desert, Chile.

3.17 Dissected pediplain in the Andes Mountains east of Copiapó, southern Atacama Desert, Chile.

3.18 Joint-controlled dissection of a quartz monzonite pediment Mojave Desert, California. Note the disintegrating surface veneer of desert varnish.

3.19 Longitudinal trench across a rectilinear pediment surface, Apple Valley, Mojave Desert, California. Alluvium nowhere exceeds a depth of two metres.

3.20 North Panamint playa, Inyo County, California.

3.21 *Centre:* playa and marginal alluvial fans, Panamint Valley, California.

3.22 *Below:* saline playa in salt-dome topography, Cerro de la Sal, Atacama Desert, Chile.

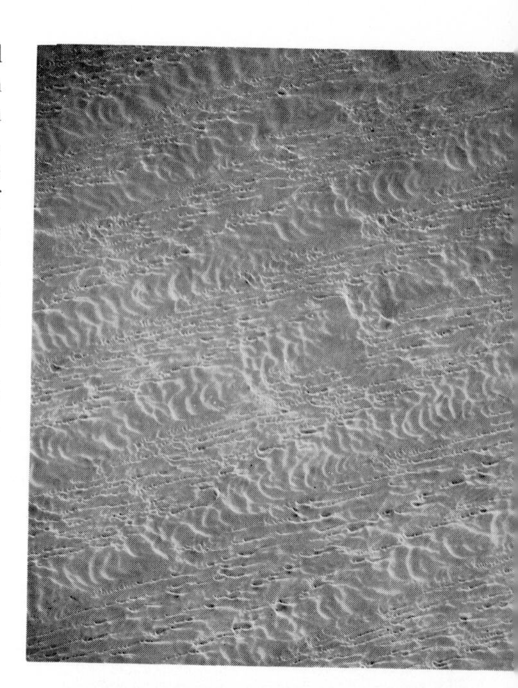

4.1 Dune and draa-patterns in the southern Ténéré Desert, Niger. There are two orders of features here: the draa-sized features are spaced at about one km apart and run ENE-WSW. Superimposed on these are siefs of fine sand running more E-W and zibar of coarse sand at right angles to the wind with linguoid forms and trends almost N-S.

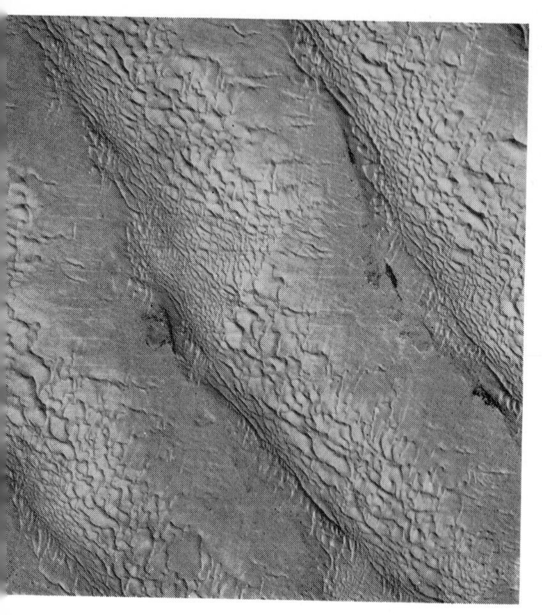

4.2 *Below: Draa* and dunes in the Idehan Oubari in Libya. The draa appear to be transverse to the wind and run NNW-SSE. A complex system of transverse and oblique dune-sized features is superimposed on the draa. Small lakes and vegetated hollows can be seen between the draa.

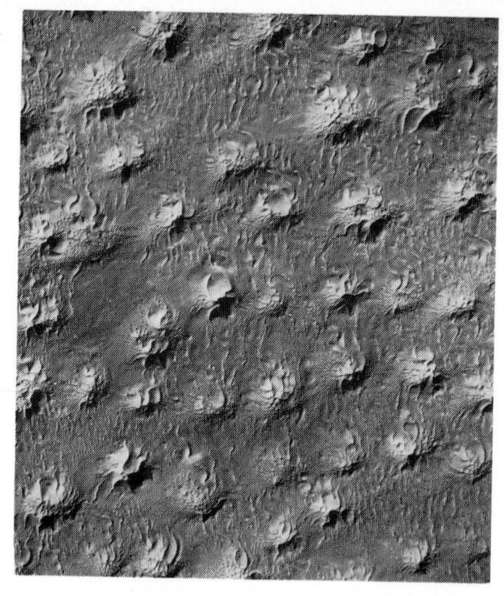

4.3 *Above: Rhourds* (sand mountains) in the Great Eastern Erg in Algeria. The rhourds are about 100–200 m high. They form at the nodes of crossing oblique and longitudinal elements. An aklé dune pattern is superimposed on the rhourds which are of draa size.

4.4 Longitudinal dunes in the Simpson desert, Australia. Photo: E. C. F. Bird.

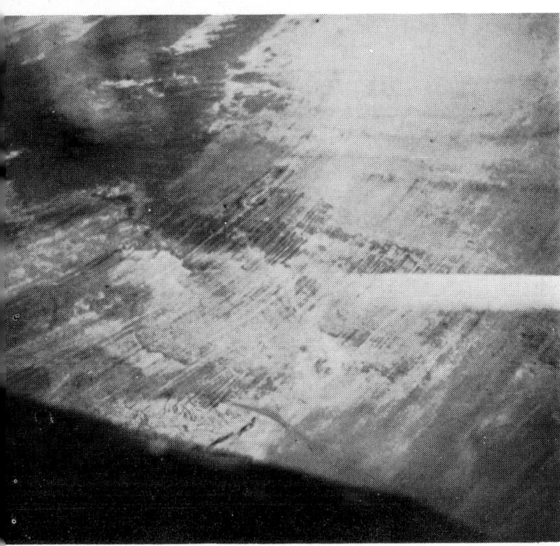

4.5 *Left:* wind-eroded grooves around the Tibesti Massif (NASA satellite photograph).

4.7 *Below:* sorting of magnetite mineral grains on ripple and dune crests in the Great Sand Dunes, Colorado, U.S.

Below: wind abrasion (fluting) ın andesite outcrop on a marine ace near Copiapó, Atacama ert, Chile.

4.8 Megaripples in coarse sand in the Ténéré Desert, Niger. Photo: D. N. Hall

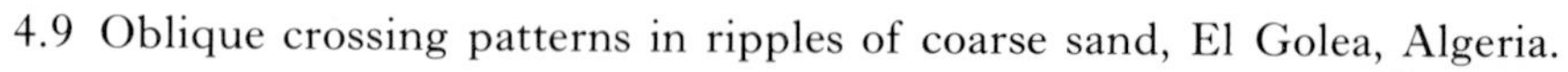

4.9 Oblique crossing patterns in ripples of coarse sand, El Golea, Algeria.

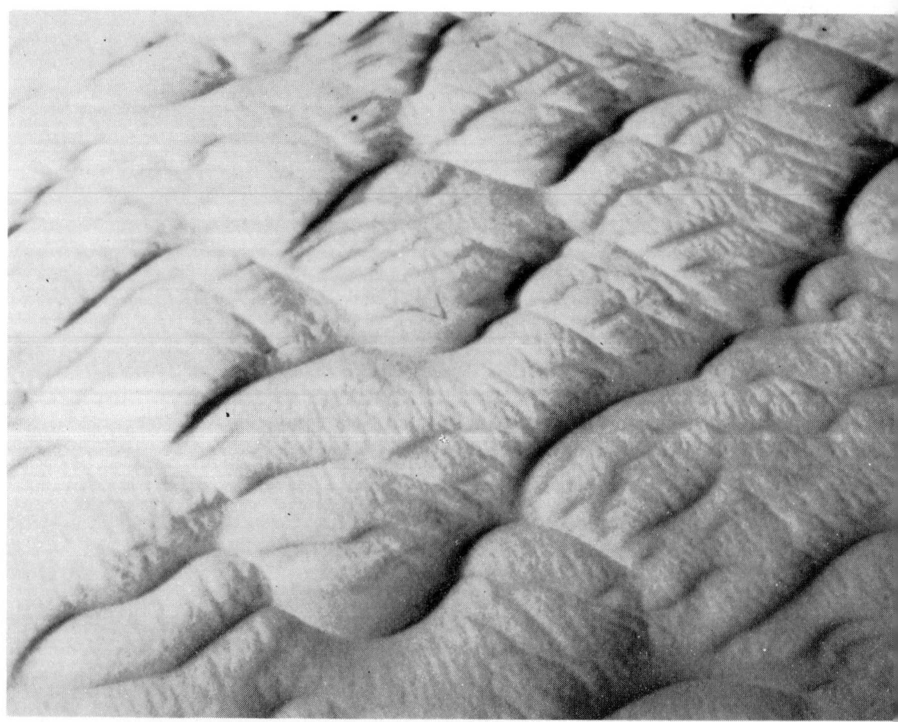

4.10 Nearly straight ripples formed by a cross wind near the crest of a dune. Note the 'smoking' of sand off the crest. Photo: I. G. Wilson

4.11 Aklé dune patterns, Utah.

4.12 A single elongated dune ridge near Hassi Messaoud in Algeria, showing the alternation of barchanoid and linguoid elements.

4.13 Barchan dune advancing across a stone pavement near La Joya, southern Peru.

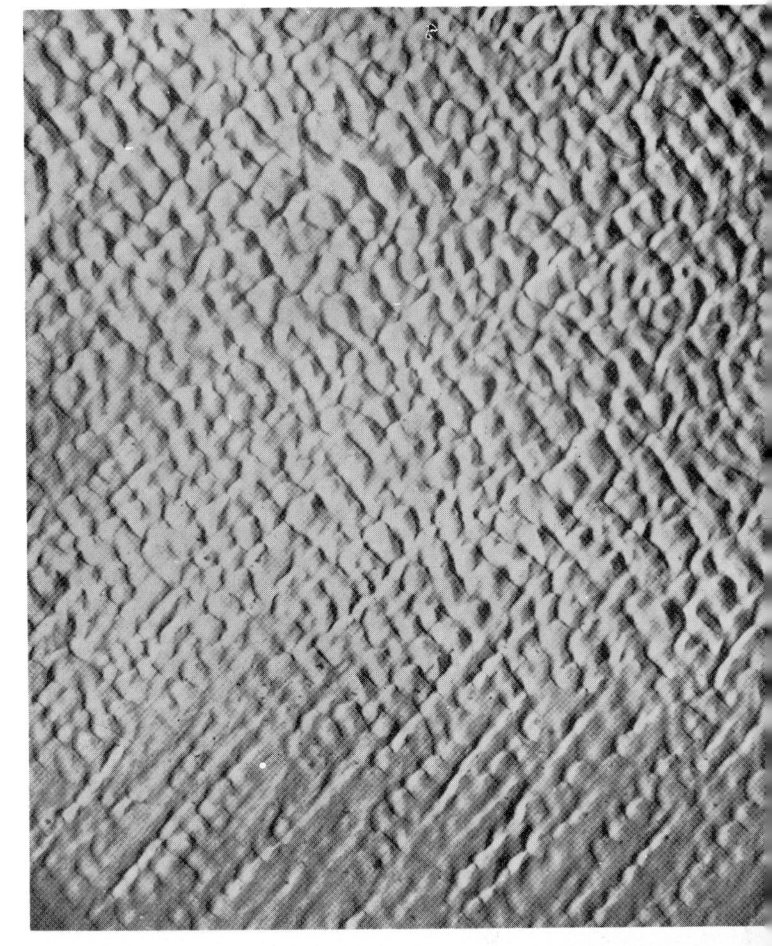

4.14 A dense reticule of simple transverse and longitudinal elements, in the Majâbat Al-Koubrâ (from Monod, 1958).

4.15 A reticule of longitudinal dunes and transverse dunes in coarser sand on the edges of the Simpson Desert, Australia. Photo: E. C. F. Bird.

4.16 A simple reticule of transverse and longitudinal elements at ripple size, near Ouargla, Algeria.

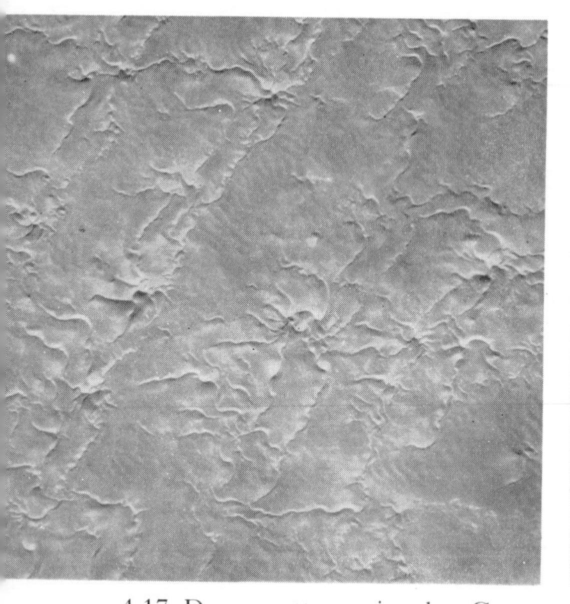

4.17 Dune patterns in the Great Eastern Erg, Algeria. There is a 'basal' pattern of transverse zibar in the coarse sand aligned from NW to SE, with two sets of oblique sief elements in finer sand running from nearly E-W and NE-SW. The oblique elements are linked to transverse barchan-like dunes.

4.18 Sief and zibar patterns in the central Ténéré Desert, Niger.

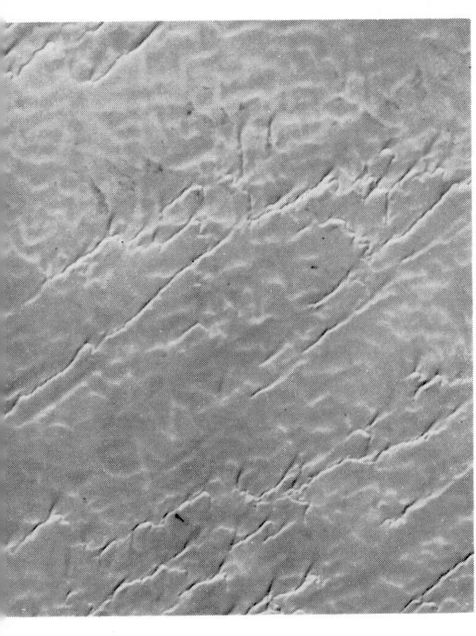

4.19 The elongation of one oblique element in an aklé pattern in the western Ténéré Desert, Niger. The aklé is aligned transverse to winds from the NNE and has prominent barchanoid elements. The elongated sief-like dunes cut obliquely across the aklé from NE to SW.

4.20 Sand-flow dune, Paso de Mayo, coastal Peru. The cutting is that of the Panamerican Highway.

4.21 Longitudinal dunes branching together in a gap between hills near Alice Springs, Australia. Photo: E. C. F. Bird.

and basin area (A)—in the following form:

$$S = 6{\cdot}69\,(H/\sqrt{A})^{0{\cdot}88} \qquad \ldots (3.32).$$

Bedrock in the drainage basin affects debris, depositional processes and sediment concentration in flows reaching the fan, and these features are all related to fan slope. Hooke's laboratory experiments exemplify these relationships. *(i)* Fan slopes vary directly with particle size. This view is only moderately supported by the field evidence collected by Denny in Death Valley (1965), but it is corroborated by Bluck (1964) who demonstrated that particles in both a mudflow and a stream deposit on an alluvial fan in Nevada decline in size exponentially away from the source area. *(ii)* Steeper slopes occur where depositional processes are associated with predominantly coarse sediments (such as debris-flow deposition and sieve deposition*), and gentler slopes occur where fluvial processes alone are dominant. *(iii)* Fans with flows of higher sediment concentration generally have steeper slopes.

(b) Segmented Alluvial Fans. Unsegmented fans are unusual in areas where fans coalesce and where they have undergone prolonged development. More commonly the piedmont zone is composed of segmented fans. Fig. 3.7 shows diagrammatically the development of segmented fans on a piedmont plain (Denny, 1967). Criteria for distinguishing fan-surfaces of different ages and segments include stone pavements, soil profiles, vegetation patterns and degree of dissection.

Many unsegmented fans (Fig. 3.7a) become entrenched near their apices with the result that the loci of deposition are moved downslope, secondary fans are built on or beyond the former segments, and the upper parts of the initial fans are no longer zones of deposition (Fig. 3.7b). The initial segments may become dissected into gullies by local runoff. In Fig. 3.7b, the head of the gully at X may be below the level of the main channel at Y so that if the left bank of the main gully is eroded, discharge in it may eventually flow down the fan gully, leading to the preferential development of a third fan segment (Fig. 3.7c). During these changes, the overall area of the fan has increased, and more and more of its surface has become abandoned by depositional processes and dominated by erosion.

* Such deposition occurs where fan material is so coarse and permeable that surface flow infiltrates before reaching the toe of the fan, a lobe of coarse debris is deposited where water can transport material no further, and the debris acts as a sieve by allowing water to pass while retaining sediment (Hooke, 1967).

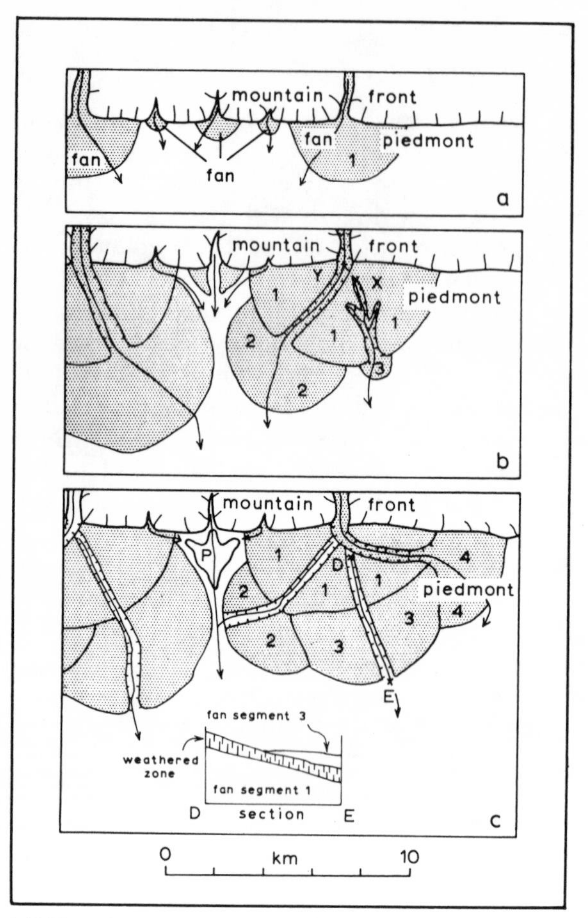

3.7 Development of segmented fans on a piedmont plain (after Denny, 1967).

3.3.3 PROCESSES AND DEPOSITS

The lithological character and quantity of the material in fans depends in large measure firstly on the type of bedrock in the catchment basin and the nature and rate of its destruction, secondly on its alteration during transport, and thirdly on alterations of the material after deposition.

The source of debris is the slopes and channels of the catchment basin. Beaty (1963) argued that material accumulates from the valley-sides in the channels for much of the time, and is occasionally flushed out by severe floods.

Although alluvial fans all result from the deposition of material carried from the mountains, personal observations of fan floods (Beaty, 1963), laboratory experiments (Hooke, 1967) and analysis of fan sediments (Bull, 1962 and 1963) demonstrate that the processes of debris movement are varied, but may range from viscous *debris-flows* to *streamflows*. These processes all tend initially to follow established channels. Sheetflow of material appears to be limited on fans, although sediments may locally overtop the established channels and widening of deposits occurs, especially downslope of the channel ends.

The various processes may often be related through time during a single flood. Eyewitness reports of alluvial fan flooding in the White Mountains of California and Nevada (Beaty, 1963) provide an illustration. Two and a half hours after a heavy thunderstorm in the mountains, masses of debris advanced downslope in a series of waves along a low front of boulders and mud; the lower ends of this debris-flow consisted of silts and fine sands with cobbles and small pebbles which constituted a rapidly moving mudflow; after the debris-flows had halted, streamflows continued for up to 48 hours and dissected the newly-laid sediments and older deposits.

Debris-flow deposits result from dense and viscous flows, and often comprise long, narrow strips that follow former channels; in cross-section they may be slightly concave, with well-defined lateral ridges of coarse debris along parts of their margins (Beaty, 1963). The deposits are poorly sorted, with coarse particles embedded in fine material; they are unstratified; and the boundaries of recent flows are sharply defined, and often have a lobate shape in plan (Hooke, 1967).

Bull (1962, 1963 and 1964*a*) successfully discriminated between mudflow and water-laid sediments in alluvial fans of California. Mudflow deposits, which largely follow defined stream channels but may overtop banks and spread out as sheets, are characterized by abrupt margins (Plate 3.13), a high proportion of clay and poor sorting, and some large 'rafted' boulders'. Horizontal orientation of flat, coarse fragments and graded bedding distinguish fluid mudflows from the more viscous flows which have no graded bedding and some larger rock fragments orientated in vertical planes. Bubble cavities are common and result from air being trapped beneath the mudflow or being entrained in it—a feature which may actually decrease the viscosity of the flow.

Clumps or trains of boulders and armoured mud balls occur on some of the fans described by Bull (1964*a*) and they were probably transported in mudflows or slightly more fluid flows. Studies of mud pebbles in flash flood environments along wadis in the Jordean Mountains and on the southern coastal plain of Israel (Karcz, 1969) indicate that mud-crack controlled collapse of wadi banks during dry periods creates screes of mud pebbles and boulders which become incorporated in subsequent floods.

Water-laid sediments are derived from flows containing much less debris: a higher proportion of water allows much better sediment sorting. Bull (1963) distinguished two types of water-laid sediment: *(i)* those from shallow flows which continually filled small channels and changed position frequently—these deposits were generally well-sorted and often showed cross-bedding or lamination; and *(ii)* water-laid sediments in major channels which were usually coarser and more poorly sorted. Water-laid deposits often have no clearly discernible margins. Mudflow deposits and water-laid sediments represent the two extremes; Bull also identified intermediate types of deposit.

3.3.4 FAN ENTRENCHMENT

The question naturally arises: what causes channel entrenchment and the moving of loci of deposition? There are several hypotheses, and they are not wholly compatible. The basic point is that changes in the erosion area cause changes in the depositional area by causing changes in the connecting stream and channel.

(a) Hypotheses related to temporary changes at present. Denny (1967, p. 85) maintained that entrenchment is 'the normal consequence of large variations in flood discharge'. This inherently reasonable hypothesis implies that modification to fans during the most frequently recurring floods consists of aggradation, but the exceptional, large event with greater competence causes fan-head entrenchment. Beaty (1970) also inclined to the view that entrenchment is a contemporary phenomenon associated with extreme events.

Entrenchment may occur during the later stages of a flood, when high waterflows continuing after initial debris-flow deposition dissect the older deposits (Beaty, 1963).

Hooke (1967) maintained that the feeder channel is often incised into the fan head area because in this zone water is able to transport on a lower slope the material deposited by earlier debris-flows, i.e. the hydraulic gradient required to move debris-flows is higher than that for water transport because the former have higher viscosity and finite yield strength. Hooke recognized, however, that some fan head entrenchment is too deep to have been produced by this process alone.

(b) Tectonic effects. Denny (1967) described how faulting parallel to a mountain front may lower the valley floor with respect to the mountain

(creating a small scarp perhaps), cause drainage incision upslope of the scarp, and move the locus of deposition to the new scarp foot.

In the San Joaquin Valley, Bull (1964*b*) recognized segmented fans in which each segment had a relatively rectilinear longitudinal profile, and in which segments sloped less steeply away from the mountains. The youngest and steepest sloping segment was adjacent to the mountains. He concluded that the fan-surface had been steepened on several occasions as a result of stream-channel steepening caused by intermittent uplift of the mountains. Hooke (1967) observed similar segmentation of fans in eastern California. Tectonic uplift might be responsible in some areas for channel incision on alluvial fans. Hooke (1967) recorded, for example, that fans on the west side of southern Death Valley, California, had been deeply incised as a result of eastward tilting of the valley.

(c) Climatic changes. An alternative explanation of fan entrenchment is that it results from a long-term change in discharge conditions which, in turn, is a consequence of climatic change or, perhaps, removal of vegetation by overgrazing. It is possible to conceive of a variety of climatic changes which could lead to the desired change from deposition to entrenchment: increasing storm frequency, increasing storm intensity, increasing total precipitation, and decline of total precipitation with increased storm intensity, are but a few.

That climatic change has occurred during fan formation in many areas is beyond dispute. Lustig (1965) cited several features associated with fans in Deep Springs Valley, California, which might be indicative of climatic change, amongst which were: great movement of the loci of fan deposition; misfit trenches near fan apices; paired terraces continuous with former fan-surfaces; desert varnish on abandoned fan-surfaces; and greater estimated tractive forces within present active channels than on the fan-surfaces. In discussing fan-head trenching in Deep Springs Valley, Lustig pointed out that the phenomenon is so widespread that a regional explanation is required. He indicated that in most instances estimated tractive force* within the trench is greater than on adjacent fan-surfaces. 'Because modern floods are confined within channels, the tractive force associated with the modern process must be greater than that which accompanied transport in the past. If this is true, then one or more of the three component variables that comprise tractive force must be greater today than in the past' (Lustig,

* Estimated approximate tractive force $= dS$, where d $=$ maximum particle size (a substitute for depth of flow), S $=$ bedslope (a substitute for energy gradient). Another commonly used variable, the specific weight of the transporting medium, is omitted.

1965, p. 175). Of the possibilities, increased density of flow seems most likely; and this may result from a change in climate.

Lustig went on to outline a climatic cycle of fan development. In a fan-building period (Fig. 3.8a) aggradation is general, and is associated with heavier precipitation than at present, higher water: sediment ratios and lesser tractive forces. In the following fan-trenching period (Fig. 3.8b), precipitation is more localized and less frequent, water: sediment ratios are low, mudflows are more frequent and tractive forces are great, so that the former aggradation surface is dissected.

3.3.5 CONCLUSION: AGE, RATE AND CONDITION

From the material we have reviewed, it is clear that studies of alluvial fan systems have provided some valuable, precise and testable ideas on the nature and operation of fluvial systems in deserts. But they have also provided controversy, and gaps remain in our knowledge. Perhaps most striking is the unresolved controversy which relates to the contemporary conditions of alluvial fans in California and Nevada. The argument is between those who believe fans to be in a steady state of one kind or

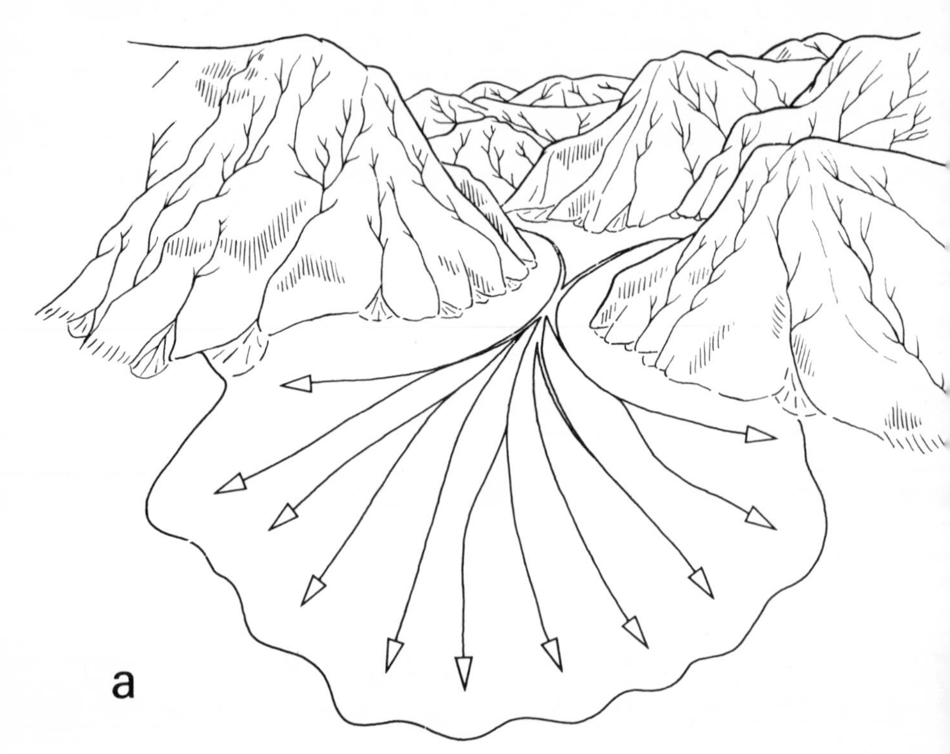

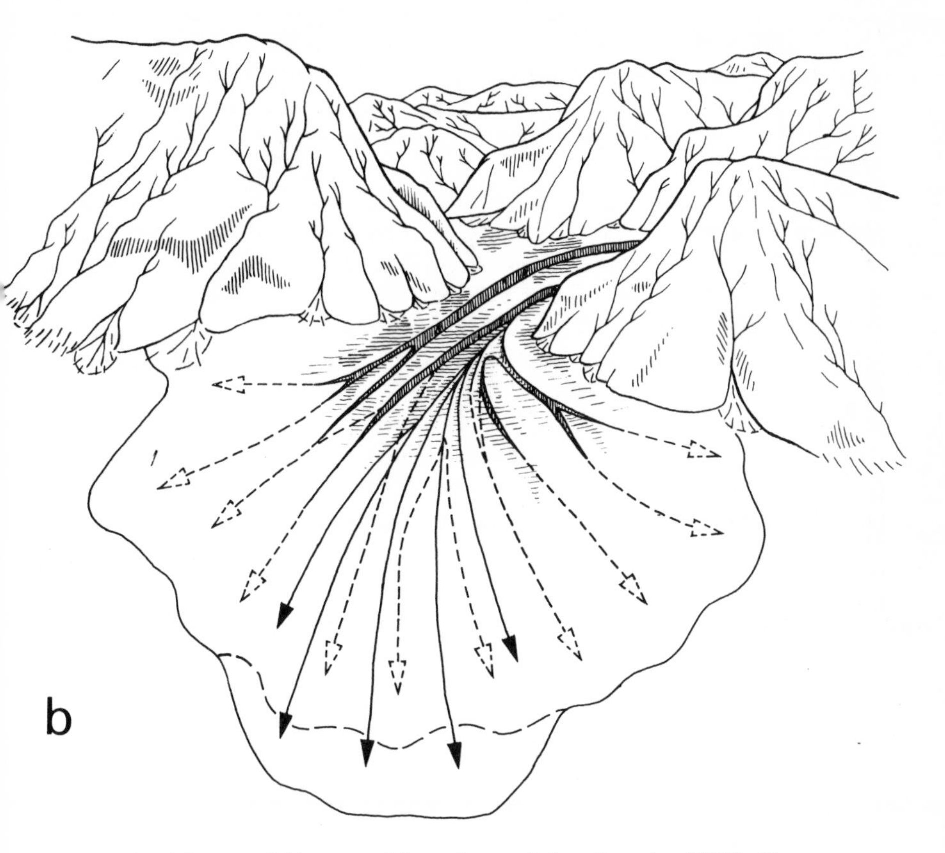

3.8 Fan-building and fan-trenching phases (after Lustig, 1965). For explanation, see text.

another (e.g. Denny, 1967; Hooke, 1968); those who recognize a contemporary condition of dissection (e.g. Hunt and Mabey, 1966); advocates of a cyclic interpretation (e.g. Lustig, 1965), and those who identify fans which are 'young' and actively growing (e.g. Beaty, 1970). At present the chronological evidence necessary to resolve some points in this controversy is rarely available. An exception is the study by Beaty (1970) in which he argued that, as the Milner Creek fan in the White Mountains of California may rest on the 700,000-year old Bishop tuff and contains approximately 0·83 billion m^3 of debris, the fan may have accumulated sediment at an average net rate of 2,134 m^3 p.a. (approximately 7·6 to 15·2 cm per 1,000 years); he also suggested that the fan is still growing—that is, after a period of 700,000 years, it is still not in an equilibrium condition.

3.4 Pediment Systems

3.4.1 CRITIQUE AND CONTROVERSY

Pediments have attracted more study and controversy, and have sparked the imagination of more geomorphologists, than most other landforms in deserts. The relevant body of literature is enormous, and it is currently burgeoning. Disagreement often begins with the problem of definition or even with the confused etymology of the word and its ambiguous architectural overtones. (The O.E.D. suggests that 'pediment' may be a corruption of pyramid, or may be derived from the Latin *pes* (*ped-*) meaning 'foot'.) And disagreement continues throughout all subsequent phases of enquiry. Here is a subject dominated by almost unbridled imagination; a subject in need of organization and direction.

Views of pediments vary widely. To some, pediments pose problems; to others, there is nothing magical about them (Logan, 1956). Birot (1968) regarded the development of pediments as the central problem of arid-land geomorphology; whereas to Lustig (1969, p. D67) 'the only real "pediment problem" is how the reduction or elimination of mountain mass occurs'.

A review of pediment literature reveals several important difficulties and weaknesses in much previous work.

(a) Definition. It is quite clear that only the most general qualitative definition can command wide acceptance; and it is equally clear that the reason for this is that the morphology of pediments is enormously variable. Definitions are almost as numerous as authors, and much controversy arises from discussion in common terms of different phenomena by different investigators. There are even discrepancies between thoughtful, general definitions. Compare, for example, 'pediments are erosional surfaces of low relief, partly covered by a veneer of alluvium, that slope away from the base of mountain masses or escarpments in arid and semiarid environments' (Hadley, 1967, p. 83) and 'les pédiments sont des surfaces rocheuses ou glacis rocheux, possédant une faible pente (inférieure a 5°), généralement localisés au pied de massifs montagneux' (Mammerickx, 1964*b*, p. 359). Discrepancies between these two definitions relate to 'erosional', a 'veneer of alluvium', the presence of a mountain mass, slope and climatic environment.

(b) Precision. Few studies present precise data which are in a form that is both verifiable and of use to subsequent investigators: most data are general and, at times, ambiguous. Two examples will illustrate this point. References to 'gently-sloping' pediments abound, and are of little value

if the reader does not know the limits of the 'gently-sloping' category. Perhaps more confusing is the slightly more precise statement to the effect that 'pediments in area X slope at between six and one degrees'—do these figures represent the range of slope for single pediments, or the range within which all pediment slopes occur?

(c) Explanatory speculation. There has been too great a dependence on and concern with a few deductive, largely unverifiable and certainly unverified explanatory hypotheses. Foremost amongst these is that of Lawson (1915), who was mainly concerned with mountain-front retreat and contemporaneous extension of piedmont plains, and that introduced by Gilbert (1875), which emphasized the role of lateral planation by streams.

(d) Processes. These hypotheses, and several others, make many gross assumptions concerning processes. Sheetflooding, a common *deus ex machina*, has rarely been observed on pediment surfaces; backweathering of mountain fronts is more honoured in the deduction than in the observation; and lateral stream planation is not a phenomenon commonly witnessed on pediments. Coupled with gross deductions concerning process is the common cause-and-effect error arising from relations between a deduced process and the visible landform. Clearly, for instance, sheetflooding cannot produce a planar surface because a planar surface is necessary for sheetflooding to occur. Again, widespread weathering and occasional removal of weathered debris is unlikely to produce a pediment surface; it is more likely to maintain, probably at a lower level, a pre-existing form. In short, there is frequently confusion between pediment-forming, and pediment-modifying processes.

A more justifiable assumption concerning processes is frequently ignored: processes have almost certainly changed in many areas—in nature, magnitude, frequency etc.—as a result of climatic changes during the course of pediment evolution.

(e) Predilection for pediments. Attention is often exclusively focused on the pediment rather than upon the erosional/depositional system of which it is a part. In this respect, pediment literature contrasts markedly with that on alluvial fans.

In view of these criticisms, and the appalling volume of literature, we propose to restrict our discussion to what we consider are more certain, promising and verifiable aspects of pediment study. For more conventional surveys of this field, the reader is referred to articles by Dresch (1957), Hadley (1967), Tator (1952 and 1953), and Weise (1970).

3.4.2 DISTRIBUTION OF PEDIMENTS

If we accept that pediments are, on a liberal definition, gently sloping surfaces developed on bedrock, then it is clear that they are a world-wide phenomenon. Indeed, King (1953) maintained that most of the world's landforms are transformed into either pediments or their collective counterparts, pediplains. And the innocent might be forgiven for recognizing similarities between pediplains and peneplains (Dury and Langford-Smith, 1964). Pediments have also been recognized within the stratigraphic column (Williams, 1969). Some have argued that the occurrence of pediments is related to particular climatic conditions. In the south-western United States, for instance, Corbel (1963) recognized three geomorphological zones, *(i)* where mechanical weathering is predominant, *(ii)* where chemical weathering is most important, and *(iii)* an intermediate zone where both occur. It is in the latter zone, where climate is 'regularly irregular' between the zones of winter snow and summer floods, that pediments are formed. Such assertions are interesting but unproven.

More commonly, pediments are said by definition to be desert phenomena. It would be tedious to reproduce a list of references which reveals that pediments have been described in every desert; it is simpler to state that there is no desert without a pediment literature.

Of greater geomorphological significance is the location of pediments within deserts. There is no doubt that pediments most commonly occur at an intermediate location between watershed and base-level, and usually between a mountain and an alluvial plain. We refer to such pediments, which are characteristic of basin-range country (e.g. Cooke, 1970*a*), as *apron pediments*. A common variant of this type is the *pediment dome* (e.g. Davis, 1933; Mabbutt, 1955), in which adjacent pediments are not surmounted by a mountain. The second common location for pediments is adjacent to major drainage lines, often exoreic streams (e.g. Tuan, 1962); such pediments we call *terrace pediments*.

But pediments by no means occur throughout all desert basins. For example, Lustig (1969*a*) identified a singular distinction between the distribution of pediments and alluvial fans in basin-range country. He argued that if pediments originate by mountain-mass reduction, 'the average mountain mass should be significantly smaller in certain areas that are characterized by the existence of pediments' (Lustig, 1969*a*, p. 67). His morphometric analysis of basin and range parameters in the western United States showed that areas occupied by smaller ranges (in terms of width, height, relief, area and volume), such as south-western Arizona and south-eastern California, are indeed those characterized by pediments. By contrast, in areas with larger ranges, alluvial fans are common and pediments are rare. In short, it could be that the distribution of pediments and alluvial fans is in part determined in the basin-range context by morphometric relationships which, in turn, may be a reflection of erosional history.

Although there is little doubt that many pediments are erosional features because they are cut discordantly across structures and rocks of different lithologies, some believe that pediments are preferentially developed on particular rock types. Coarsely-crystalline rocks, and notably granite, are certainly commonly associated with pediments (e.g. Logan, 1964; Mabbutt, 1955*a*; Tuan, 1959; Warnke, 1969). In a detailed study of the Sherman erosion surface of Colorado and Wyoming, Eggler, Larson and Bradley (1969) convincingly demonstrated that the surface is associated exclusively with a granite facies which disintegrates rapidly and totally to gruss as the result of the expansive effects of altering biotites; other granites and crystalline rocks give rise to parkland-tor and more rugged topography. That is to say, the surface is closely related to bedrock lithology, and in this case to a particular type of granite. But there is no doubt that pediments occur on rocks of other lithologies. In southern Arizona, for instance, they are found on alluvium, indurated sedimentary deposits, metamorphic, and several contrasted types of volcanic rocks, and elsewhere (especially in the Sahara) they have been described on alternating 'hard' and 'soft' rocks or generally weaker rocks (the *glacis d'érosion* of Dresch, 1957 and Birot and Dresch, 1966), and on sandstone, schist, mudstone, shale and limestone.

If pediments eschew locational definition in climatic and lithological terms, they also appear to be unrelated to the broad pattern of endogenetic processes for they are found both in unstable basin-range deserts and in stable 'shield' deserts. In detail, however, tectonic activity may have an important role to play.

3.4.3 FORM OF PEDIMENTS

(a) Search for a system. Unlike alluvial fans, pediments are not features that can easily be considered as parts of clearly circumscribed, functioning systems. Often the surface occurs in more than one drainage basin. In addition, it may be impossible to assume that present drainage networks on a pediment were associated with its formation. In other instances, there may be several isolated pediments in what is essentially a uniform plain within a single drainage basin. And elsewhere, the pediment may legitimately be considered as a functional component of a drainage basin. Nevertheless, for the purposes of precise description and improved explanation, it is desirable and useful to distinguish a pedi-

ment system. In the context of the basin-range topography of the Mojave Desert, Cooke (1970*a*) suggested a beginning might be made by using a unit (called a pediment association) which includes the pediment, the mountain area tributary to it, and the area of alluvial plain to which it is tributary. The unit is not entirely satisfactory, even in this context, but it is useful; elsewhere, alternative solutions might be devised.

(b) Gross morphometry of pediments. (In this section, we shall be referring to apron pediments.) Analysis of quantitative pediment data is confined to a few articles (e.g. Cooke, 1970*a*; Corbel, 1963; Dury *et al.*, 1967; Mammerickx, 1964*a*). Such analyses depend in the first instance on the definition of variables; in the following account we accept the definitions of the authors concerned. One scheme of definition which may be of general value is shown in Figure 3.4. Most of the working definitions are explained by Cooke (1970*a*). But it should be emphasized that a critical point of definition, which can affect most calculations, is the lower limit of the pediment on the piedmont plain; Cooke defined it as the line at which the alluvial cover becomes continuous.

(i) Descriptive data. Pediments vary enormously in area. Cooke found in the western Mojave Desert, California, a range of values from 0·78 to 35·1 km^2, and a mean value of 5·98 km^2. Pediments in the neighbouring Sonoran Desert, Arizona, are generally larger: Corbel (1963) cited a range from 2·5 to 620 km^2, and a mean value of 100 km^2. Similar variation is apparent in other deserts: in the Atacama Desert, Chile, for instance, both very small and very large pediments have been described (e.g. Cooke and Mortimer, 1971; Clark *et al.*, 1967; Galli-Oliver, 1967).

Also of importance is the fact that the proportion of an area occupied by pediments greatly varies. To contrast the Mojave and Sonoran deserts again (Table 3.5), it is clear that the proportion of the area in pediment is greater in Arizona than in California. The ratio of mountain area to plain area also varies. In the western Mojave Desert, the ratio is 0·36, and 9·4 per cent of the plains' area is composed of pediments; McGee (1897) suggested that the ratio in southern Arizona is 0·25, and that 40–50 per cent of the plains' area is composed of pediment.

Slope is an important feature of pediments because it is more susceptible to analytical evaluation than most pediment properties (see below), and it is also variable, usually within the range 0–11 degrees. Table 3.6 shows a sample of pediment-slope data from south-western North America. The data are derived from study of several pediments in each of the areas mentioned. Slope also usually varies along the profile of a single pediment: generally, the profile is concave upwards and slope

TABLE 3.5 *Comparison of Landform Areas in Part of the Mojave Desert and Parts of Arizona, U.S.A.*

	Average percentages of landform areas	
	Western Mojave Desert	*Parts of Arizona*
Mountain	28·9	10
Piedmont	71·1	90
Pediment	6·7	30
Pédiment zone couverte	—	10
Alluvial plain	64·4	50

SOURCE: Cooke, 1970*a*; Corbel, 1963.

TABLE 3.6 *Pediment Slope Data from Arid and Semi-arid Areas of South-western North America*

Area	*Range of longitudinal pediment slope (degrees)*	*Mean pediment slope (degrees)*	*Reference*
Western Mojave Desert	30′–11°	2°35′	Cooke, 1970*a*
Papago Country, Arizona	30′–2°12′	—	Bryan, 1925, p. 95
Ajo Region, Arizona	48′–3°18′	1°–1°40′	Gilluly, 1937, p. 332
South-east Arizona	1°30′–4(?)°	—	Tuan, 1959
10 pediments in Arizona	30′–4°	2°	Corbel, 1963, pp. 53–54
Mojave and Sonoran Deserts	1°30′–5°40′ (means range) 30′–7°	—	Mammerickx, 1964, p. 421; Balchin and Pye, 1955, p. 171
Ruby-East Humboldt Mountains, Nevada	West flank: 48′–3°48′ East flank: 4°54′–5°25′	—	Sharp, 1940, p. 357
South-west U.S.	30′–7°	2°30′	Blackwelder, 1931*b*, p. 137
	6′–2°42′	—	Bryan, 1936, p. 770
	7′–8°5′	2°11′–3°15′	Leopold, Wolman and Miller, 1964, p. 494
	15′–10°	—	Tator, 1953
Cape Region, Baja, California	1°–5°	—	Hammond, 1954, p. 79

varies within the range 0–11 degrees; sometimes, however, profiles may be rectilinear (e.g. Cooke and Mason, 1973).

(ii) Analysis of morphometric data. Analysis of morphometric data is only in an inchoative stage, and here we shall confine our commentary to a consideration of pediment-slope data through the examination of several simple explanatory hypotheses. Unless otherwise stated, pediment slope is defined as the slope of the line joining the highest and lowest points of a pediment profile along the line of pediment association length (Cooke, 1970*a*).

1 Pediment slopes in an area are positively correlated with the relief and length dimensions of the pediment associations in which they occur. This hypothesis is substantiated in the western Mojave Desert, where the correlation of pediment slope with pediment association relief/length ratio gave an r value of 0·4416 (P = 0·001). In this region, pediment association relief results largely from tectonic movements and, as pediment association relief is positively correlated with pediment association length (r = 0·6728; 0·001 > P), it seems reasonable that the pediment association relief/length ratio may be considered as an indirect expression of the effects of tectonic movements. To explore this interpretation further, a second hypothesis is proposed.

2 The slope of pediments associated with faults is significantly steeper than the slope of pediments not associated with faults. This hypothesis is confirmed by data from the western Mojave Desert (Table 3.7).

TABLE 3.7 *Relations of Pediment Slope and Faults*

	Pediments	*No.*	*Mean Slope*	*Range*	*Standard Deviation*
1	Associated with faults	31	2°55′	39′–5°23′	1°12′
2	Not associated with faults	22	2°10′	31′–3°50′	48′

t = 2·7; significance level (P) ≈ 0·01.
SOURCE: Cooke, 1970*a*.

Thus, although the evidence is not conclusive, it does appear that in one area at least tectonic activity may be a significant factor in determining the general slope of pediments.

3 On the assumption that pediment slope is a function of hydraulic variables in a fluvial system, additional hypotheses may be formulated. Firstly, it might be hypothesized that pediment slope is inversely

related to catchment basin area, on the deductive grounds that pediment slope should be inversely related to water discharge, that discharge should be proportional to catchment basin area, and therefore that gentler pediment slopes should be associated with larger catchment basins. The last of these assumptions was not substantiated in the western Mojave Desert by Cooke (1970*a*; $r = -0{\cdot}2335$; $P > 0{\cdot}1$), or in Arizona and California by Mammerickx (1964*a*).

Secondly, it might seem likely that pediment slope will vary along a single profile according to the nature, and especially the size, of debris transported across it. At Middle Pinnacle, near Broken Hill, Australia, Dury (1966*a*) demonstrated that two surveyed pediment profiles were adequately described by logarithmic curves to the base 10, that particle-size decreased in an orderly fashion downslope, and that rates of particle-size decrease were less than the rates of slope decline. Cooke and Reeves (1972) considered slope and particle relationships in a study of pediment profiles in the Mojave Desert. They showed that with distance along surveyed pediment traverses, slope usually declines in an orderly fashion, but particle-size reduction varies both with distance and in its relationship to slope change. The size of the largest particles generally decreases most rapidly downslope, but this rate of decline is usually less rapid than the rate of slope decline. They attributed the variability of particle-size change downslope to variations along the profiles in the relative proportions of transported and *in situ* debris. In addition, they described differences in the nature of slope changes between mountain front and pediment on different lithologies: in quartz monzonite areas there is often a marked contrast between debris sizes on mountain fronts and on pediments, which may account for the distinct break of slope between the two landforms in these areas; on other rock types, where the particle-size contrast is less marked, the change of slope is less abrupt.

If pediment slope is in part related to debris-size, the broader possibility exists that pediment slope is significantly related to rock type. Despite some assertions to this effect, the idea remains to be rigorously tested and Mammerickx (1964) concluded that bedrock lithology is not a decisive factor in determining pediment slope.

4 Finally, it might be hypothesized that pediment slope decreases with pediment length, in the deductive belief that as a pediment extends simultaneously in the horizontal and vertical planes its relief : length ratio will decline. At present this hypothesis does not appear to be justified. Firstly, both Mammerickx (1964) and Cooke (1970*a*) identified only weak relationships between pediment slope and pediment length in the south-western United States. Secondly, Dury *et al.* (1966) in a short

study on pediments in New South Wales, demonstrated that pediment steepness (expressed numerically as some function of the angle between the horizontal passing through the lower pediment limit where best-fit groundslope is one degree, and the inclined line drawn from this point to the point where groundslope is eight degrees) fails to correlate significantly with pediment length.

(c) Details of pediment topography. In this section we do not pretend to be comprehensive; we merely identify some of the important characteristics of certain pediments which appear to us to have received too little attention and which substantiate our contention that the morphology of pediments is enormously variable. In particular circumstances, the occurrence of some of the features we describe may be of critical diagnostic value in explaining the origin of pediments.

Although many published accounts may give a contrary impression, a pediment which is a clean, smooth bedrock surface is rare indeed. In most cases, the pediment is a complex surface, comprising patches of bedrock and alluvium, in places capped by weathering and soil profiles, punctuated by inselbergs, and scored by a network of drainage channels.

The patchiness of bedrock and alluvium may result partly because some material is temporarily in transit across the surface and partly from the incomplete removal of formerly more extensive alluvial covers. Inselbergs vary in number and dimensions: often they appear to become larger and more numerous towards the upper limit of the surface; and frequently they occur either on more resistant bedrock or on major interfluves.

Rather surprisingly, little attention has been given to the occurrence of weathering and soil profiles on pediments, although their evidence may be critical to the interpretation of pediment development. On the Apple Valley pediment, California, Cooke and Mason (1973) have described a soil profile which has been eroded in places and buried in others and the weathering profile in quartz monzonite beneath the bedrock surface. Mabbutt (1966) sketched the weathering features on granite and schist pediments in central Australia; and Twidale (e.g. 1964) has recorded evidence of deep weathering at the upslope margins of pediments in South Australia. The relationships between pediments and zones of supergene enrichment of copper minerals in the southern Atacama Desert has also been studied (e.g. Clark *et al.*, 1967; Segerstrom, 1963). Other examples of studies considering such features are few.

Another important yet neglected feature is the presence of cut-and-fill features on pediments. Channels 1–3 m deep and now filled with alluvium have been described by Cooke and Mason (1973) in the Apple Valley pediment, California; Dury (1966) recorded a buried channel at Middle Pinnacle, New South Wales; and in southern Arizona there are numerous examples (e.g. Fig. 3.9). The presence of buried channels indicates that the relations between erosion and sedimentation in the pediment zone have changed during the period of pediment development, probably as a consequence of changed environmental circumstances. The filling of channels and other depressions in bedrock by alluvium is commonly responsible for the general smoothness of many pediments.

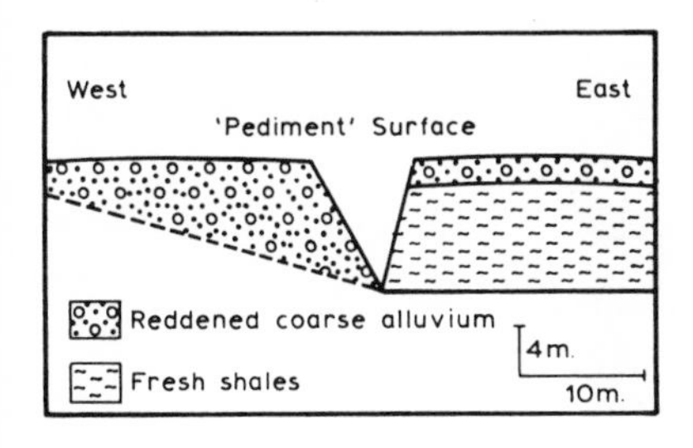

3.9 Section across a 'pediment' surface at Arivaca, southern Arizona.

Closely related to buried channels are pediment drainage nets. These, too, have rarely been considered. There are three common types. *(i)* Channels occurring in the upper part of the piedmont plain, which commonly form a distributary system and die out lower down the surface. Such channels often straddle the piedmont angle, and they are deepest at intermediate positions on their longitudinal profiles. *(ii)* Channels occurring on the lower part of the piedmont plain, which are generally deepest at the lowest point in their longitudinal profiles, and usually form part of a drainage system that has been rejuvenated on one or more occasions by lowering of base-level. Such systems may cover the whole pediment. When drainage in this type of net is rejuvenated it often leads to the destruction of the pediment surface. For instance, as Fig. 3.10 shows, drainage channels on a pediment near Hodge, California, are in general accord with the slope of the pediment, but a number of low-order channels follow strike directions normal to pediment slope and thus disrupt its longitudinal continuity. *(iii)* On relatively undissected surfaces, often between areas characterized by types *(i)* and *(ii)*, drainage nets may consist of complex and frequently changing patterns of shallow rills.

These drainage nets are similar in pattern and location to those on alluvial fans, and they may perhaps be explained in similar terms. Type *(i)* is probably generated by drainage in the catchment area behind the

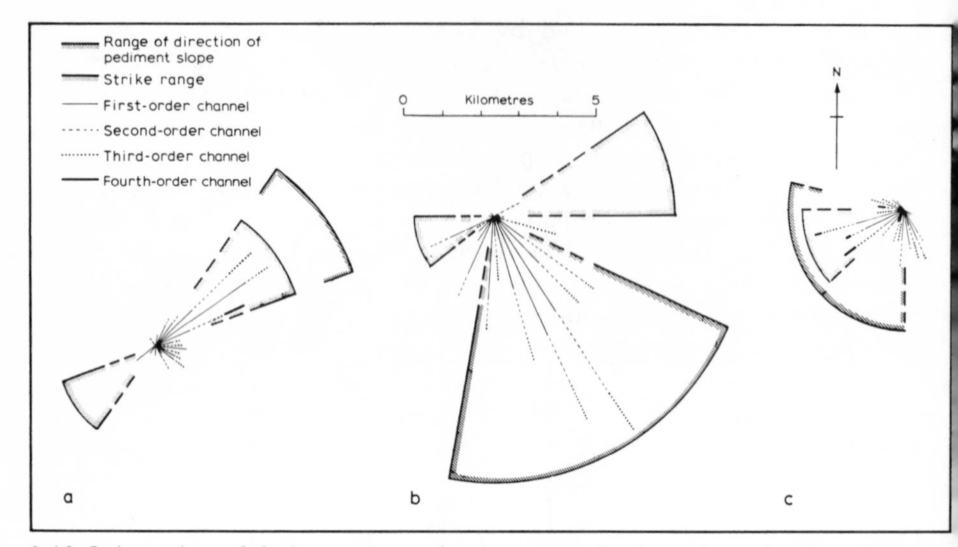

3.10 Orientation of drainage channels, the range of orientation of rock strike and pediment slope on three pediments near Hodge, Mojave Desert, California.

pediment, type *(ii)* may result from runoff on the pediment surface itself, and type *(iii)* probably arises from rillflow, perhaps characteristic of declining sheetfloods, in the intermediate zone. Drainage incision may reflect adjustments to climatic or tectonic changes, or changes in the nature of waterflow within the system. Such changes could have accompanied pediment formation, or they could be younger and lead to pediment destruction. Doehring (1970) has suggested that on high quality topographic maps, pediments can be differentiated from alluvial fans because drainage texture becomes significantly finer upslope on pediments but not on alluvial fans (see also Miller, 1971).

Where bedrock is exposed at the surface, the latter is often discordant to the structure of the former. For example, Mabbutt (1966) described the longitudinal profile of a schist pediment in which the surface *generally* slopes at right angles to the strike; in *detail*, the surface comprises discontinuous ridges of schist parallel to the strike and up to 0·6 m high, separated by flatter, debris-mantled zones. In contrast, surface and structure occasionally accord where, for example, the surface coincides with semi-horizontal joint planes in quartz monzonite.

Downslope, a pediment often disappears beneath an alluvial cover, where it is known as a *suballuvial bench*. There has been much speculation on the nature of this feature. One persistent deduction is that it ought to be convex (e.g. Lawson, 1915; Davis, 1933). The limited available evidence does not confirm this idea. The suballuvial bench of the Apple Valley pediment, California, is a continuation of the rectilinear exposed surface, and its slope is only two minutes steeper than that of the over-lying alluvium. In a gravimetric and seismic survey of the suballuvial bench of Cima Dome, California, Sharp (1957) was unable to identify any convincing convexity of the profile. And a seismic-refraction survey of suballuvial profiles at Middle Pinnacle, Australia, did not reveal any convexity in the suballuvial bench (Dury, 1966; Langford-Smith and Dury, 1964).

The upslope margin of pediments is frequently clearly demarcated as the boundary between mountain and plain. In plan it may be linear, or crenulate with embayments into the mountains where major lines of drainage emerge; where two embayments meet across a divide, breaking up the continuity of a mountain mass, a *pediment pass* is formed (e.g. Howard, 1942; Warnke, 1969). The transverse profile of the boundary may be relatively flat and horizontal or, where there are embayments, undulating.

The boundary often separates slopes of very different inclinations; the angle produced by the two slopes is called the *piedmont angle*. Two comments are appropriate on the form of the boundary and the piedmont angle, features central to an understanding of pediment evolution. Firstly, the piedmont angle may be as little as 90 degrees, but usually it is much greater. The abruptness of the junction has been exaggerated, often because it appears to be more abrupt than it actually is when viewed from a distance across the piedmont plain. In many cases, the longitudinal profile forms a continuous logarithmic curve, uninterrupted by a break of slope. Secondly, there is definitely no simple single, universally applicable explanation of the feature, as some authors have suggested. We shall return to the problem of origin below (section 3.4.5).

3.4.4 GAMUT OF PEDIMENT PROCESSES

Here we are concerned with identifying the spectrum of pediment processes, fully recognizing that some are localized in their effect, and that a particular feature may be produced by different processes or combinations of processes in different areas. Studies of pediment processes fall broadly into three groups: a small body of empirical observations, a larger group of casual observations, and a very large body of inferences drawn from the evidence of landforms and deposits. One of the main obstacles to the effective investigation of pediment processes is that they are not effectively or easily monitored in short periods over a large area; thus observations of real value for testing explanatory hypotheses are few. One solution, as we shall explain below, is to study small-scale terrain analogues. A further problem is that in some areas pediments may be fossil features that were created under different conditions in the past.

(a) Empirical observations. Empirical observations of pediment processes comprise firstly studies of erosion and sedimentation on small-scale analogues of the mountain-plain system and secondly, several attempts to record processes operating in full-scale systems.

(i) Small-scale analogues. Higgins (1953) described small hills and pediment-like surfaces, separated by sharp breaks in slope, developed in late Pliocene volcanic breccia near Calistoga, California. Although these features are not strictly small-scale analogues of true mountains and pediments, their study is instructive. Two rainstorms were observed in the area, and their effects were recorded by changes to bands of white dolomite sand arranged across the surfaces. Finer material in the bands was dispersed several metres downslope on the undissected surfaces, and even further where the bands crossed a channel. Movement of debris, Higgins concluded, was largely by 'slopewash', a process which includes sheetwash, raindrop impact and splash, and is locally more pronounced in the channel.

The Badlands National Monument, South Dakota, is an outstanding area of rapidly changing badland-slope and miniature-pediment topography, developed on Oligocene clays, shales, sandstones and volcanic ashes of the Chadron and Brule formations. Smith (1958) surveyed in detail landforms on the Brule Formation and concluded that the badland slopes retreat by disaggregation, slumping and especially sheet and rill erosion; that the pediments are extended by spreading sheetwash close to rill mouths at the badland-slope/pediment junction and that sheetwash is the principal process of transporting debris across the pediments. As sheetwash declines following a rainstorm it changes from continuous layer flow into subdivided rillwash. Detailed though Smith's observations are, his conclusions are mainly inferences drawn from inspection of form and deposit.

Fortunately, Schumm (1956 and 1962) has monitored changes in the same area by means of 15 8·4 mm—diameter, 46 cm—long reinforcing rods inserted into the ground until they were flush with the surface. He observed exposure of these rods and the retreat of badland slopes between 1954 and 1961. Two of the stakes were buried; the remainder were exposed by 10–75 mm. Erosion was greatest on the badland slopes; and on the pediment it decreased away from the badland-slope/pediment junction (Fig. 3.11). Two important facts are clear. Firstly, the badland

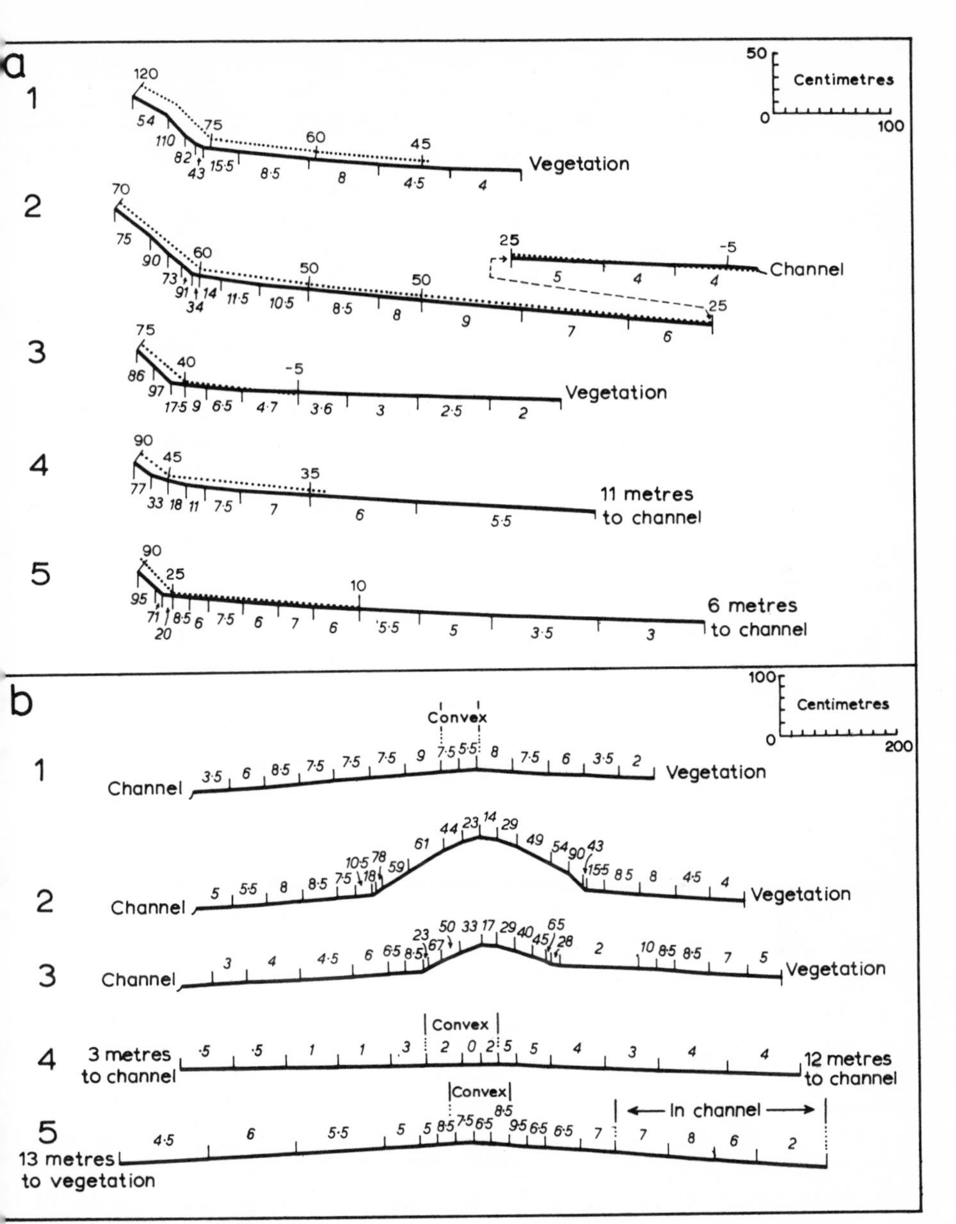

.11 Slope profiles in Badlands National Monument, Dakota. (*a*) Profiles in May, 1961. Numbers below profiles are slopes in per cent; those above profiles are erosion depths in mm at stake positions; dashed lines represent profile in July, 1953. (*b*) Profiles illustrating headward extension and coalescence of miniature pediments: the profiles were measured at the same time in different locations. Numbers above profiles give slope in per cent. (After Schumm, 1962.)

slopes have retreated by 6–12 cm, leaving behind extensions of the pediments which are steeper than the older parts of the pediments. And secondly, the pediment surfaces are re-graded as they are extended. Smith's view that sheetwash is important on the pediment is confirmed by Schumm who showed that it both re-grades and transports debris across the surface. Schumm elaborated on one important point: the transition from steep badland slope to gently-sloping pediment is not accompanied by deposition at the boundary; sediment is removed across the pediment to a zone of deposition further downslope; and at the same time, the pediment is re-graded. The reason for this is almost certainly that the velocity of flow is of the same order of magnitude on both badland slope and pediment, despite their different inclinations. This argument may be illustrated by reference to the Manning equation for mean velocity of flow which states

$$V = \frac{1{\cdot}49 D^{\frac{2}{3}} S^{\frac{1}{2}}}{n} \qquad \ldots (3.33)$$

where V = mean velocity of flow in feet per second, D = mean depth of flow (or hydraulic radius), S = slope, and n = roughness coefficient. If it is assumed that hydraulic radius = mean depth of flow, and that depth of flow is the same on hillslope and pediment, D can be dropped from the equation. Mean relative velocity (Vr) is defined as

$$Vr = \frac{1{\cdot}49\ S^{\frac{1}{2}}}{n} \qquad \ldots (3.34)$$

Then, if $S = 0{\cdot}8$ on the hillslope, and $0{\cdot}1$ on the pediment, and $n \simeq 0{\cdot}06$ on the hillslope and $0{\cdot}02$ on the pediment, Vr would be 24 on the pediment and 23 on the steep, rough hillslope. *i.e.* Decrease in roughness on the pediment apparently compensates for decrease in slope. Furthermore, D is likely to be greater on the pediment, and thus pediment flow velocities would be proportionately greater. This slope-surface roughness contrast is one that is common in full-scale pediment systems, especially those developed on coarse-grained igneous rocks, and Schumm's argument may apply to them. For example, in the western Mojave Desert (using Cooke's data), quartz monzonite mountain fronts have a mean slope of $28^\circ\ 40'$ and a roughness of, say, 0·06; quartz monzonite pediments have a slope of $2^\circ 31'$, and a roughness of, say, 0·02. Given these facts, Vr on mountain fronts is approximately 18, and that on pediments 15; the relative velocities of flow on the two landforms are probably very similar.

(ii) Processes in full-scale systems. Measurements of processes in full-scale pediment systems are rare. Cooke and Reeves have monitored

the movement of painted natural surface particles and easily identified objects designed to represent the various shapes of natural particles (e.g. cubic wooden blocks, marbles and metal washers), and recorded changes in the position of washers on nails driven into the ground, at several sites along surveyed mountain-front and pediment profiles in the western Mojave Desert over a four-year period. Although their results do not provide evidence of the nature of pediment evolution, it is clear that the following processes are at work in this area: on the mountain fronts, concentrated and unconcentrated flow, animal activity, and slight mantle creep; and on the pediment, concentrated and unconcentrated wash, and animal activity. The main evidence of unconcentrated surface-wash was the lowering of washers on nails, and accumulation of debris in sediment traps.

(b) Processes: inferences and casual observations. Since the oft-quoted and vivid description by McGee (1897) of a sheetflood (or of something which has come to be interpreted as a sheetflood), the phenomenon has been considered as an important process in developing pediments (Davis, 1938; Rich, 1935). Several comments are appropriate here. Firstly, there certainly are occasions when water may flow across relatively undissected desert surfaces in a sheet (Plate 3.14). Secondly, sheetflow of flood proportions is extremely unusual, and interpretation of early travellers' graphic descriptions of runoff in terms of sheet-flooding has rightly been questioned (e.g. Lustig, 1969*a*). Thirdly, whatever the effect of sheetflow, it cannot create the surface, for the surface is a prerequisite for its establishment; it can at best merely modify the surface by removing debris and perhaps by undercutting steep slopes (e.g. Lefèvre, 1952). Fourthly, because records of sheetflow are rare, knowledge of their effects is limited. One of the few informative studies was made by Rahn (1967) who demonstrated from observations of four floods in south-west Arizona that runoff from heavy storms on the bajada may be of a sheetflood character whereas runoff from mountain storms is more commonly in the form of a streamflood; and that the flow is usually supercritical. Discharge resulting from the four storms varied from 260 cfs. to 9·0 cfs.

That streamflow occurs on pediments is a fact more commonly verified by observation, although measurements of discharge are nevertheless scarce. Anastomosing drainage patterns on undissected surfaces suggest streamflow, although it is possible that many of these ephemeral patterns result from the translation of sheetflow to streamflow as a runoff episode closes (e.g. Davis, 1936). A problem would exist in the interpretation of channel data with respect to pediment formation, even if such data were available, because it is not clear in most instances whether the channel system created, or is creating, the pediment or whether the system was established on the pediment after its formation.

The consensus view is certainly that the channel system is related to pediment creation. In particular, many authors place emphasis on the possibility that a stream flowing across a pediment may swing from side to side and progressively erode the surface by lateral planation. Proponents of this process are numerous: Gilbert (1875 and 1877) is credited with its discovery; Johnson (e.g. 1932) with its promotion; Sharp (1940), Howard (1942) and others with its elaboration; and more recently, Rahn (1966) and Warnke (1969) with its re-invocation. (These examples are taken from the American literature; the idea is commonly encountered in other languages.) Evidence used to identify lateral planation on pediment margins is circumstantial, and includes that of rock fans, and that of inselberg slopes which are steeper where channels impinge against mountain/piedmont junctions (Rahn, 1966). In addition, Sharp (1940) in one of the more convincing statements on lateral planation, suggested that evidence for the process in the Ruby-East Humboldt Range, Nevada includes 'The facts that terraces cut largely by lateral planation along adjacent streams coalesce to form partial pediments; that the largest and smoothest areas of pediment are adjacent to the largest streams; that the surfaces indent the mountain front along the streams and are restricted in areas of hard rocks . . . Of the several features cited, the formation of partial pediments by coalescing terraces, obviously cut by lateral planation as shown by meander scars and related features, is considered the most convincing argument favouring the lateral planation theory' (Sharp, 1940, p. 363).

It is well known that lateral planation often occurs in the flood-plains of perennial rivers and it could be that the process is most often responsible for fashioning pediments where perennial flows cross piedmont plains from mountain areas. Or perhaps the process is most effective where relatively unresistant rocks are involved (e.g. Denny, 1967; Sharp, 1940). On the other hand, the effectiveness and indeed the presence of the process has been denied by some (e.g. Lustig, 1968); empirical observations have not confirmed its efficacy (e.g. Schumm, 1962) and some evidence cited in its favour in basin-range situations is, to say the least, unusual. As Lustig (1969*a*, D65–D66) observed: 'No doubt, channels will migrate with time on these (pediment) surfaces, and some erosion will therefore be accomplished, but the part of any pediment that abuts the mountain front cannot be explained in this manner. The hypothesis virtually requires that streams emerge from a given mountain range, and, on occasion, turn sharply to one side or the other to "trim back" the mountain front in interfluvial areas. Such stream paths, nearly perpendicular to a sloping surface, would defy the laws of gravity and have not been observed except in those areas where drastic tilting has occurred.'

In some areas, and notably in Australia, emphasis has been placed on the role of weathering processes in fashioning pediments (e.g. Mabbutt, 1966; Twidale, 1967). Mabbutt (1966), in describing pediments of central Australia, demonstrated that granite pediments are characterized by weathering mantles up to 1·7 m thick beneath thin and mobile alluvial covers, and that these mantles are more stable on middle-pediment slopes, where the degrees of clay weathering and horizon differentiation are greater. The rock face in contact with the mantle is more uneven than the surface, indicating differential weathering at depth and, perhaps, reduction of surface irregularities by alluviation. In addition, subsoil weathering is not only more effective than surface weathering but is also more important in trimming back the hillslope and extending the pediment. In places on schist pediments, surface irregularities reflecting lithological differences are associated with an uneven, chemically weathered suballuvial floor, but in contrast to granite pediments, rock breakdown is more rapid at the surface: 'ground-level trimming' is the name given by Bryan (1925*a*) and adopted by Mabbutt for this process in which 'the mantle imparts a general levelling to an otherwise uneven and heterogeneous bedrock surface, directly through its base-level control of ground-level sapping, and indirectly through its provision of an even surface of attack upon weathering outcrops by rainwash and sheetflow' (Mabbutt, 1966, p. 89).

Subsurface weathering is likely to be particularly pronounced at the mountain-pediment junction because of the natural concentration of water there. Reasons for this concentration include the facts that runoff tends to disperse as it reaches the junction and percolation may be increased (especially if a debris veneer is present), and that more runoff-producing rains occur in the mountains and the runoff often travels only a short distance on to the plains (Twidale, 1967). Twidale (e.g. 1968) identified zones of especially intense weathering at piedmont junctions and related these to a sequence of pediment-surface development (Fig. 3.12); but such zones are by no means a universal phenomenon.

Two further comments arise from a recognition of the presence of surface weathering phenomena. Firstly, the relative rates of weathering and fluvial erosion processes require careful examination. Does the presence of soil profiles indicate a stable surface, or a balance between rates of weathering and rates of surface lowering? Secondly, it is important to know the relationship between soil-profile development and

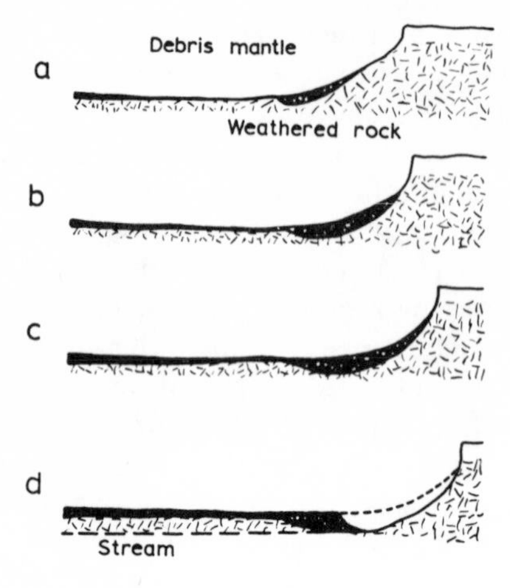

3.12 Pedimentation, scarp retreat and preferential weathering at the junction of pediment and scarp (Twidale, 1968).

pediment extension. If, for example, a pediment grows at the expense of a mountain mass, its surface is clearly diachronous, being youngest at the mountain front. Is this fact reflected in the pattern of pediment soils, or are the weathering features much younger than the pediment?

Corbel (1963) concluded from a study of pediments in Arizona that chemical processes are relatively, and often absolutely, more important in the piedmont zone than in other areas, especially where a pediment is thinly covered with alluvium. He estimated, on the basis of rather few data, that in Arizona the rate of pediment erosion is 36 m^3/km^2/p.a. (of which 70 per cent is chemical), and in California it is said to be 6m^3/km^2/p.a. (of which 85 per cent is chemical). These estimates emphasize the potential importance of chemical processes in pediment development. Taken together with the comments of others mentioned above, it would seem that such processes deserve greater attention.

3.4.5 MODELS OF PEDIMENT DEVELOPMENT

It is clearly unrealistic to seek a single comprehensive hypothesis of pediment development because the landform varies enormously from place to place, and the effective processes are variable in both space and time. Furthermore, detailed diagnostic evidence of pediment evolution is rare, and much interpretation rests on impression, intuition and imagination; and some essential evidence may never be reclaimed. In this section we present a possible general framework within which pediment evolution may be considered, and briefly outline some of the more attractive hypotheses.

(a) A general model. In most localities, the pediment forms one area within a system generally comprising a mountain area, a pediment, an alluvial plain and a base-level plain (Fig. 3.4). As Corbel (1963) has indicated, the pediment is often intermediate between zones of bedrock and alluvium, erosion and sedimentation and, perhaps, uplift and subsidence. The critical boundary in this system is that between areas where erosion is dominant and areas where deposition is preponderant. More often than not, this boundary coincides with the mountain/piedmont junction and there is no pediment; elsewhere, the boundary may be on the piedmont plain, and there is a pediment. The location of this boundary will be determined in large measure by the relationships between the rates of debris supply and removal. We can envisage an equilibrium condition, where the boundary is stable and the two rates are approximately equal (e.g. Denny, 1967). In these circumstances, the actual position of the boundary will be determined by local circumstances—it could be anywhere along the profile. If the equilibrium is upset, so that supply changes with respect to removal, the boundary will migrate in the appropriate direction until a new equilibrium is established. The movement of the boundary may or may not be accompanied by a change in form, e.g. dissection. In areas with pediments it appears that erosional forces are adequate to remove available debris efficiently from at least part of the piedmont plain.

Within this general framework, we may envisage three situations in which pediments may evolve. Firstly, erosional processes may become ascendant over those of debris supply, so that the alluvial margin migrates downslope exposing a pediment in its wake. This we shall call the *exhumation hypothesis*. Secondly, the pediment (or the piedmont plain) may be extended at the expense of the mountain mass within the erosional zone. Here there are several explanatory hypotheses which we can survey under three headings: the *parallel retreat hypothrsis*, in which mountain fronts are said to retreat parallel to themselves, leaving incipient pediments at their bases to be modified subsequently by weathering and runoff; *the drainage basin hypothesis* which admits to at least two major zones of change in a drainage basin—along the principal river valleys and along the mountain front between and transverse to them (Lustig, 1969*a*); and the *lateral planation hypothesis*, in which lateral swinging of streams is said to be responsible for mountain mass reduction. A third situation may be one in which processes of debris supply equal or exceed those of removal, and weathering plays an important role in fashioning the pediment. This we may call, following Mabbutt, the *mantle-controlled planation hypothesis*.

In addition, we may recognize four phases of pediment evolution common to most of the explanatory hypotheses: initiation, extension, modification and destruction. Initiation of a pediment is probably associated with the creation of relief, and this can only be achieved either directly by tectonic movements or indirectly by drainage incision consequent upon changes in endogenetic and/or exogenetic condition. Pediment extension may be accomplished by one or more of the processes we have previously described and which are enshrined in the hypotheses we have identified. Once formed, a pediment may be modified in several ways: by fluvial regrading (sheetflow, streamflow or lateral planation), and by mantle-controlled planation. Processes of pediment destruction are of two kinds: processes of aggradation which lead to burial of the surface; and processes of dissection which destroy its continuity. Both of these changes may be accomplished by base-level changes, tectonic changes, or climatic changes.

(b) Brief exposition of evolutionary hypotheses. Below we briefly review the major hypotheses of pediment evolution. Where possible we have merely quoted the relevant author. We have not tried to evaluate their relative merits, because of the difficulties we have repeatedly emphasized.

(i) Parallel retreat hypothesis. Lawson (1915) viewed the geometry of mountain and plain development in basin-range country as follows. A mountain front, however it may have been initiated, undergoes weathering and erosion, achieves a characteristic slope (determined by lithology) and retreats roughly parallel to itself (Fig. 3.13a). As the mountain front retreats, alluvium accumulates at its base, and a bedrock bench is produced beneath it. The nature of the mountain front-piedmont plain profile is determined by many features which include the initial geometric framework; processes and rates of denudation; structure and lithology; and tectonic conditions.

In the 'ordinary case' envisaged by Lawson, of a piedmont plain with a rising base-level, the mountain front retreats at a uniform rate, but the alluvium increases in thickness at its foot at a decreasing rate because successive alluvial increments are spread over a progressively larger alluvial plain (Fig. 3.13a). As the piedmont plain continues to develop, however, the smaller quantities of alluvium derived from the mountain may be insufficient to keep the extending piedmont plain covered, and a bedrock bench may be formed subaerially (a pediment). Ultimately, the mountains may be completely eliminated, and only piedmont slopes may remain.

Since it was introduced, Lawson's imaginative hypothesis has been critically examined, modified, embraced and rejected (e.g. Davis, 1930; King, 1962; Tuan, 1959). In general, if not in detail, its assump-

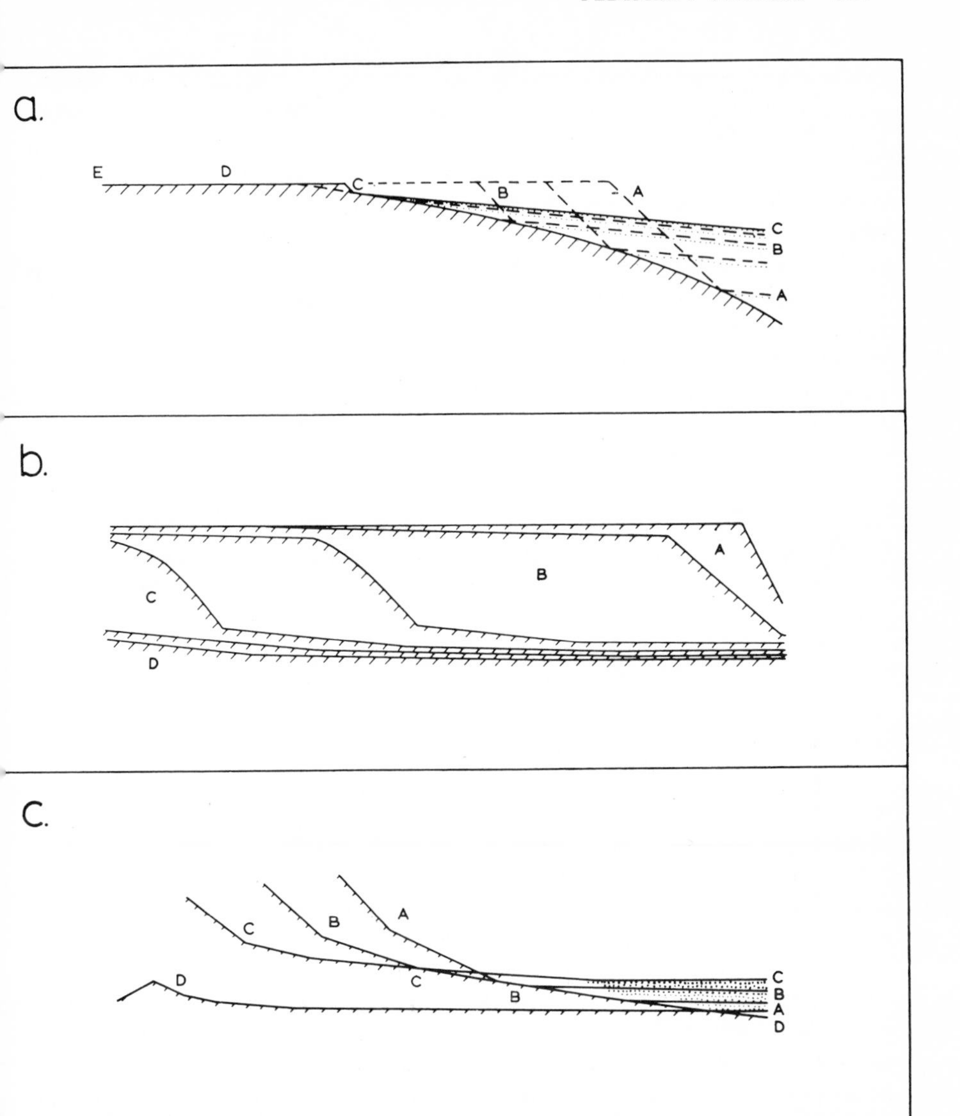

.13 Hypothetical sequences of mountain and piedmont plain development in the desert. (*a*) Lawson's scheme in an area with a continuously rising baselevel (after Johnson, 1932); (*b*) stages in the evolution of a granitic mountain area in the Sudan (after Ruxton and Berry, 1961); (*c*) successive profiles of regrading allied to lateral planation of pediments (after Johnson, 1932).

tions accord well with observations of pediments in many basin-range deserts; its principal failing appears to be that the role of major drainage courses from the mountains is underplayed. Ruxton and Berry (1961) formulated an interesting extension of the ideas of Lawson and others in explaining the evolution of granite landforms in arid, semi-arid and savanna areas of the Sudan (Fig. 3.13b). They suggested that once an equilibrium profile is established, it continues to develop by parallel retreat of both the hillslope and the 'plinth', until eventually downwearing becomes of greater importance than backwearing. No progressive accumulation of basin sediments is involved in the area, but central to the hypothesis is the presence of a regolith zone along most profiles which comprises a superficial layer of migratory debris and beneath it a sedentary weathering zone in which material is prepared for removal.

Rates of mountain-front retreat and pediment extension are largely a matter of speculation. Melton (1965*a*) reckoned pedimentation in southern Arizona may have proceeded, assuming a uniform rate, at about 0·1 cm per year. This compares with King's (1953) estimate of 0·1–0·2 cm per year.

A problem closely allied to the parallel retreat hypothesis is that of the origin and development of the piedmont angle. We have already pointed out that the piedmont angle varies greatly from place to place; and that the boundary between mountain and plain may be anything from a gradual change of slope to an abrupt break. At times the mountain front is extremely steep, and occasionally it appears to have been over-steepened. There appears to be general agreement—but little proof—that the piedmont angle is maintained as the landforms evolve. A corollary of this view is that the mountain front retreats parallel to itself (Balchin and Pye, 1955). This appears to be established in certain places where change is rapid (e.g. Schumm, 1956), and is eminently probable in arid and semi-arid areas, where there is unimpeded removal of weathered debris from the base of mountain fronts, and slope inclination is adjusted to environmental conditions. A further corollary is that as the mountain front retreats it *automatically* creates a pediment surface which can later be modified (e.g. Büdel, 1970; King, 1953). We may conclude by briefly noting explanations of the origin and maintenance of the piedmont angle. *(a)* Lateral planation by streams (Johnson, 1932; Rahn, 1966); *(b)* uplift of a mountain block along a fault or fold; *(c)* change from one type of bedrock lithology or structure to another Denny, 1967; Rich, 1935; Twidale, 1964 and 1967) and, allied to this, maintenance of steep mountain fronts by resistant cap-rocks (e.g. Everard, 1963; Fair, 1947); *(d)* difference between processes operating on mountain front and pediment (e.g. change from rillwash to sheetwash from turbulent to laminar flow; from gravity-controlled slopes to fluvially-controlled slopes); *(e)* change from one type of debris to another (e.g. Cooke and Reeves, 1972); *(f)* maintenance or even oversteepening by ground-level trimming (Mabbutt, 1966) or locally intense weathering followed by exhumation (Twidale, 1962); *(g)* 'unloading' and sheet jointing especially in granitic rocks (e.g. Twidale, 1962); and *(h)* concentrated subsurface erosion at the hillfoot due to lateral flow or water in the weathering profile (Ruxton and Berry, 1961, Bocquier, 1968).

(ii) Drainage basin hypothesis. This hypothesis represents a logical extension of Lawson's view in the context of modern drainage basin studies. It has been succinctly propounded and encapsulated by Lustig (1969*a*, p. D66–67) whose account we quote here:

> It has been proposed that pediments result from parallel slope retreat of the mountain front . . . The writer's observations in several deserts of the world suggest that few, if any, escarpments are not dissected by prominent drainage systems . . . The existence of drainage basins in the mountain ranges is of central importance to the various pediment arguments. These basins are the loci of the most effective erosional processes that operate on mountain ranges . . . The mountain fronts may well retain some characteristic slope angle that reflect rock strength, structure, weathering characteristics, and other variables, and they may retreat at this angle. This does not prove, however, that ranges are primarily reduced by parallel retreat of escarpments . . .
>
> . . . There is no question that processes of subaerial and suballuvial weathering occur on pediments today, nor that fluvial erosion also occurs. A pediment must exist prior to the onset of these processes, however, and in this sense the origin of pediments resides in the adjacent mountain mass and its reduction through time . . .
>
> To say that pediments result from the reduction of adjacent mountains through time is to give voice to a seemingly obvious and intuitive argument . . . Each range has some local base level in an adjacent basin, and it cannot be eroded below this level. Given stability for a sufficient period of time, the consequences of mountain reduction must inevitably include the production of a pediment . . . The only real 'pediment problem' is how the reduction or elimination occurs.
>
> Observation shows that the course of any master stream channel from a given drainage basin from the mountains on to the pediment surface and thence to the basin floor has no sharp change of slope . . . The interfluvial areas, however, generally do exhibit a marked change

in slope, at least within a narrow zone parallel to the mountain front. The reason for the existence of such a zone is precisely that it is an interfluvial area; the dominant process that operates on a mountain front is not fluvial.

Qualitatively, it can be argued that two basic processes are operative on a given mountain mass. The steep slopes of the mountain front are interfluvial areas that are subject to weathering. Runoff on these steeply sloping surfaces is of short duration and is not concentrated. The runoff serves largely to remove the finer weathered debris that is transportable. Larger particles generally remain in place until they are reduced in size by weathering. The rates of mountain front retreat are basically unknown, but by any reasonable assessment they are slow in relation to rates of processes that are operative in drainage basins . . .

(iii) Mantle-controlled planation hypothesis. This hypothesis may most appropriately be reviewed by quoting Mabbutt (1966, pp. 90–91). The remarks apply only to central Australia, but they may well be relevant in other areas:

. . . it is concluded that grading on these granitic and schist pediments has largely acted through the mantles, whereby the smoothness of depositional profiles has been transmitted to suballuvial and part-subaerial pediments. The action of the mantle is most direct on granitic pediments, in that suballuvial notching and levelling proceed through weathering in the moist subsurface of the mantle. On schist pediments control of levelling by the mantle is less direct, in that its upper surface is the plane of activity of ground-level sapping and of erosion by rainwash and sheetflow. In this way, continuity of grade is established across a succession of separate schist outcrops. On schist pediments, where trimming is more rapid, a single depositional cycle may leave a widespread erosional imprint; on massive granitic rock, where weathering advances on a broad front and much more slowly, each cycle of mantling may produce only minor effects. The processes described are most active near the hill foot, where mantling and stripping alternate more frequently; the lower parts of both schist and granitic pediments in central Australia appear to be largely and more permanently suballuvial . . .

It might be objected that the processes described are tantamount to insignificant secondary modification, under stable conditions, of a pedimented landscape left by other primary agents. However, this would lose sight of landscape evolution in these areas since the Tertiary. Relics of a weathered Tertiary land surface survive as laterite-capped low platforms on plains adjacent to all study areas

> and occur close enough to the hills in the Aileron and Bond Springs areas for the former piedmont profiles to be reconstructed. The profiles show that the present plains were already the sites of lowlands on the Tertiary surface, topped by the hills of today, and that subsequent evolution has involved mainly the breaching of the duricrust and the etching away of a layer of soft weathered rock . . . In most sites the hill base was set back to a structural boundary from which there can have been little subsequent retreat . . . What has been involved in pedimentation has been slight back-trimming at the hill base, partial weathering of an irregular, exhumed weathering front, and the imposition of continuous, concave profiles . . .

This scheme is genetically allied to that of Thomas (1966), who found in Northern Nigeria that the main processes working in the inselberg landscape were etching and stripping, and to that of Ruxton and Berry (1961) noted above.

(iv) Lateral planation hypothesis. In Douglas Johnson's (1932, pp. 656–58) words:

> The essence of the theory is that rock planes of arid regions are the product . . . of normal stream erosion. A peculiarity of the stream erosion theory is the relative importance attached to lateral corrosion . . .
>
> Every stream is, in all its parts, engaged in the three processes of *(a)* vertical downcutting . . . *(b)* . . . aggrading, and *(c)* lateral cutting . . . If we imagine an isolated mountain mass in an arid region, it will be evident that the gathering ground of streams in the mountains . . . will normally be the region where vertical cutting it at its maximum . . .
>
> Far out from the mountain mass conditions are reversed . . . Aggradation is at its maximum . . .
>
> Between the mountainous region of apparently dominant degradation and the distant region of apparently dominant aggradation there must be a belt or zone where the streams are essentially at grade . . . Lateral corrasion is again dominant not only in fact but in appearance . . . Heavily laden streams issuing from the mountainous zone of degradation are from time to time deflected against the mountain front. This action, combined with the removal of peripheral portions of interstream divides by lateral corrasion just within the valley mouths, insures a gradual recession of the face or faces of the range.

As the processes described by Johnson proceed, the pediment profile is progressively regraded (Fig. 3.13c). We have commented on the acceptibility of the lateral planation process above (section 3.4.4).

(v) Exhumation hypothesis. This hypothesis differs from the others in that it assumes a mountain-piedmont plain relationship to be established; indeed it could have been established in terms of any of the hypotheses described above. All that is required is a transfer of the alluvial boundary across the piedmont plain to expose the bedrock pediment. Evidence for such movement of the boundary may include truncation of soil and weathering profiles, and the presence of alluvial outliers. The exhumation will normally be accomplished by running water, which may be concentrated in channels, or may merely be unconcentrated wash. General causes of such changes may be: changes in the geometric relations of mountains and plains in the system; climatic or tectonic changes; or base-level changes. Mabbutt (1966), Paige (1912) and Tuan (1962) are amongst the several proponents of this hypothesis.

(c) Sequences of pediments. Most of our discussion has been concerned with the nature and evolution of a single pediment in an active system. In many areas, however, several pediments may coexist as a 'geomorphological staircase' and, although they may have been initiated at different times, it is possible that they are all continuing to be extended. Sequences of pediments have been described by Twidale (1967) in South Australia, by Butzer (1965) at Kurkur in Egypt, by Cooke (1964*b* and *c*) in Chile, by Rich (1935) at Book Cliffs, Utah, by Sharp (1940) in the Ruby-East Humboldt Range, Nevada, and by several authors in Arizona and New Mexico.

There are, for instance, magnificent flights of terrace pediments along the valleys of the San Pedro, Rio Grande and other major rivers in Arizona and New Mexico. The principal explanatory hypothesis states that the succession of surfaces results from valleyside planation graded to successively lower base-levels of the downcutting rivers in late Tertiary and Quaternary times (e.g. Bryan, 1926 and 1936). Tuan (1962) considered the causes of the suite of surfaces in the San Pedro Valley (Fig. 3.14) in some detail. He showed that the simple 'base-level hypothesis' was not strictly applicable in this area. More probably, he claimed, it was tectonic movements and their effects on the fluvial system which led to the succession of surfaces. Evidence for tectonic activity, and contrary to the base-level hypothesis, includes the recognition of structural deformation during the period of landform development, the irregular arrangement of basin surfaces, their differential dissection and exhumation, and the straightness of scarps between upper and lower surfaces.

The explanation of pediment sequences lies in the recognition of the processes responsible for drainage incision. The range of possible

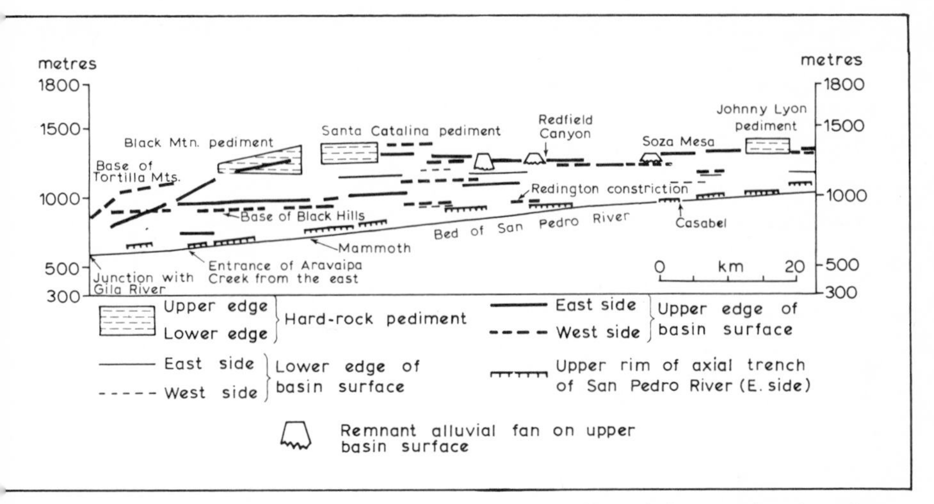

.14 Longitudinal profile of the lower San Pedro Valley, Arizona, showing the elevation of piedmont surfaces on both valley flanks (after Tuan, 1962).

processes is actually quite limited and is related ultimately to changes of climate or sea-level, or to crustal activity. Specification of the process or processes active in a particular area is often a matter of controversy and speculation.

3.5 Playa Systems

3.5.1 INTRODUCTION

The lowest areas within enclosed desert drainage basins are often marked by almost horizontal, largely vegetation-free surfaces of fine-grained sediments. Similar surfaces may also occur at intermediate locations along a desert drainage network. Such base-level plains are common and distinctive desert features, and they have acquired many different local names. General terms include *nor* (Mongolia), *pan* (South Africa), *sebkha* (North Africa), and *dry lake* and *playa* (North America). The variety of local terminology is evident in the Sahara, where terms such as *sebkra* (*sebjet*, *sebchet*, *sebkha*, *sabkhah*, *sebcha*), *zahrez*, *chott* and *garaet* are used for similar features in different areas (Coque, 1962; Smith, 1969). Here we shall use *playa* as a general term for base-level plains in desert drainage basins. Different types of playa have acquired specific names. A clay-silt playa, for instance, is known as a *clay pan* in Australia, a *takir* in the U.S.S.R., a *khabra* in Saudi Arabia, and a *qu* in Jordan; and a saline playa is called a *kavir* in Iran, a *salar* in Chile, and a *tsaka* in Mongolia (Neal, 1969).

In the following review we draw heavily on recent studies of American and Saharan playas, and on our own observations in the Americas, emphasizing those features which appear to be of more general significance. Reports by Krinsley (1968 and 1970) on Iranian *kavirs*, and by Krinsley, Woo and Stoertz (1968) on Australian playas do suggest that the features we describe are of general significance, but in some instances this remains to be substantiated.

Playas have received much attention, for four main reasons. Firstly, they are sites of mineral wealth—notably salts and especially chlorides, sulphates, carbonates, nitrates and borates—and their economic potential has engendered a large literature (e.g. Gale, 1914; Smith, 1966). Secondly, the record of playa deposits has been valuable in determining the nature of climatic and ecological change in desert basins during the Quaternary (e.g. Martin, 1963; Morrison, 1965; Smith, 1968). Thirdly, horizontal smooth surfaces are attractive for high-speed vehicular movement and military activity (e.g. Neal, 1965*a* and *b*, 1968 and 1969). Finally, playas form unusual, distinct and varied geomorphological and hydrological environments (e.g. Coque, 1962; Motts, 1965).

The distribution of playas has now been mapped in many deserts. There are over 1,000 playas recorded in North Africa; of some 300 playas in the United States, at least 120 are the relics of Pleistocene lakes (e.g. Feth, 1961*a*; Snyder, Hardman and Zdenek, 1964); Bowman (1924) described the pattern of predominantly moist *salars* in the Atacama Desert and the Andes; and Gregory (1914) located over 190 salt-clay pans in Western Australia. The occurrence of playas is principally controlled by climate (Tricart, 1954), as we shall describe below, but as most playas are located within enclosed drainage basins the causes of enclosure are clearly of some interest. Enclosure of drainage basins most commonly results from tectonic movements (e.g. creation of basins or blocking of established drainage lines by faulting, folding or general subsidence), aeolian processes (e.g. formation of hollows by deflation, sand deposits blocking drainage-ways), or volcanic activity (e.g. creation of craters and calderas, blocking of drainage by volcanic effusions). Tectonic movements are of regional significance in the basin-range deserts of North America, the Atacama, Iran and Inner Asia; volcanism is significant in Atacama; and aeolian activity is reported to be important in creating enclosed basins in the Sahara, and especially in the eastern Sahara (Smith, 1969). Many basins, of course, arise from several causes, and some have developed over long time periods.

Very few data are available for relating size and shape of playas to drainage basin conditions. As we mentioned in our discussion of alluvial fans (section 3.3), mutual adjustment of playa area and fan area might be envisaged. Similarly, playa area might be related to other morphometric properties of drainage basins. Using approximate map data, a preliminary analysis by Cooke showed that for 38 playas in the Mojave Desert, California, playa area is positively correlated with drainage basin area ($r = 0{\cdot}6645$, significance level—0·1 per cent), and playa area is negatively correlated with the ratio of drainage basin relief to basin area ($r = -0{\cdot}4055$, significance level—1.0 per cent). Such an analysis does tend to confirm the notion that playa area is related to drainage basin conditions, and this line of enquiry could profitably be pursued, especially with reference to changing playa dimensions and hydrological circumstances. The actual areas of playas vary greatly, from a few square metres to over 9000 km^2. Lake Eyre in Australia is one of the largest, with an area of approximately 9300 km^2 (Twidale, 1968). The shape of playas, despite short-term variability, may reflect the origins of, and the contemporary processes at work within, playa basins. For instance, a straight margin may be imposed by faulting (e.g. Krinsley, 1970; Neal, 1965*a*, p. 113), or a jagged, irregular boundary may be determined by the pattern of deposition at the termini of a complex system of drainage lines.

3.5.2 DISTINCTIVENESS AND CLASSIFICATION

(a) Playa environments. Playas are pre-eminently receptacles for sediment and water, and their nature is in large measure determined by their sedimentary and hydrological properties.

Playas are characterized by fine-grained clastic and non-clastic sediments. Elastic sediments are mainly derived from surface runoff by deposition either from flows of water or from standing water bodies. The clastic material is usually fine grained, comprising clay, silt and granular particles. Non-clastic sediments, composed mainly of saline deposits, come largely from ground-water, although it may be difficult to determine precisely the proportions resulting from ground-water and surface runoff. Playas are distinctively areas of evaporite formation, as we have discussed in section 2.5.3., and in our discussion of solonchak soils (section 2.3.2).

The accumulation of fine-grained sediments has three main geomorphological consequences. Firstly, many playas—especially those with smooth, hard surfaces—tend to be relatively impermeable, and thus the accumulation of surface runoff is encouraged. Secondly, because of the shrinkage properties of the sediments, desiccation phenomena are common (section 2.5.1). And thirdly, aeolian activity is probably relatively more effective in these areas of fine-grained sediments than elsewhere.

Water on playas results from surface runoff, direct precipitation, or ground-water discharge. Ground-water discharge at playa surfaces may be from capillary rise, or, in areas where the water-table intersects the surface, directly from the ground-water zone. Surface runoff is removed by infiltration and evaporation; ground-water is lost by evaporation and evapotranspiration. All playas exist in areas where annual evaporation is considerably greater than annual precipitation, but as the ratio is reduced, and the supply of surface- and ground-water is increased, so the period in which water is at the surface is extended and eventually perennial lakes may be formed (Langbein, 1961). Such lakes without outlets ('closed' lakes) can only exist where precipitation on the lake and inflow to it are equal to or less than the rate of evaporation. The local relations between present climate and basin floor conditions, and the hydrological balance, are influenced, of course, by many variables. The geometric, soil and vegetation characteristics of the drainage basin are all important. The rate of evaporation in any one area depends not only on climatic conditions but also on other factors such as water salinity (Harbeck, 1955) and the geometry of the water body. Further complications arise if, for example, part of the water supply is from outside the arid region, or from 'fossil' ground-waters which have moved long distances; or if, as is often the case, the playa is inherited from prior pluvial conditions and is now only inundated by infrequent floods. Nevertheless, dry playas and perennial 'closed' lakes may be viewed as the two extremes of a continuum representing the conditions of enclosed-basin floors in arid lands.

Motts (1965) has suggested a simple, qualitative classification of playas based on the relative amounts of ground-water and surface inflow (Table 3.8). Examples of the different types occur in many deserts. For example, the 'hard, dry playas' of the western United States and the *garaets* of Tunisia are supplied largely by surface runoff (1), whereas the 'self-rising moist playas' in the former area and the *chotts* of the latter are fed mainly by ground-water discharge (5). The relations between ground-water divides and topographic divides, and the position of ground-water discharge within ground-water divides were criteria used by Motts (1965) to classify further the ground-water discharging playas.

(b) Surface features. Within the general context of playa sedimentology and hydrology we consider next the variety of surface features. Adjectives commonly used to describe playa surfaces are hard, soft, wet, dry, rough, smooth, flaky, puffy, salty or non-saline (Neal, 1969). Several combinations of adjectives frequently recur. For example:

TABLE 3.8 *Hydrological Classification of Playas*

1		2	3	4	5
Increase of surface-water discharge →			increase of ground-water discharge → increase of surface-water discharge ←		All ground-water discharge
All surface-water discharge		Relatively small amount of ground-water discharge and large amount of surface-water discharge	Approximately equal proportions of ground-water and surface-water discharge	Relatively large amount of ground-water discharge and small amount of surface-water discharge	
1A	1B				
Small amount of surface-water discharge	Large amount of surface-water discharge				

SOURCE: Motts, 1965.

hard, dry, smooth surfaces; and soft, friable, puffy surfaces. And often a particular surface type is associated with a specific group of environmental conditions. Thus hard, dry smooth surfaces commonly occur on playas dominated by surface runoff and silt-and-clay-sized sediments. Many of the surface properties can be precisely measured. Langer and Kerr (1966), for instance, used a Knoop indenter to measure the penetration resistance of bulk playa samples. But such quantitative data are few, and thus in the following discussion we rely on the use of rather general qualitative terms. Such terms do not make for precision, but they do facilitate comparison and generalization and they permit simple hypotheses to be outlined.

In a detailed survey of playas in Iran, Krinsley (1970) recognized seven major playa surface types and determined their extent and frequency (Table 3.9). This analysis emphasizes the relative predominance of salt-crust and clay-flat surfaces in Iran—a feature common to playa areas throughout the world.

Motts (1965) classified playa surfaces according to the position of the water-table and the nature of ground-water discharge, as these hydrological factors influence the sedimentology and ultimately the geomorphology of playa surfaces (Table 3.10). This classification provides a suitable basis for describing playa surfaces, although each category is not necessarily exclusive because different hydrological processes may produce similar surface features. We examine each of the six categories below, basing our general comments mainly on descriptions of United States playas by Motts (1965), Neal (1969), and Langer and Kerr (1966).

TABLE 3.9 *Extent and Frequency of Playa-surface Types in Iran*

Surface type	*Area (km²)*	*Percent of total playa area*	*Number of playas (partly or wholly of particular surface type)*	*Percent of all playas*
Salt crust	27,624	41	50	83
Clay flat	23,724	35	24	40
Wet zone	6,702	10	14	23
Fan delta	3,229	5	6	10
Swamp	2,889	4	3	5
Intermittent lake	1,912	3	2	3
Lake	1,172	2	2	3
TOTAL	67,252	100	60	100

SOURCE: Krinsley, 1970.

TABLE 3.10 *Classification of Playa Surfaces*

i Total surface-water discharging playas where the water-table is so deep that no ground-water discharge occurs at the playa surface
ii Playa surfaces where discharge occurs at the surface by capillary movement
iii Playa surfaces where discharge occurs directly at the water-table
iv Playa surfaces where discharge occurs by phreatophytes and other plants
v Playa surfaces where discharge occurs by springs
vi Combinations of all of these

SOURCE: Motts, 1965.

(i) Total surface-water discharging playas. The surfaces of these playas are characteristically smooth, hard, commonly dry, and composed of fine-grained clastic sediments. Over half of their material may be of clay, and there may be some calcium carbonate accumulation, but the proportion of salines is often low. They are relatively impermeable and have relatively high bearing strength. Because there is a downward component to water movement when the surface is flooded, a soil profile may be developed below the surface and this may have a distinctive zone of caliche development. The surface of some playas in this category, such as that of Sabzever Kavir, Iran (Krinsley, 1968), may show traces of a distributary drainage network. Aggradation is by

deposition from surface floods or by aeolian deposition. The quantity and quality of incoming sediment will depend largely on the geologic and climatic environment within the basin. Motts (1965) suggested that in many playas of the western United States, for example, more clay and less silt were deposited in times of more humid climate than is the case at present because of greater chemical weathering. The actual surface type which develops, however, is in part related to the conditions of sedimentation. Different surface types could develop from essentially similar sediments entering the basin, depending on the nature of the major salines at the time of deposition (Kerr and Langer, 1966). For example, waters with high carbonate content may have deflocculated clay particles, leading ultimately to the formation of homogeneous, fine-grained, compact crusts (Kerr and Langer, 1966).

High shrinkage coefficients for surface and subsurface material in playas of this type mean that desiccation phenomena are common (section 2.5.1). Neal (1969) noted that the surface glaze on these playas is perhaps related to a high degree of fine-particle orientation.

(ii) Capillary-movement playas. Capillary movement of water to the playa surface will vary mainly according to the relations between the opposing forces of evaporation and capillary pressure.* The former may vary seasonally or over longer time periods. The latter increases as the radii of particles in playa sediments decreases (Langer and Kerr, 1966). In places, artesian pressure may significantly supplement capillary pressure (e.g. Neal and Motts, 1967). Rapid capillary discharge is often associated with soft, friable puffy ground—the 'self-rising ground' of some authors—and with salt-thrust polygons. Slow capillary discharge may yield a thin salt crust. In either case, the surface tends to be saline, soft, fairly permeable and loosely compacted, and it may have a microrelief of up to about 15 cms. In comparison to the hard, dry crusts of type *(i)* (Table 3.10) the surfaces of capillary-movement playas in the United States appear to have a lower clay content, and a higher proportion of salines and larger-sized particles. Moisture content of the surfaces, and hence the surface character, may vary greatly during the year.

(iii) Playas with direct ground-water discharge. Salt crystallization characterizes playa surfaces where groundwater discharges seasonally or perennially on to the surfaces. There may be thin salt crusts or

* Langer and Kerr (1966) define capillary pressure, Pc, as

$$Pc = \frac{Ts}{g}\left(\frac{1}{r_1}+\frac{1}{r_2}\right),$$

where Ts is the surface tension between liquids and particles of radius r_1 and r_2.

thicker salt 'pavements'. In Australia such salt layers may be up to 46 cms thick (Krinsley, Woo and Stoertz, 1968). Pressures of salt crystallization tend to cause surface disruption, creating salt ridges, salt thrust polygons, and extremely irregular microtopography. The surfaces are generally wet, soft and sticky; and they may be 'puffy' in places. Solution phenomena—such as small pits—may also occur.

(iv) Playas with phreatophyte discharge. Although many playas are without vegetation, loss of water by evapotranspiration through phreatophytes and other plants can occur where the water-table is sufficiently close to the surface for it to be tapped by plant roots and where the water is of low enough mineralization to allow plant survival. Features associated with playa plants are ring fissures (section 2.5.1), slight surface subsidence, and phreatophyte mounds. Evapotranspiration may lead to local lowering of the water-table beneath plants and, as a result, there might be slight subsidence of the ground. Phreatophyte mounds arise from the accumulation of fine-grained wind-blown material and from the precipitation of salts near to the base of plants. They may be up to five metres high. A sequence of phreatophyte mound development is described by Neal and Motts (1967). Initially the plant grows to maturity near to the playa surface. Aeolian debris begins to accumulate around it and salts may form a surface crust on the mound. As the mound becomes larger, the plant continues to survive at the top of the mound by extending its root system. There will come a time when the plant is so high that its roots can no longer draw water from the water-table, and it will die. The mound may resist erosion because of the salt crust, but once the crust has been dissolved by rainwater, the aeolian deposits beneath will soon be degraded by wind, rainwater, streams, or even by lacustrine erosion.

(v) Playas with spring discharge. Springs occur where the water-table or piezometric surface is higher than the playa surface. If the piezometric surface should rise, the springs may become foci of evaporite deposition, producing spring mounds. Spring mounds can theoretically grow to the level of the piezometric surface (Motts, 1965; Neal and Motts, 1967).

(vi) Combination of surface types. Any single playa surface may include the characteristics of more than one of the above groups. For example, a hard, smooth clay surface may grade into a puffy crust, and a puffy crust may merge into a salt pavement. Such a spatial gradation would probably be accompanied by a decrease in the argillic component and an increase in salt content (Langer and Kerr, 1966), and by an increase in surface moisture. Similar changes may affect a single part of

the playa surface as conditions, especially hydrological conditions, change through time.

(c) Wind activity. The accumulation of material in playas is related mainly to evaporation processes and to surface runoff. The latter may also be a locally significant process of erosion. But as most playas occur in the lowest areas of enclosed basins, solid material can only be removed from them by the wind. The mechanisms of wind erosion and dune formation will be fully discussed in Part 4. Here it is only necessary to state a few points specifically pertinent to playas.

Firstly, not all playa sediment is equally susceptible to deflation. For example, hard, dry crusts with a high clay content are very much more resistant to deflation than surface materials rich in salts (Coque, 1962). In addition, the susceptibility of playa sediment to deflation varies with its moisture content, which is itself often quite variable areas, in depth, and through time. The effective lower limit of deflation in playas is the water-table.

Secondly, removal of playa sediment by the wind does not necessarily mean that there is net erosion from the drainage basin. Some material may be carried in suspension out of the basin, but most will be deposited either on the playa (e.g. in phreatophyte mounds), or around its margins, perhaps in the form of sand dunes (section 4.3). Some of the marginal deposits may be returned to the playa. Indeed, we may envisage a cycle within a closed basin in which clastic and saline materials are supplied to the playa by rainfall, runoff and ground-water, are removed by deflation, and are subsequently returned to the playa by runoff or from recharged groundwater (see section 2.3.2d. *(vi)*).

Thirdly, wind action has been invoked to explain rather unusual 'stone tracks' on some playa surfaces in the United States. The tracks, which often change their direction, are up to about three centimetres deep, several centimetres wide, and vary in length from a few metres to a recorded maximum of 270 m. At one end of each track there is usually a stone or some other object, embedded in the surface and weighing up to 200 kg. Clearly the tracks have been created by the scraping of the stones across the playa. Two principal mechanisms of stone movement have been suggested. Some have argued that loose stones are blown by strong winds across wet playa surfaces (Clements, 1952; Kirk, 1952). An imaginative alternative hypothesis (Stanley, 1955), tested in detail on Racetrack Playa, California, is that the stone tracks are produced by wind-blown ice-floes dragging protruding stones across the wet playa surfaces. This hypothesis is strongly supported by the identical signatures of groups of tracks, which imply that the scraping stones were held in fixed positions relative to one another during movement. Only ice could provide the planar body required to maintain those positions. A significant feature of Stanley's analysis is his demonstration that mathematical prediction of tracks produced in a rotating floe closely corresponds to observed patterns. The occurrence of ice-ramparts on playa shores and reports of ice-floes on playas in winter lend support to this hypothesis. As Stanley emphasized, stone tracks occur on only a few playas; it could be that they are absent from many because there are no suitable stones or because there are no ice-floes. In some playas, such as North Panamint Playa, California (Cooke, 1965), stones are scattered across the surface but are not associated with tracks. In such cases, tracks may not have been formed, or they may have been obliterated by subsequent changes in the distribution of fine surface sediments. Although the ice-floe hypothesis is convincing, the wind-and-wet-surface mechanism still deserves attention. The theoretical discussion and experimental work of Sharp (1960) showed that the force produced by the drag of the wind must be sufficient to overcome the frictional resistance of the mud, and that the wind velocities required to move scrapers over wet mud are higher than those which commonly occur naturally. These observations, and the fact that there is no evidence at present of stone tracks in playa surfaces where ice-floes never form, suggest that the wind-and-wet-surface mechanism may be relatively unimportant. All are agreed, however, that it is wind which causes the stone movement.

3.5.3 PLAYA CHANGES

Short-term changes in the character of playas are mainly associated with seasonal and annual variations in the availability of surface and subsurface water. Both seasonal and annual variations are evident, for example, in the Chott el Djerid, Tunisia, where there is surface-water in most, but not all winters, and the inundated area varies from year to year (Coque, 1962). Tricart (1954) argued that typical *sebkhas* only occur in regions with contrasting seasonal precipitation where the water evaporates sufficiently in the dry season to cause the emergence of extensive, saline beaches. Over a shorter period, a single storm may alter significantly the pattern of playa surface conditions, as Neal and Motts (1967) showed for the Pago Playa, New Mexico (Fig. 3.15).

The commonest short-term playa changes include alterations in the distribution of saline and non-saline areas of desiccation and salt phenomena, and especially of crust types. For example, at Harper Playa, California, Neal (1968) attributed the conversion of soft, dry friable surfaces to hard, dry crust, over four years, to the dissolution of

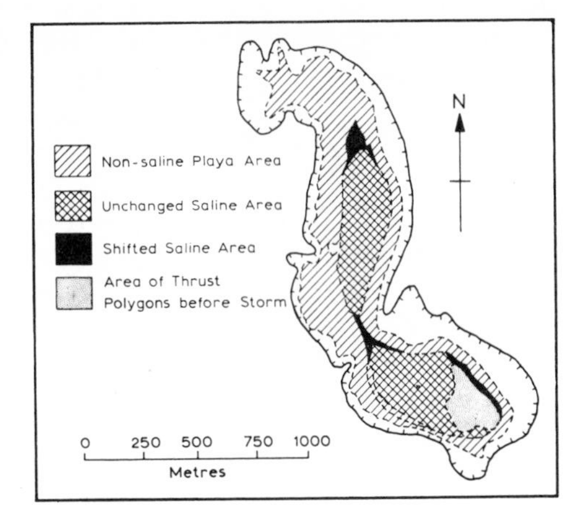

3.15 Pago Playa, Estancia Valley, New Mexico, showing shift of saline contacts after a storm in 1964 (after Neal and Motts, 1967).

surface evaporite minerals during flooding. An additional type of change may occur where a playa is underlain by water-soluble evaporites, and the water-table falls. Downward percolating water may here dissolve salts and cause surface subsidence in forms ranging from vertical-sided collapse sinks to smoothly rounded dolines (Neal and Motts, 1967). In the Aghda Playa of Iran, sink holes and collapse depressions are due to ground-water discharge reaching the surface and leaching the salt from the clayey-silt deposits (Krinsley, 1970). Although most surface-water of the Chott el Djerid is evaporated in summer, in some areas circular holes in the surface remain filled with water. These holes, called *aïoun*, have nearly vertical sides, are up to four metres deep and five metres wide, and are surrounded by aureoles of salt efflorescence (Coque, 1962). The origin of *aïoun* is not clear, but we might speculate that they result from solution subsidence.

Progressive changes over a period of years may lead to the creation of some distinctive surface features. Changes of vegetation, surface runoff, and water-table levels in the western United States during the last century have had significant effects on playas (Neal and Motts, 1967). The main consequences of falling water-table, for example, appear to have been the formation of giant desiccation polygons and stripes, and the creation of relict spring mounds; solution and subsidence of surfaces underlain by evaporites, the local replacement of soft, puffy surfaces by hard, dry compact playa surfaces; and playa expansion.

Many authors have seen particular significance in the alternation of 'wet' and 'dry' conditions on playa surfaces. 'Wet' periods are generally times of sediment accumulation. In 'dry' phases, erosion, largely in the form of deflation (but also by running water around playa margins), is ascendant. It is thus possible to envisage a regular alternation of erosional and depositional episodes. The net effect of such an alternation would clearly depend on the relative rates of the erosional and depositional processes, and the frequency and duration of the episodes. This is an attractive hypothesis for, although evidence on rates of processes and frequency and duration of episodes is very limited, it is clear that some playas are characterized by erosional forms such as deflation hollows, striae and dissected lacustrine sediments (e.g. Blackwelder, 1931*b*), whereas others are without such phenomena and seem only to have undergone a net accumulation of sediment.

Over the much longer timespan of the Quaternary there is ample evidence of change in some playas. The two main categories of evidence are erosional and depositional features marginal to the playa, and the stratigraphic record of the deposits beneath the playa. Marginal features include strandlines, tufa deposits, spits, bars, deltas, abrasion terraces, cliffs and overflow channels (Plate 1.9). Such phenomena clearly represent formerly more extensive lakes, but they only have chronological value if they contain or can be related to datable material, such as artifacts, mollusca, wood etc. Often more illuminating is the stratigraphic record which may contain valuable evidence of environmental conditions during the evolution of the playa, together with datable materials.

The detailed work of Smith (1968) in Searles Basin, California (Fig. 3.16) provides an outstanding illustration of studies of playa evolution. The Searles Basin is extremely sensitive to regional climatic changes. When inflow of water from the Sierra Nevada increased, a lake was formed and 'mud layers' were deposited. At times the lake overflowed into Panamint Valley. When the flow diminished, the lake waters were evaporated, and a salt lake or salt pan was created. Fortunately in this area the sensitivity of the basin to climatic changes is associated with a wealth of evidence of such changes in the past. Lacustrine sediments and shoreline information allow reconstruction of high lake-levels. Marginal sediments can be related to subsurface sediments. Subsurface saline deposits indicate the nature of low lake-levels, and some of these deposits can be correlated with interpluvial soils in the basin. Gaps in the depositional sequence are identified by unconformities and fossil soils. The succession has been dated by the C14 method for ages less than 40,500 B.P., and by estimation of ages over 45,000 B.P. from rates of mud sedimentation determined within the C14-dated beds. Analysis of the mineral assemblages in the saline deposits allows estimation of the temperatures of deposition, and permits inferences concerning season of saline deposition and, finally, the temperature/precipitation character-

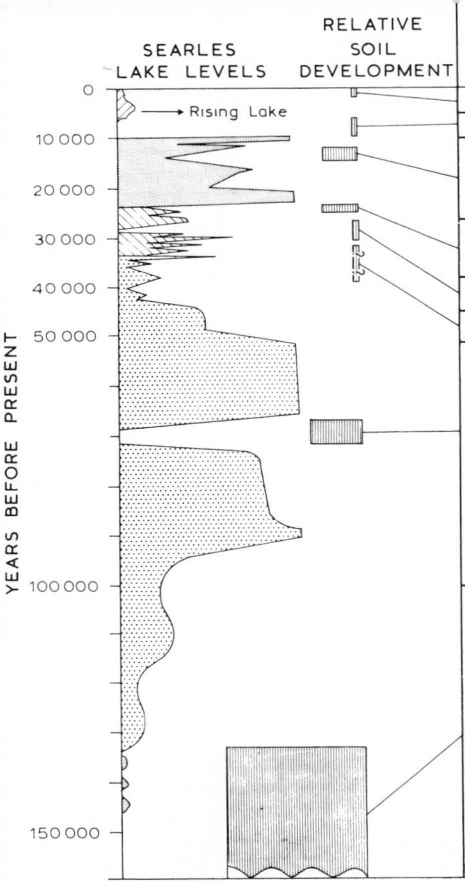

SEARLES LAKE LEVELS

RELATIVE SOIL DEVELOPMENT

YEARS BEFORE PRESENT

0
10 000
20 000
30 000
40 000
50 000
100 000
150 000

→ Rising Lake

TYPE OF SOIL	OF SALINE DEPOSITION (°C)	SEASON OF SALINE DEPOSITION	SEASONAL CHARACTERISTICS			
			SUMMER		WINTER	
Calcic	?	Summer ?	Hot	Dry?	Cool	Moist?
Calcic	25° — 40°	Summer	Hot	Dry	Cool	Moist
Calcic, Cca up to 6ft. thick	(No Salines)	None	Hot	Dry?	Cool	Moist?
Noncalcic	15° — 20°	Fall, Winter, or Spring	Warm Moist		Cool	Dry
Calcic?	20° — 40°	Summer	Hot	Dry	Cool	Moist
?	5° — 15°	Winter	Warm	Wet	Cold	Dry
Calcic to Noncalcic Cca up to 4ft thick	5° — 15°	Winter	Warm	Wet	Cold	Dry
Calcic Cca up to 6ft thick	10° — 20° ↑ Gradual Change ↓ 20° — 40°	Fall Winter or Spring / Summer	Warm / Hot	Wet / Dry	Cool / Cool	Dry / Moist

3.16 Diagram relating lake history, degree of development of fossil soils, crystallization temperatures and season of correlative salines, and inferred climatic characteristics of selected interpluvial intervals in Searles Basin, California (after Smith, 1968).

istics of summer and winter. The changing lake-level in the Quaternary and associated information on playa conditions are summarized in Fig. 3.16.

Many less detailed studies have been made of the Quaternary history of playas in this and other deserts (e.g. Coque, 1962; Morrison, 1965; Twidale, 1968), and it is not appropriate to review these studies in detail here. Nevertheless, although the evidence varies from place to place, it is beyond doubt that many playas were formerly occupied by more extensive and more permanent lakes in one or more periods during the Quaternary—occasions often contemporaneous with glacial stages elsewhere—and that the present forms are inherited from climates which were wetter (e.g. Flint, 1963) or colder (e.g. Galloway, 1970), or both. In some areas, the removal of large water bodies has resulted in isostatic recovery which may be reflected in warped shorelines and other phenomena. In the area of former Lake Bonneville, which covered some 50,000 km^2, the Bonneville shorelines have been raised in a broad domical uplift of about 64 m (Crittenden, 1963). But not all playas have related evidence which points towards more extensive lakes. In Western and South Australia (Neal, 1969) and in parts of the Sahara, for instance, no marginal features indicating more extensive ancient lakes have been identified. Throughout this section we have noted that diversity of playas, in terms of their origin, hydrology, mineralogy, processes and surface forms. We may conclude by emphasizing that diversity also characterizes the age and evolution of playas, even within a single desert. Desert climates permit playas to be formed; local conditions determine the form they take.

PART 4

Aeolian Geomorphology in Deserts

4.1 Introduction

4.1.1 PREFACE

(*a*) *Extent.* Large areas of aeolian sand are the most distinctive of desert landforms and with the striking, if not so widespread, signs of deflation they have understandably led many geomorphologists to over-emphasize the role of the wind in deserts. Since there has been some reaction to these ideas in the recent literature, we need to introduce the discussion of aeolian forms with a short review of their extent.

We have seen in section 2.1 that for the world's deserts as a whole estimates of between one-quarter and one-third sand cover seem to be reasonable. Most of this sand occurs not in isolated dunes but in large sand bodies known as sand seas (the *ergs* of the northwestern Sahara). In a world-wide survey, for example, Wilson (1970) estimated that 99·8 per cent of all 'active' aeolian sand was found in ergs whose area was greater than 125 km^2. Further large amounts of aeolian sand are to be found in 'fixed' ergs on many desert margins (e.g. Grove and Warren, 1968; Fig. 4.1.) Estimates of areas covered by wind-eroded forms are difficult to make, but we can say with confidence that they dominate the topography of some areas (Fig. 1.9, and Plate 4.5) and it is clear that much of some desert surfaces has suffered deflation to a greater or lesser extent.

(*b*) *Regularity.* A preliminary glance at the patterns of any aeolian forms will show them to be regular (Plates 4.1., 4.2., and 4.3.) although the forms created on loose sand are usually more complex than those eroded in hard, cohesive materials, but both show repetition of size, spacing and form. The serious study of these patterns dates from the late nineteenth century when it was realized that they belonged to a larger group of regular patterns now known collectively as bedforms (e.g. Bashin, 1899; Cornish, 1914; Exner, 1920; Harlé and Harlé, 1919). Bedforms formed beneath water are perhaps better understood than those formed by the wind, and it has become common to use sub-aqueous analogies to explain aeolian forms (Allen, 1968; Escande, 1953; Rim, 1958). A full bibliography of desert dune studies can be found in Warren, (1969).

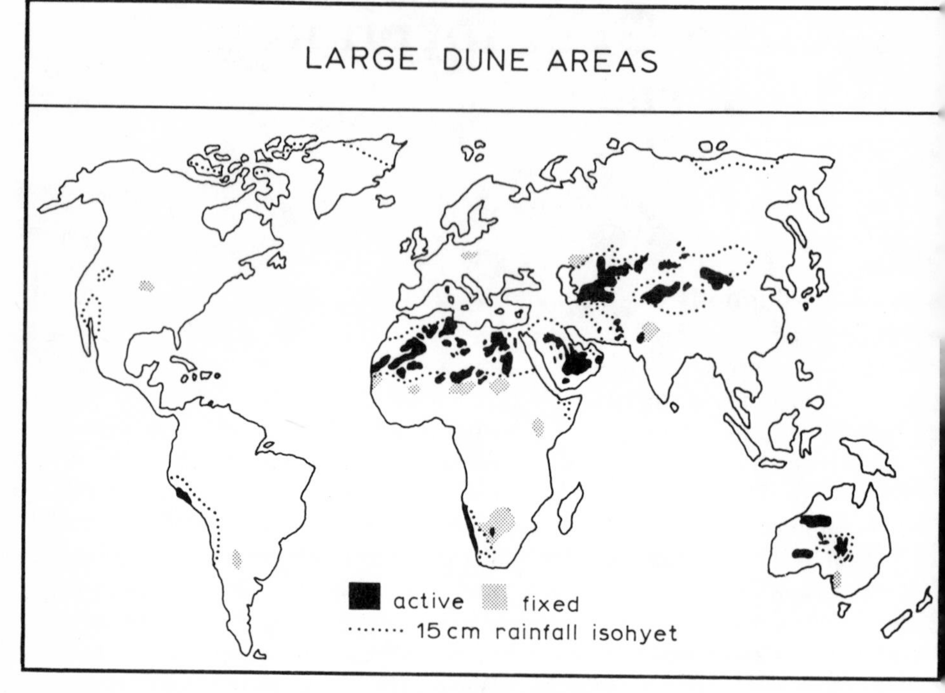

4.1 World map of principal fixed and active ergs.

Bedforms can be defined as 'a regularly repeated pattern on a solid surface which has formed in response to the shearing action of a fluid' (Wilson, 1972*b*). They are the result of an interaction between the flow patterns of the fluid and the bed in which there is development towards a condition of dynamic equilibrium. Bedforms are more common in loose granular sands than in cohesive beds.

(*c*) *Bedform Hierarchy*. Bedforms are usually arranged in an hierarchy of superimposed forms so that in any one area there are distinct size classes with very few forms of intermediate size (e.g. Allen, 1968*b*). In sub-aqueous bedforms the existence of an hierarchy has been recognized for many years (e.g. Gilbert, 1914), although it is still a matter of controversy (the arguments are reviewed by Allen, 1968*a*). The sub-aqueous forms most closely analagous to aeolian bedforms are found beneath shallow seas and here the existence of three groups in the hierarchy—namely ripples, small dunes and large sand waves—seems to be well established (e.g. Houbolt, 1968).

The accompanying air photographs (e.g. Plate 4.3) show clearly that aeolian bedforms also have an hierarchical arrangement. Ripples are nearly always found on the backs of dunes, and dunes are themselves commonly found on the backs of larger features, known in north Africa as *draa*, and an hierarchy can also be found in erosional forms. Since this is a fundamental point in their understanding it should be established from the outset of a discussion on aeolian forms.

The hierarchy of aeolian bedforms has been the subject of conjecture among desert geomorphologists for many years. Some early writers favoured the idea that aeolian ripples were simply nascent dunes and that, given time, they would grow into dunes (Cornish, 1914; Hedin, 1904). It was more common to draw a distinction between ripples and dunes (e.g. Bourcart, 1928; Högbom, 1923; King, 1916). In particular Bagnold (1941) emphasized this distinction claiming that ripples and dunes were formed by completely different mechanisms. Structures larger than dunes were not known to most of the early workers in Europe, but when explorers penetrated the Sahara and the Asian deserts a new group of larger phenomena were observed. Most of the early desert geomorphologists assumed that these features were simply too big to be dunes. Frere (1870) suggested that the large ridges in the Thar Desert in India were parallel horsts and graben, and Chudeau (1920) and others supposed similar features in the Sahara to be hills covered with sand. It was widely recognized, however, that the large features were distinct from the smaller dunes that covered them. Later studies, as we shall see, attributed large features to aeolian erosion, to stronger winds in the past, or to great age (sections 4.3.3.*h.* *(i)* and 4.3.4).

Queney (1945 and 1953) appears to have been the first to propose that different orders of contemporary structures existed, and that the distinctions between ripples and dunes could be extended to larger-sized features. He distinguished between four groups in the Idehan Marzūq in Libya: *(i)* ripples *(petites rides)* of fine sand with wavelengths of between 6 and 30 cm, and a common wavelength of 10 cm; *(ii)* large ripples *(grandes rides)* of coarse sand with wavelength between 50 and 300 cm; *(iii)* dunes; and *(iv)* larger features with wavelengths of the order of one kilometre.

The most complete statement of the hierarchical arrangement of aeolian bedforms has been made by Wilson (1970; 1972*b*). His size limits and terminology are listed in Table 4.1. Except for the fourth order, this hierarchy appears to apply to purely erosional as well as to depositional bedforms. The hierarchical arrangement applies to transverse, longitudinal and oblique forms alike.

TABLE 4.1 *Hierarchical Classification of Dunes*

Order	*Name*	*Wavelength*	*Amplitude*
First	draa	300 to 5,500 m	20 to 450 m
Second	dunes	3 to 600 m	0·1 to 100 m
Third	aerodynamic ripples	15 to 250 cm	0·2 to 5 cm
Fourth	impact ripples	0·5 to 2,000 cm	0·01 to 100 cm

SOURCE: Wilson, 1970.

These groups overlap in their size ranges, but when wavelength is plotted against grain-size, as on Fig. 4.2, they can be seen to occur in three quite distinct zones. As the grain-size of the sand increases so the wavelength of the feature also increases. A greater grain-size means that a stronger wind is needed to move the sand, so that grain-size controls effective windspeed. We can therefore say, in addition, that as wind-speed increases, so wavelength of features increases. But at any one grain-size or windspeed there are at least three distinct groups of feature. In fine sands we could find aerodynamic and impact ripples, dunes and draa, but on nearby coarser sand we might find ripples as big as the fine-sand dunes, or dunes as big as the fine-sand draas.

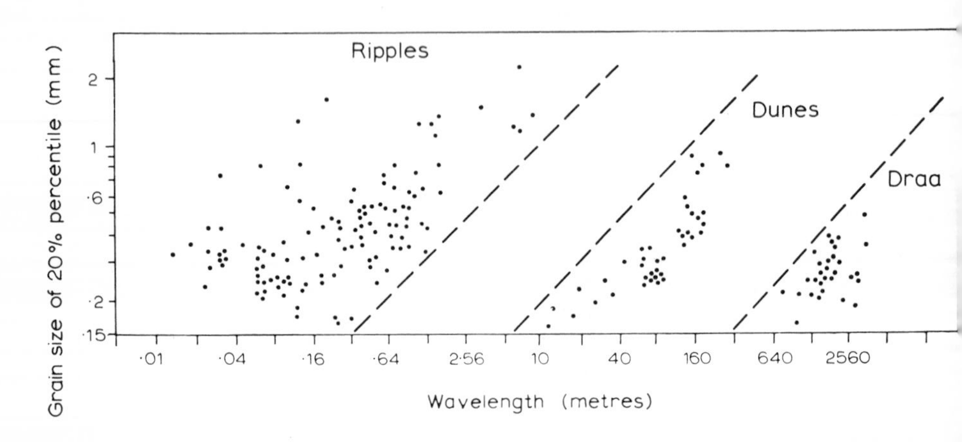

4.2 Plot of the wavelengths of aeolian bedforms against the grain size of the 20th percentile of their sands (after Wilson, 1972*b*).

4.1.2. WINDS

Aeolian bedforms are the result of processes occurring at the interface between the bed and airflow. This section briefly describes some of the relevant characteristics of the wind. Later sections will describe some relevant properties of the bed.

The patterns in the wind are by definition vital to the character of aeolian bedforms. Winds have three main causes of variability (viewed at a large scale), namely speed, direction and turbulence.

(a) Speed. Windspeed is of primary importance in determining the amount of sand that can be moved (section 4.3.1.*b*). Windspeed measurements in meteorological records, are mere summaries of a very complicated picture, but we can view mean horizontal windspeed as the dominant unidirectional motion when the superimposed eddies of turbulent motion have been filtered out.

Winds are inadequately measured in deserts. Not only are the lengths of record short, but the systems of recording are often unsatisfactory and most stations are at sites where readings are often distorted by topographical or man-made obstacles, or by vegetation. However they usually are the only records available in most deserts, so that they must be used with adjustments. Bagnold (e.g. 1941 and 1953*b*) adapted some of his formulae for use with simple wind records, and Dubief (1952) extended Bagnold's earlier method by comparing the resultant winds arrived at using the Bagnold formulae with the records of sand storms *(vents de sable)* in the Sahara; he found that there was usually a close correspondence between the resultant directions of the two, so that he then used the more widely available records of sand-storms to discover sand-movement directions over the whole desert.

Windspeeds in meteorological records appear to have a log-normal distribution: there are a large number of gentle winds, and progressively smaller numbers of stronger winds.

(b) Direction. We have mentioned the broader aspects of wind direction in section 1.3.3. In dune studies we are interested in the amounts of sand blown from each direction and this can be derived by using some of the formulae set out in section 4.3.1.*b*. There is, of course, great variety in the patterns of sand movement in deserts. Wilson (1971*b*) has used the data of Dubief (1952) to construct a map of potential movement across the Sahara (Fig. 4.3). It can be seen that there are two distinct flows separated by a sand divide. We shall return to the significance of these patterns in later sections.

(c) Turbulence. Turbulence is a property of the wind which is infrequently measured, because it is difficult to measure and to model mathematically, but turbulence nonetheless is very important. It is

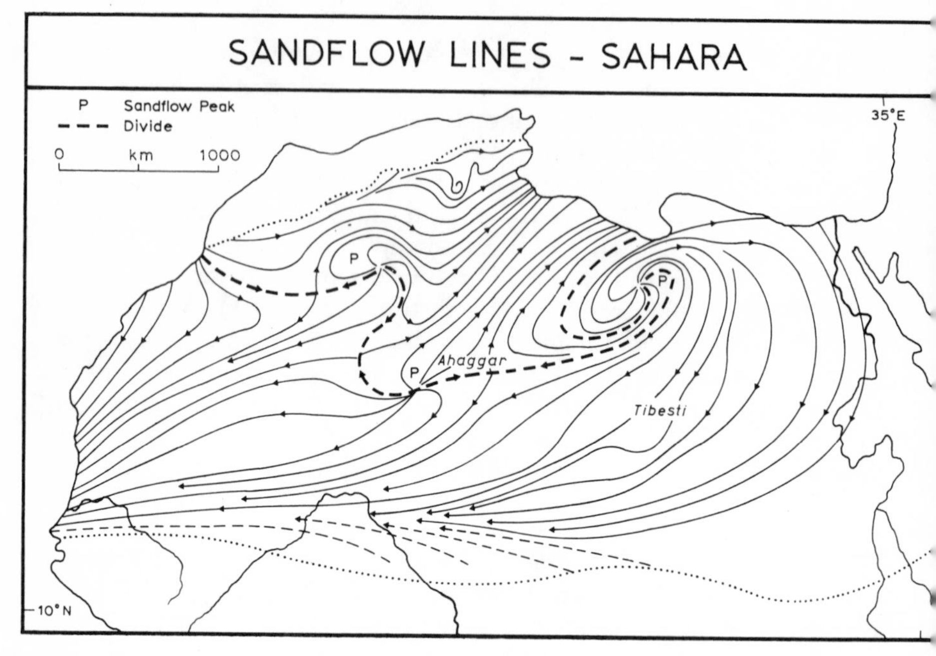

4.3 Sandflow lines in the Sahara constructed from wind data (after Wilson, 1971*b*).

important to the entrainment of sediment and Bourcart (1928), von Kàrmàn (1953), Queney (1953), Kampé de Feriet (1953) and Dubief (1953) believed that turbulence was an additional factor that should be considered with windspeed in calculating the amount of sediment in movement, although Kawamura (1953) and Bagnold (1953*c*) put forward arguments to show that turbulence could only be important in the entrainment of particles finer than sand.

The second important aspect of turbulence is its possible effect on bedforms. In this respect it is important to establish what are the characteristics of turbulence as measured by meteorologists. At the end of a highly complex statistical treatment of turbulence, Lumley and Panofski (1964) presented an 'extremely tentative' description of the geometry of motion in the wind. They concluded as follows:

(a) In nearly all conditions (except when the atmosphere is thermally very stable) there is a 'background' pattern of small, nearly isotropic eddies.

(b) In thermally stable air with light winds there is little turbulence apart from such isotropic eddies; 'what there is has scales more or less

independent of height, is elongated in the direction of the wind, and has small vertical extent.'

(c) 'There is some evidence that, with sufficient wind, eddies tend to be of the corkscrew variety with axes parallel to the mean wind . . . Superimposed on [this] small scale turbulence are much larger, essentially horizontal, eddies that have horizontal dimensions of hundreds of metres and up. These eddies do not decrease [in size] with height' (p. 210).

(d) In neutral, thermally stable conditions the latter type of eddy is of much less importance; in these conditions there are eddies of many sizes near the ground but the smaller ones disappear with height and only larger ones remain.

(e) Under convective conditions (to be expected in hot deserts) there are more or less equally-spaced upward moving plumes of air (thermals), and corresponding areas of downdraughts, and superimposed on these movements there are mechanically-induced eddies that diminish in size with increasing height. If there is only light windshear the plumes are equally spaced in all directions, but stronger winds bend them in their direction of movement, and this leads to the merging of several plumes in lines parallel to the wind.

Corkscrew motion parallel to the ground is known as Taylor-Görtler movement (e.g. Allen, 1968). Thermal turbulence as described under *(e)* has been suggested as an explanation for the equal spacing of sand 'mountains' in the Algerian Sahara (Clos-Arceduc, 1966; Cornish, 1914; Folk, 1971*b*; Gabriel, 1965) and for the equal spacing of draa ridges parallel to the wind (Bagnold, 1953*a*; Folk, 1971*b*). As we have noted, the conditions for thermal turbulence are likely to be ideal in deserts. Hanna (1969) has presented detailed evidence for this type of flow between longitudinal dunes.

It appears to be assumed by most authorities that there is a whole range of eddy-sizes, but Sutton (1953) quoted figures to show that about two-thirds of eddy energy is consumed by eddies with periods of less than five seconds, and that a common period is about 10 to 20 cycles per second, although the characteristic eddy spectrum at any one place will depend on local roughness factors. With normal windspeeds, Sutton's figure would mean a wavelength of about 25 cm. Chepil and Siddoway (1959) observed a decrease in eddy size with height and a direct relationship between drag velocity and eddy size. Regularity has also been observed by van der Hoven (1957) in his studies of the cyclic variation in windspeeds from several American stations. The first of the two distinct peaks in his analysis seems to be due to frontal movements; the second is at about one cycle per minute. With normal windspeeds this gives a wavelength of eddies which might correspond to *draa* sized features. Shaw (1942) and Chepil and Siddoway (1959) have also noticed very regular fluctuations in windspeed. Longitudinal regularity was noted by Kuettner (1969) to be very common in the atmosphere. He recorded many cases of bandstructure and well-developed cloudstreets in trade-wind areas. Common spacings of cloudstreets were between 5 and 10 km. They were associated with fast winds and upward movement beneath the clouds and compensating downward movement between the clouds. Cloudstreets usually occur when there is heating of the atmosphere from below.

The discussion of regular transverse wave motion in the atmosphere is dominated by considerations of the flow of a stratified atmosphere over an obstacle such as a hill. There is a very large literature on this kind of motion and it seems to be well understood mathematically (e.g. Corby, 1954). The passage of the air over the hill produces a series of wave-like motions above, and fairly regularly spaced rotors near the ground. In other words a series of regularly spaced zones of faster and slower and reversing flow are found to the lee of the hill. The appearance of this kind of motion depends on the height and shape of the obstacle in relation to the vertical temperature and wind profile. For example an obstacle about 200 m high might initiate wave motion with windspeeds greater than 5 m/sec. As with thermal turbulence, regular motions in the lee of obstacles are very likely in the atmospheric conditions that are found in most hot deserts.

In summary it can be said that there is some meteorological evidence for both transversely and longitudinally spaced regularity in the turbulence of the atmosphere. We return, in section 4.3.2, to the ways in which this motion interacts with growing bedforms.

4.2 Wind Erosion

4.2.1 WIND EROSION PROCESSES

(a) Introduction. Sediment removal by wind has been studied very extensively because of its importance in the context of soil erosion of agricultural lands, but less attention has been given to this important subject by desert geomorphologists. The work on wind erosion in the Great Plains of the United States and Canada by Chepil and his colleagues is of great significance, and much of it is directly relevant to the study of wind erosion in deserts. The following remarks are largely derived, unless otherwise stated, from Chepil's work and from the earlier work by Bagnold (e.g. 1941). More detail can be found in Bagnold's (1941) classic discussion of sand movement and in Chepil and

Woodruff's (1963) comprehensive review paper, which also contains an extensive bibliography.

(b) Eroding action of the wind. Wind erosion is related to the properties of the wind near the surface, and to the properties of surface materials. There is some disagreement among the authorities as to exactly what mechanism in the wind is responsible for sediment entrainment. Kampé de Feriet (1953), von Kàrmàn (1953) and Queney (1953) all maintained that vertical movements involved in turbulence were important in lifting particles from the surface, and de Félice (1956) suggested that rotatory 'magnus' effects must also play a part. Bricard (1953) and Walther (1951) observed that very noticeable electrical effects can accompany dust storms and Walther and Gabriel (1965) emphasized the possible importance of electric charges in sand entrainment. However, Bagnold (1941 and 1953*b*) suggested that electrical effects and turbulent upward movements were only of sufficient power to lift particles smaller than sand, and Chepil and Woodruff (1963) made similar observations. They maintained that entrainment of sand particles was attributable only to the surface drag of the wind.

As the wind passes over a stable surface it is retarded at its base by friction. There is a very thin layer at the base in which the velocity is zero, the thickness of the layer depending on the roughness of the bed. It is of the order of 1/30th of the particle diameter on a granular bed. Above this layer the wind velocity increases as in Fig. 4.4. If z is the

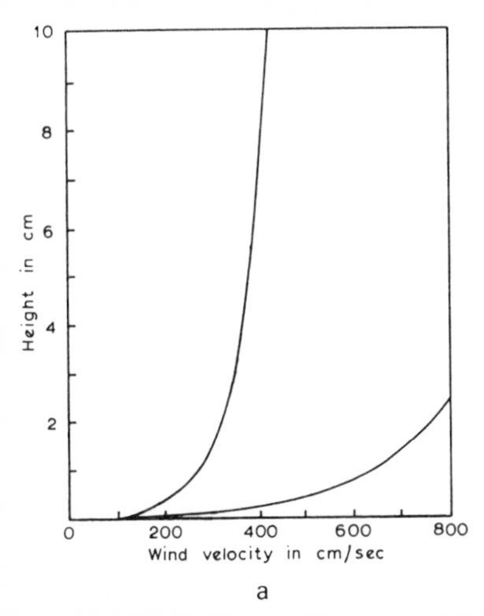

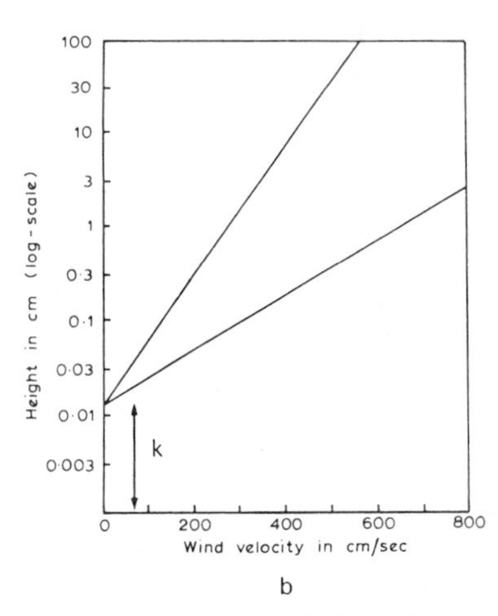

4.4 The velocity pattern of winds near the ground (adapted from Bagnold, 1941): (*a*) on an arithmetic height scale; (*b*) on a logarithmic height scale.

height at which the wind measurements are taken, and the velocity there is found to be u, and k is the height at which the velocity is zero, the slope of this line is defined by the Prandtl/von Kàrmàn equation:

$$U_* = \frac{K u}{\log {}^z/_k} \qquad \ldots (4.1)$$

where K is the Kàrmàn constant which can change according to the temperature gradient (von Kàrmàn, 1953), but it is usually taken to be about 0·04. The quantity U_* is referred to as the drag velocity, and it has been found to be proportional to the slope of the line as on Fig. 4.4b. It is related to the drag exerted on the wind by friction at the surface τ, and to ρ which is the density of air (e.g. von Kármán, 1953):

$$U_* = \sqrt{\frac{\tau}{\rho}} \qquad \ldots (4.2)$$

A number of empirical relationships have been worked out to define the Prandtl equation (e.g. Horikawa and Shen, 1960; Williams, 1964). As the windspeed rises so does the drag on the bed and thus U_* also increases; in other words, the slope of the line in Fig. 4.4b becomes more pronounced. As the shear of the wind increases, loose particles on the surface will be subject to increasing stress, and will eventually begin to move. The critical velocity at which particle movement begins was referred to as the fluid threshold by Bagnold (1941). It can be defined in the following way:

$$U_{*t} = A\sqrt{\frac{\sigma - \rho}{\rho} gD} \qquad \ldots (4.3)$$

where U_{*t} is the threshold drag velocity, σ is the specific gravity of the grains, ρ is the specific gravity of air, g is the gravity constant, D is the grain diameter and A is a coefficient which, for particles above 0·1 mm diameter, was found to be 0·1.

The fluid threshold can be seen to vary with grain diameter. The relationship in equation 4.3 actually only holds for quartz particles bigger than 0·1 mm. Below that value the threshold velocities rise again. Fig. 4.5 illustrates the effect of winds on particles of different sizes. Bagnold (1941) found in wind-tunnel experiments that sediments composed only of smaller material were harder and harder to move as grain size decreased. This is due to increased interparticle cohesion

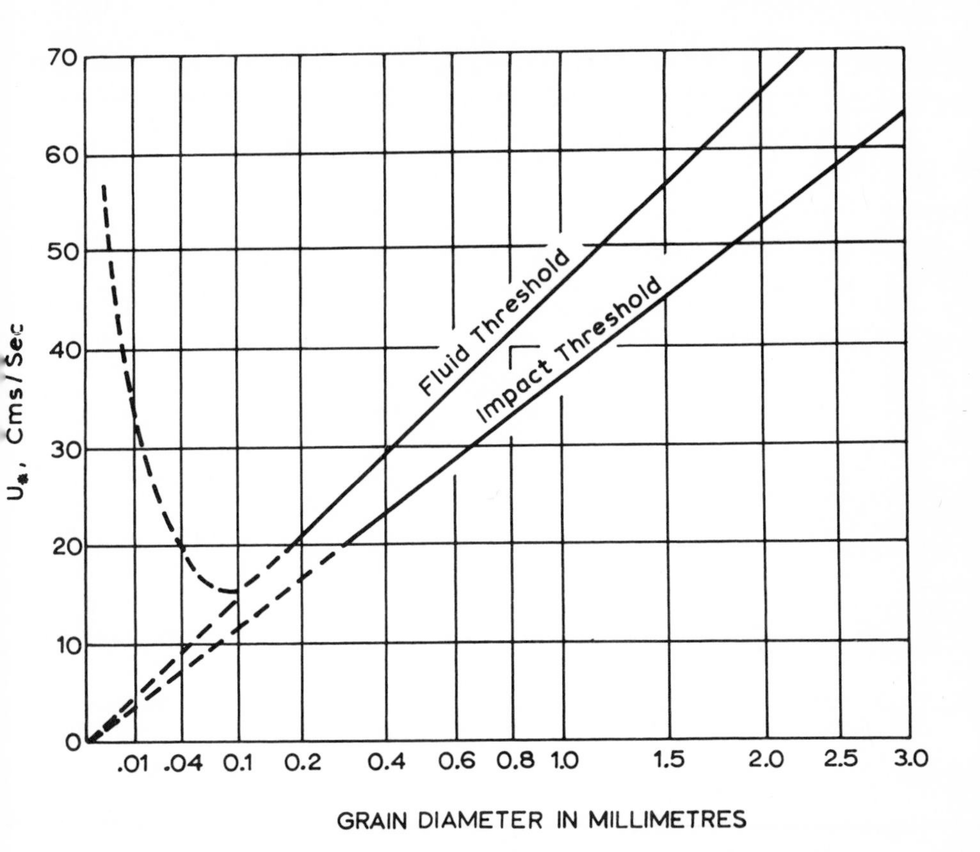

4.5 Variation of threshold velocity with grain size (after Bagnold, 1941).

with weak chemical bonds, to greater moisture retention, and to lower values of surface roughness (e.g. Smalley, 1964).

When particle movement starts the bed is bombarded with grains and this initiates further movement, so that sediment movement can be maintained at lower velocities than were needed to initiate movement. There is therefore a new threshold value which Bagnold (1941) called the impact threshold. Its value can be defined with the following formula:

$$U_{*t} = 680\sqrt{d} \log 30/d \qquad \ldots (4.4)$$

where U_{*t} is the threshold wind velocity and d is the grain diameter.

These relationships between particle diameter and threshold velocity have been repeatedly confirmed (e.g. Belly, 1962; Chepil and Woodruff, 1963; Horikawa and Shen, 1960). The threshold velocity has been found to be slightly affected by other factors such as particle shape (Williams, 1964). Chepil found that very few particles with equivalent diameters* exceeding 0·5 mm (actual diameters 0·84 mm) are eroded. In general the potential erosion of a soil increases as the percentage of soil fractions greater than 0·84 mm in diameter declines (Woodruff and Siddoway, 1965). Bagnold (1953*b*) proposed a general figure of 16 km/hr (as recorded by regular meteorological observations) as the threshold velocity for most desert sands.

At the threshold of grain movement, three types of pressure are exerted on the surface particle: *impact* or *velocity pressure* (positive) on the windward area of the particle; *viscosity pressure* (negative) on the leeward area of the particle; and *static pressure*, a negative pressure on the top of the particle, caused by the so-called Bernoulli effect (which arises from pressure reduction where fluid velocity is increased, as at the top of the particle). *Drag* on the top of the particle is due to the pressure difference against its windward and leeward sides, and *lift* is caused by decrease of static pressure at the top of a particle compared with that at the bottom. The values of drag and lift required to initiate movement are, of course, affected by the character of the grains. Once the particle has been entrained, drag and lift change rapidly. The process of grain transport is considered more fully below (section 4.3.2).

Once movement has started, small particles move in suspension, medium-sized sands move by saltation (bouncing) and coarser particles move as surface creep (rolling).

(c) Properties of the surface relevant to its wind erodibility. So far our discussion of surface materials has only been concerned with individual loose particles. But such particles are frequently aggregated in various ways to produce erosion-resistant structural units. Chepil and Woodruff (1963) distinguished four major types: primary (water-stable) aggregates; secondary aggregates, or clods; fine material among clods; and surface crusts. Primary aggregates are held together by water-insoluble cements of clay and organic colloids; clods are held together in a dry state by cements comprising mainly water-dispersible particles smaller than 0.02 mm in diameter; cohesion between clods is provided largely by water-dispersible silt-and-clay sized particles; and surface crusts arise from raindrop impact and clay-particle reorientation as described in section 2.4.2 *a*.

The binding agents for these dry structures include chiefly silt, clay and organic matter, and it is the properties of these agents that determine

* equivalent diameter = $P_e D/2{\cdot}65$ where P_e = the bulk density, D is the diameter and 2·65 is the material density of the particles.

the mechanical stability of the structures. Chepil and Woodruff (1963, p. 262) observed that 'the relative effectiveness of silt and clay as binding agents depends somewhat on their relative proportions to each other and to the sand fraction. The first five per cent of silt or clay mixed with sand is about equally effective in creating cloddiness, but the quality of the clods is different. Those formed with clay and sand are harder and less subject to abrasion by windborne sand than those formed from silt and sand. For proportions greater than five per cent and up to 100 per cent the silt fraction creates more clods, but these are softer and more readily abraded than those formed from clay and sand. The greatest proportion of non-erodible clods exhibiting a high degree of mechanical stability and low abradability is obtained in soils having 20 to 30 per cent of clay, 40 to 50 per cent of silt, and 20 to 40 per cent of sand.'

Decomposed organic matter (humus) on and in soils is another cementing substance. Humus is derived from plant and animal residues by the decomposer micro-organisms, and includes their secretory products. The humus content of desert soils is very low (section 2.3.1).

Calcium carbonate tends to decrease cloddiness and mechanical stability because of its lightness and mechanical instability. As we have described (section 2.3.2*d.v*), many dry-land soils contain calcium carbonate, and when it is near to the surface, the erodibility of the soils may be high.

Large structural units act as non-erodible or less-erodible obstacles to wind erosion. If wind erosion is to proceed beyond the removal of existing loose particles, the structural units must be broken down by weathering, raindrop impact or wind abrasion. Soil structures vary in their resistance to these forces. Their susceptibility to wind abrasion varies inversely with their mechanical stability and this in turn is a function of interparticle cohesion (Smalley, 1970). In a dry state, primary aggregates are generally most stable, whereas clods, crusts and fine material among clods are progressively less stable. Abrasion by other particles carried by the wind may cause the progressive breakdown of soil structure as erosion continues.

Non-erodible particles (including stones) seriously restrict the progress of wind erosion, since the amount of material removed is limited by their height, number and distribution. As erosion proceeds, the height and number per unit area of non-erodible particles increases until ultimately they completely shelter erodible material from the wind, and a 'wind-stable surface' is created, (section 2.4.2*a*).

Other characteristics of the surface, such as soil moisture and surface roughness, are also important to wind erosion. Only dry soil particles are readily erodible by the wind; soil moisture promotes particle cohesion and restricts erodibility. The rate of soil movement varies approximately inversely as the square of effective surface soil moisture (Woodruff and Siddoway, 1965). The soil moisture at any particular time, of course, is determined by the properties of the soil and the particular weather conditions. Surface roughness is determined by vegetation characteristics and other roughness elements such as particle size. A rough surface is more effective in reducing wind velocity than a smooth one, and is thus less susceptible to erosion.

Vegetation cover influences the nature of wind erosion in several ways, Firstly, the quantity of vegetation, as represented by the proportion of covered ground, is a reflection of the extent to which the surface is exposed to erosion. Secondly, vegetation tends to increase both elements of surface roughness, and it hence tends to reduce wind erosion. Olson (1958) found that grassy vegetation on coastal dunes could increase surface roughness by up to 30 times its value over bare surfaces. The taller and denser the vegetation, the more effective its protection.

Of the numerous variables in the wind erosion system, some are permanent, others change. The characteristics of the wind, the structural units, the organic residues, the soil moisture and the vegetation may all change over short periods, and especially seasonally. The texture of the surface material, on the other hand, tends to be fairly constant.

(d) Initiation and progress of wind erosion. Wind erosion may begin when the equilibrium of the system is disrupted by a change in one or more of the component variables. The changes may be in precipitation, temperature, wind velocity, the soil aggregates, the surface roughness, vegetation cover etc.

For saltation to start, it is necessary that the fluid threshold velocity of the most easily erodible loose surface material is reached. If, as is often the case, there is a hard surface crust which has to be broken, then the initial velocity must be higher than that required to move loose particles. The impacts from saltating particles initiate movement of other particles if the impact threshold velocity of the latter is achieved. The impact threshold velocity is lower than the fluid threshold velocity for grains of increasingly greater size and density (Fig. 4.5). As erosion progresses across a surface, the quantity of debris in motion increases until it is at the maximum sustainable by the wind. The relationship of wind velocity to the amount of sand movement is discussed in section 4.3.1.*b*. Surface abrasion also increases and this tends to destroy soil structure and to increase the supply of erodible particles.

The quantity of material X (in tons per acre) removable from a given area, may be expressed in terms of drag velocity as follows:

$$X = a\,(U'_{*})^5 \qquad \ldots (4.5)$$

where coefficient a varies with many factors such as the size distribution of the erodible particles, the proportion of fine dust in the soil, the proportion and size of the erodible particles, position in an eroding belt and amount of moisture in the soil. The variable X may be described in a dimensionless form as an erodibility index (Chepil and Woodruff, 1963).

(e) The wind erosion equation. The major factors involved in wind erosion have been expressed in terms of a functional equation:

$$E = f\,(I', K', C', L', V) \qquad \ldots (4.6)$$

where E = erosion in tons per acre per annum, I' = soil and knoll erodibility index, K' = soil ridge roughness factor, C' = local wind erosion climatic factor, L' = field length (or equivalent) along prevailing wind-erosion direction and V = equivalent quantity of vegetative cover. The derivation of the elements in this equation, and ways of solving the equation for predictive use, are described by Woodruff and Siddoway (1965).

Yaalon and Ganor (1966) used the climatic factor, C, to delimit the relative wind erosion conditions in Israel, as defined in the equation:

$$C = \frac{\bar{u}^3}{(P\text{–}E)^2} \qquad \ldots (4.7)$$

where $\bar{u}$ is average annual wind velocity in miles per hour at a standard height of 10 m, and $(P\text{–}E)$ is Thornthwaite's moisture index. The climatic index is based on the fact that rate of soil movement varies directly as the cube of wind velocity (section 4.3.1.*b*), and inversely as the square of effective moisture, which is taken to be proportional to the $P\text{–}E$ index. The base point for determining values of the climatic index is the annual average value of it at Garden City, Kansas. Values at other stations are expressed as a percentage of this figure, so that for any given locality

$$C' = 100\,\frac{\bar{u}}{(P\text{–}E)^2}\Big/2.9 \qquad \ldots (4.8).$$

Fig. 4.6 shows the pattern of wind erodibility in Israel as revealed by this index in 1958 and 1960/61. Yaalon and Ganor observed that the boundaries of the climatic wind erosion index coincide with the arid and semi-arid climatic zones in Israel.

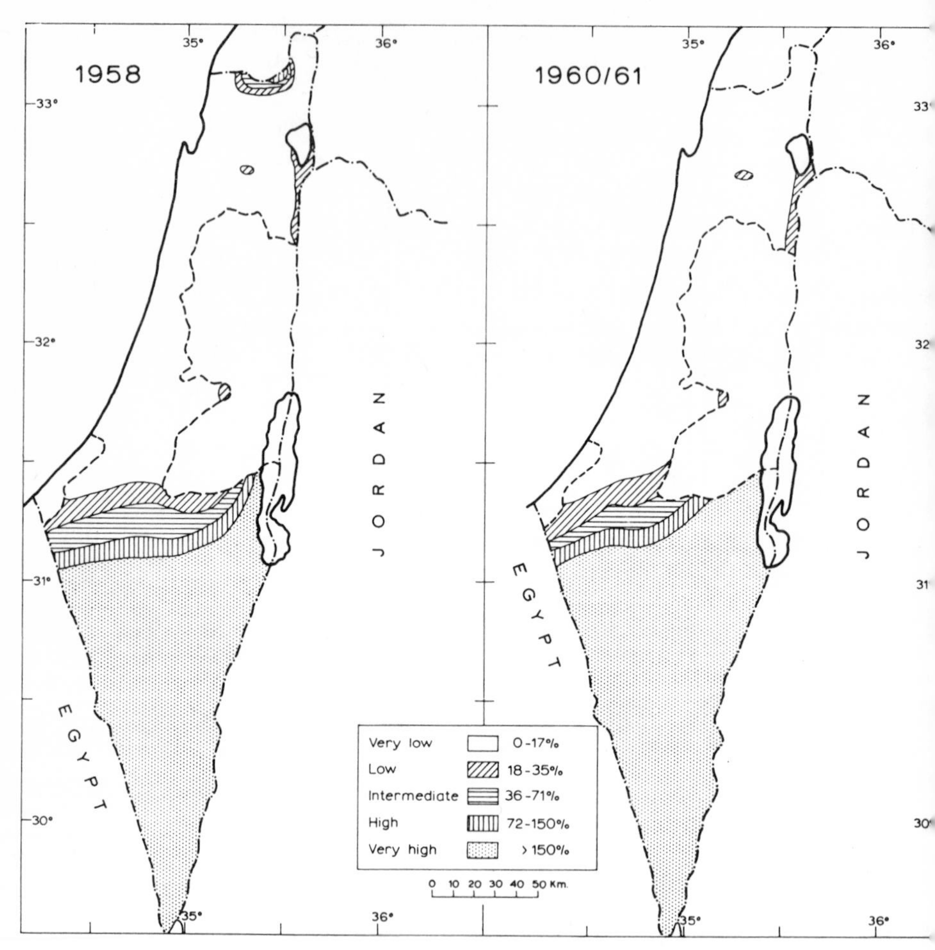

4.6 Wind erodibility in Israel, 1958, and 1960/61 (after Yaalon and Ganor, 1966).

(f) Large-scale approaches to the wind erosion problem. Evidence for and estimates of wind erosion over long periods can be obtained by considering regional patterns of sand production and accumulation. In Peru, detailed calculations of the life expectancy of dunes and sand-flow divergence led Lettau and Lettau (1969) to conclude that there must be active deflation of sand within the dune field they were studying. Some of their figures appear on Fig. 4.22. The calculated rate of erosion of sand was 21·8 mm in 100 years.

Wilson (1971*b*) has also arrived at estimates of long-term wind erosion by considering the occurrence and growth of ergs in the Sahara. He defined the mean long-term deflation rate for all particle sizes as $\overline{E}$, and the mean sand deflation rate as $\overline{E_s}$; this latter can be defined in the following way:

$$\overline{E_s} = \frac{\overline{Q\sigma}}{x} \qquad \ldots (4.9)$$

where $\overline{Q\sigma}$ is the potential sandflow rate as calculated from wind records (Bagnold's Q, section 4.3.1.*b*); $\overline{Q\sigma}$ is only reached quickly over bare sand; x is the saturation distance or the distance needed for the wind to be saturated after passing on to a sand patch. Using the results of Bagnold's experiments (section 4.3.1.*b*) it can be calculated that a light sand-free wind with a velocity of 4·9 m per sec. will move 0·015 gm per cm per sec. which gives a figure of $\overline{E_s}$ in the order of 66,200 kg per m^2 p.a. This figure, of course, only applies to short distances over bare sandy surfaces. Wilson suggested that the actual figures for sand-free areas between sandflow peaks and ergs in the Sahara (Figs. 4.3 and 4.43) may be of the order of 0·023 kg per m^2 p.a. of sand.

Chepil and Woodruff (1957) outlined a method for estimating soil loss from observations of dust concentration in the atmosphere. By relating measured dust concentrations and height-concentration relationships to commonly recorded measurements of visibility and windspeed, they hoped to be able to arrive at figures for dust transport and so for soil loss.

4.2.2. WIND ABRASION, DEFLATION AND EROSION PHENOMENA

(a) Introduction. Our discussion of the aeolian bedform hierarchy (section 4.1.1) suggested that there are three or four distinct size categories which are expressed in granular material as ripples (aerodynamic and impact), dunes and draa. Wind abrasion and erosion or deflation phenomena also appear to occur in three groups, corresponding to these three in the hierarchy. The distinctions between the groups are not well established, but a division of the published research into the three categories offers few problems of overlap.

(b) Small-scale features: ventifacts and rock-surface etching. The nature of small-scale abrasion features produced on the surface of stones and on rock faces by sand-charged winds depends on the characteristics of the rocks and rock fragments, the wind and the debris it carries, and the groundsurface conditions. The features of surface stones that are of importance include their height above surface, their shape and the area

of their bases, the position, inclination and orientation of their faces and edges, their rock type and its constituent minerals and their surface texture. The wind characteristics that are of importance include direction and velocity. Features of the transported debris said to be important include the size, angularity and mineralogy of the transported grains (including especially their hardness and density), the angle of grain impact, and the density and nature of particle movement in the saltation curtain. Significant groundsurface factors include the surface roughness and the degree of stone cover, and any activity—such as sheetwash—that can move surface stones. Wind abrasion produces varied forms at different rates, but the conditions are close to the ideal when their is an adequate but not too great a supply of tough abrasives, carried in strong winds across vegetation-free ground littered with relatively soft rock fragments. Thoulet (quoted in Kuenen, 1928) established that, in general, the degree of wind abrasion is proportional to the wind velocity, and to the size and sharpness of the transported sand grains. The height above ground of maximum wear varies: it occurs where grain-size, particle concentration and velocity combine to give the greatest energy of impact (Sharp, 1964).

Wind-abrasion phenomena have been reported from most of the world's deserts (Bather, 1900), but it would be incorrect to conclude that they occur extensively in every desert. In the Atacama Desert, for example, ventifacts are common only in exposed coastal situations and in the ignimbrite fields of the high Andes. The phenomena are not, of course, confined only to contemporary deserts. They were produced, and still are produced, in glacial outwash areas (Antevs, 1928), and they have been reported from areas as diverse as Pleistocene gravel-mantled erosion surfaces in Wyoming (Sharp, 1949) and exposed areas of the Marlborough coast, New Zealand (King, 1936). In this section we shall first look at the nature and origin of ventifacts, and secondly at surface ornamentation of ventifacts and other rock-surfaces.

The terminology of ventifacts is rather confused. Here we shall merely refer to an original surface on a stone as a *face*, a wind-eroded surface as a *facet*, and the boundary between two surfaces as an *edge*. Ventifact shapes vary greatly, but the triquetrous pyramid (dreikanter) is particularly common; ridge-shaped types with a roof-shaped form (with two dominant facets) or an *einkanter* form (with one or two facets and one edge) are also frequent.

Where an original face is normal to the wind, the pounding of saltating grains on the face will gradually wear it away to produce a facet at right angles to the wind, and a sharp edge between the facet and the lee or sides of the pebble. According to Sharp (1949), the shape of the new facet will depend in part on the relation between the height of the original exposed face and the depth of the sand-laden wind. The depth of the sand-laden layer, and the mean height of the grains travelling by saltation are both greater over stony surfaces than over sand-covered surfaces (section 4.3.1.*b*). If a stone lies wholly within the dense bottom layer of the sand-laden wind, it will be cut uniformly and a planar surface will be formed. If, on the other hand, the stone extends through this bottom layer, a concave surface will initially be formed because the intensity of abrasion will be less on the upper part of the stone. Eventually, the surface will be reduced until it is wholly within the dense layer, and then it will become a plane.

The reasons for a particular slope angle on a wind-abraded facet are not well understood. The inclination of the original face may be progressively reduced, or a stable angle may be achieved and retained as the size of the ventifact is reduced. Some argue that a stable angle is achieved (e.g. Cloos in Kuenen, 1925), others that it is progressively reduced until the facet is horizontal (e.g. Wade, 1910), and still others envisage a rapid reduction to a certain angle and thereafter relatively slow reduction (e.g. King, 1936). Schoewe (1932) experimentally investigated the abrasion rate on faces of selenite pebbles normal to the wind and inclined at 90, 60 and 30 degrees. He showed that maximum abrasion occurs between 30 and 60 degrees, that wear on a vertical face begins slowly, proceeds more rapidly when the plane has been reduced to about 60 degrees and then wears very slowly when the plane is at an angle of about 30 degrees. Kuenen (1928) also demonstrated that some reduction of slope occurs, at least for initially steeply-sloping faces. It is possible that the angle of cutting is related in part to rock type (Kuenen, 1928), but this possibility is as yet unconfirmed. Schoewe also showed that faces parallel to the wind suffer practically no abrasion.

Heim (1887) hypothesized that where a wind blows on to an edge, as opposed to a face, the airstream is divided into two, and abrasion moulds the two subtended faces and rounds off the edge. A slight change of wind direction could lead to a resharpening of the edge. An extension of this view is embodied in Kuenen's (1928) hypothesis that the ultimate shape of a ventifact exposed to winds from different directions is dependent upon the original shape of the pebble base. Kuenen's experimental study of the effects of sandblast from 16 different directions on the shape of chalk-powder blocks confirmed this. For example, triquetrous pyramids were obtained from two blocks which had in common only the triangular shapes of their bases. Kuenen also showed that the new shape, once attained, tends to be retained, but diminishes in size as abrasion proceeds.

Schoewe (1932) repeated some of Kuenen's experiments and confirmed some of his results. But Schoewe was convinced that variable winds, such as the winds from 16 different directions postulated by Kuenen, were unlikely in nature, and that wind-faceted pebbles were therefore unlikely to be formed in the way Kuenen described. In experiments in which the same variety of shapes of original block were used as in Kuenen's experiments, but in which winds blew only from one or from two opposed directions, Schoewe produced ridge-shaped *einkanter*, so that the shape of the original pebble base was shown to be unimportant under conditions of constant winds. In addition, Schoewe produced triquetrous pyramids from *einkanter* by turning them over on to their facets and exposing their bases to wind abrasion. Since ridge-shaped *einkanter* can be developed in winds from one direction, or from two opposed directions, regardless of the shape of the original pebble base, the triquetrous form can clearly be created from any original shape provided the intermediate form is turned over, for example, by undermining. According to Schoewe, therefore, the triquetrous shape is the final form of wind-abraded pebbles and other forms only represent intermediate stages of development.

King (1936) emphasized the importance of the relation between the area of base and particle height. He suggested that for limestone fragments attacked by winds from opposite directions, those with high basal area to height ratios tend to be reduced to flat or broadly convex plates, whereas those with low basal area to height ratios yield more irregular forms. In the latter case, the particles tend to become unstable as abrasion proceeds and may be moved or even rolled over by the wind, thus exposing new faces to abrasion and leading to the formation of multi-faceted ventifacts. Eventually the size of the ventifact may be so reduced and the number of facets so great that the particle can be rolled by the wind.

The shifting of particles—regardless of their basal area to height ratio—may play an important role in the development of multi-faceted ventifacts, especially where winds are from only one direction. Shifting certainly occurs, as discordant sets of superimposed surface markings indicate (Sharp, 1949). Several stone-shifting processes have been described, including surfacewash, frost action, and wind activity. Sharp (1964) made observations of the erosion of gypsum cement blocks at an experimental site in the Coachella Valley, California, and concluded that wind activity was the most effective overturning agency in two ways: *(i)* in scouring away material round the base of a particle so that it becomes unstable and moves (e.g. Bather, 1900) often by tilting or overturning in an upwind direction and *(ii)* in overturning unstable particles in a downward direction. Such particle shifting means, of course, that fresh surfaces are exposed to wind abrasion or abraded facets are reoriented to the sand-charged winds.

Higgins (1956) resurrected the idea that the shape of the pebble base was important to the final ventifact shape. He also suggested that fine suspended matter may be important in fashioning ventifacts.

Wind abrasion can also ornament the facets it creates, whether they be on ventifacts or rock outcrops. The three major forms of ornamentation are pits, flutes and grooves (Plate 4.6). In places, surfaces may also be polished. Although these features are common, most accounts of them are general (e.g. Blake, 1855; Maxson, 1940; Powers, 1936) and they have rarely been described in detail. Perhaps the best account is by Sharp (1949) who described the ornamentation on fossil ventifacts produced by unidirectional winds in Wyoming. In this area pits were produced on surfaces facing into the wind and sloping at angles of over 55 degrees. They are up to 2–3 cm in diameter and depth, and are generally developed in less resistant parts of the rock. Projecting points between the pits are often capped by more resistant minerals, such as quartz. As the inclination of the surface declines, flutes are developed, especially on slopes of about 40 degrees. Flutes are scoop-shaped in plan, U-shaped in cross-section, usually open in a downwind direction, and indiscriminately cut across different mineral constituents of the rock. In Wyoming they are up to 15 cm long, 4 cm wide and 2 cm deep. Sometimes they have overhanging sides. On more steeply inclined surfaces the flutes are shorter and deeper. Grooves are similar to flutes, but they are longer and open at both ends. Both flutes and grooves are best developed on surfaces oblique to the wind direction, and they appear to be more prominent on coarsely-crystalline igneous rocks. Sharp (1964) also demonstrated that flutes and grooves can be formed on essentially horizontal surfaces. Tremblay (1961) described similar grooving on sandstones in central Canada. Although most pits, flutes and grooves on rock-surfaces in deserts probably result from wind abrasion, a degree of interpretative caution is necessary: running water can produce similar forms, and the vermicular patterns on limestone surfaces may be formed by solution (Maxson, 1940).

(c) Intermediate-scale features: yardangs and small hollows. The early desert explorers were very sensitive to the observation of the possible effects of the wind, and the literature contains numerous descriptions, sketches and photographs of sometimes grotesque erosional forms of 'dune' size which were attributed to the wind. Two terms are used to describe these forms: *yardangs* are forms elongated with the wind with

rounded upwind faces and long pointed downwind projections (Fig. 4.7). If the yardangs are composed of bands of unequal strength, a resistant cap-rock may be retained and the forms are known as *zeugen*. Descriptions of these features can be found in many accounts of deserts (e.g. Blackwelder, 1934; Bosworth, 1922; Capot-Rey, 1957; Krinsley, 1970; Mainguet, 1968).

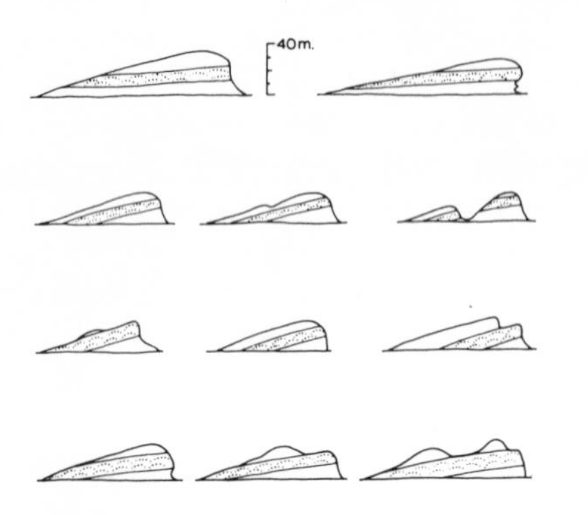

4.7 Longitudinal cross-sections of yardangs in northern Tchad (after Mainguet, 1968).

In most cases there can be little doubt that these features are wind eroded. They are aligned with the prevailing winds as recorded in meteorological stations and by nearby dunes. In Borku in northern Tchad, for example, the winds form a series of barchans on the plain near Faya whose two elongated arms indicate an unusually consistent wind direction (section 4.3.3.*b*). The yardangs in this area are unequivocally aligned with these constant winds (Mainguet, 1968). They vary in size from a few metres to one km in length and one or two to 200 m in height. As in other parts of the world, yardangs in the Broku region are usually eroded from rather soft sediments such as ancient lake deposits.

However, apart from this type of general observation, there appears to have been very little research into yardangs. There is little real diagnostic evidence on the yardang slopes themselves that they are wind eroded, since between the more intense winds that perform most of the removal there are probably periods in which rainfall or dew destroys the evidence. In some instances there are signs on the edges of yardangs that there is stream or rill erosion, and this may be important in their degradation in some instances (e.g. Blackwelder, 1934; Bosworth, 1922). But how much of their form is attributable to the wind and how much to other agencies cannot be determined from the research that has so far been published.

Many authors have described small hollows in desert surfaces and have attributed them to the local removal of surface debris by the wind. In Arizona, for example, Bryan (1923*b*) noted undrained depressions on the Kaibito Plateau and the plateau south of Glen Canyon. They are unrelated to drainage lines, and are scattered over rock-surfaces, arranged in lines along the outcrops of 'softer' beds, or found among dunes. They are up to 10 m wide, 17 m long and one metre deep, and their floors are often covered with an effective seal of slime and dust. Bryan attributed these hollows to localized wind erosion which proceeds because of a low water-table, the absence of moisture-retentive soil, and an easily weathered rock (sandstone).

The formation of small deflation hollows is controlled by the variables we have described in the wind erosion system. Optimum conditions for the creation of such hollows occur, for instance, where strong winds blow across dry, bare surfaces composed of unconsolidated sediment. But in order to create a hollow, differential erosion is necessary. Circumstances leading to such differential deflation might include locally advantageous conditions of loose-debris preparation. Amongst these are concentrations of salt and water (which might promote weathering), lack of particle-binding materials, and 'weaker' rocks. However the regularity of spacing and size of many of the features indicates that they are probably related to some regular structure in the wind.

The shape and size of deflation hollows will be determined not only by the wind structure but also by the areal pattern of erosion as expressed by the relations between above-threshold winds and erodible material. The hollows will develop to a size and shape which is in equilibrium with the prevailing conditions of wind erosion, but it is also likely that processes other than deflation (such as rill erosion) will modify them. The lower limit of erosion, the depth of the hollows, might be determined by the structure of the wind, by the shelter effect of the sides of the hollows if these pierce a hard cap-rock, or by the downward termination of the supply of erodible material. This could happen if downward erosion were to reach the water-table or a resistant horizon in the soil. Deposition of an erosion-resistant lining (e.g. of clay) on the floor of the hollow would also hinder erosion.

It should be emphasized that these remarks are speculative, as very little research has been carried out on deflation hollows. In addition, many hollows in deserts may be incorrectly attributed to deflation. The *dayas* in Algeria, for example, appear to be solution hollows (Capot-Rey, 1939).

(d) Large scale phenomena—grooves, pans and basins. Although some yardangs described in the literature are very large (Mainguet, 1968;

Bosworth, 1922) large-scale wind-erosion phenomena are dominated by grooves and large deflation basins.

Extensive areas of parallel grooving are best known from ancient periglacial environments. Lineated topography has been described from the western and mid-western United States and eastern Europe (section 4.3.1.*a*), but doubt often exists as to whether the grooves themselves are erosional or whether the inter-groove ridges are depositional. In parts of Colorado, however, there seems to be little doubt that the predominant process was erosional (Shawe, 1963; Stokes, 1964).

Until recently reports of similar features in deserts were rare. Bagnold (1933) noted giant grooves 3–15 m deep near Tekro in the northern Sudan, and Bosworth (1922) reported similar grooves in Peru, But only when air and satellite photographs became available was spectacular and widespread grooving noticed in certain areas. On the southeastern flanks of the Tibesti Massif (Plate 4.5) these features dominate the topography of some 90,000 km^2 and other smaller areas can be found nearby (Fig. 1.8.). The grooves are between half and one kilometre wide and are spaced 500 m to two kilometres apart (Hagedorn, 1968; Mainguet, 1968). They are eroded in Palaeozoic sandstones and cut almost at right angles across the fluvial drainage lines.

The regular size and spacing of these features and their consistent alignment with the curving path of the northeasterly winds around the massif leaves little doubt that the features are eroded by wind (Durand de Corbiac, 1958). The sides of the grooves are steep, whereas the floors are flat and often reveal bedrock, although they are occasionally filled with small sand dunes. The rock on the floors and sides is often varnished indicating that wind activity is slow if not inactive (Mainguet, 1968).

Other groove-like features have been noticed on the southern shores of the Algerian chotts and the northern edges of the Grand Erg Oriental (Wilson, 1970), and in a similar position on the northern shore of Lake Eyre in Australia (King, 1956). King was able to drill into sand-covered parallel ridges and discover that the sand merely mantled a ridge beneath that had been formed by the excavation of material from the surrounding areas. The relationship of these features to longitudinal dunes, which they resemble in many respects, will be discussed in section 4.3.3.*i*.

Many geomorphologists have noticed enclosed basins of various sizes in deserts and have often attributed them to wind erosion. The basins range in size from the small inter-dune hollows discussed in the previous section to basins that include large oases. Fig. 4.8 illustrates some of these features. The smaller hollows are usually shallow and are aligned with the prevailing wind direction. They are often found between dunes (Flint and Bond, 1968; Warren, 1966) or as isolated features associated with a lunette dune on the downwind margin (e.g. Bettenay, 1962; section 4.3.3.*g*). In South Africa, where these features are known as pans, they cover large areas of the 'panveld'. They range in size from a few hundred square metres to 300 km^2 in area and seven to ten metres deep (Wellington, 1955). Many authorities agree that material on the basin floor is loosened by weathering, perhaps by crystallizing salt, and then removed by the wind (e.g. Tricart, 1954). It appears that the basins may be associated with semiarid rather than arid climates (e.g. Bettenay, 1962).

The origin of much larger enclosed basins in deserts is difficult to establish. Perhaps the most celebrated of them are the hollows containing the oases of the Western Desert in Egypt. Since there are no obvious fault systems surrounding these basins they have often been attributed to wind erosion (e.g. Ball, 1927). In recent years many more of these basins

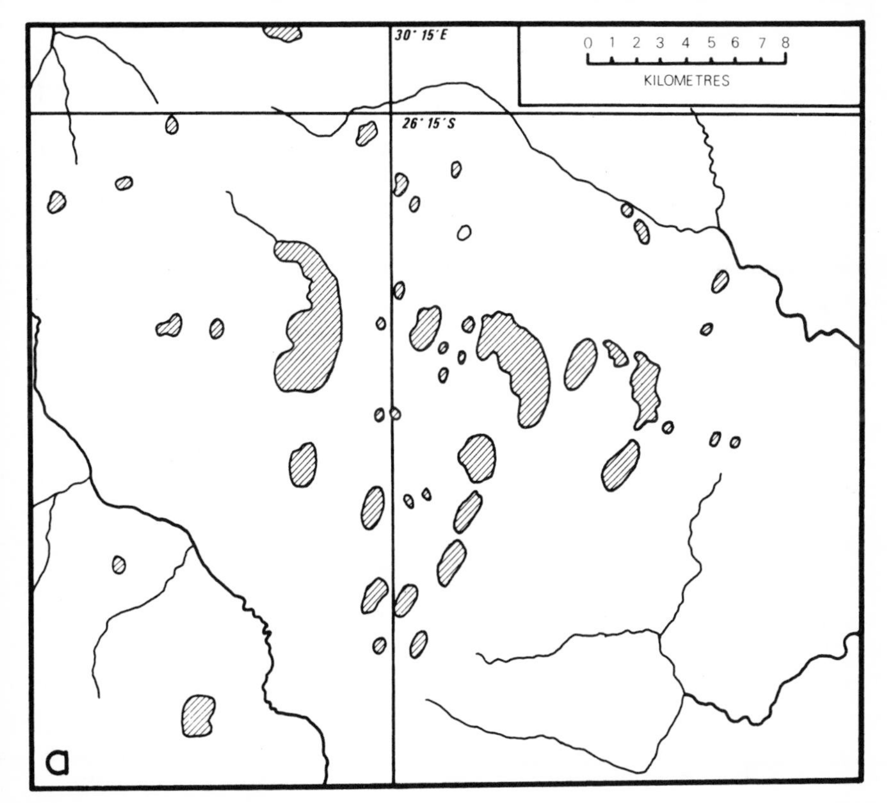

Fig. 4.8a

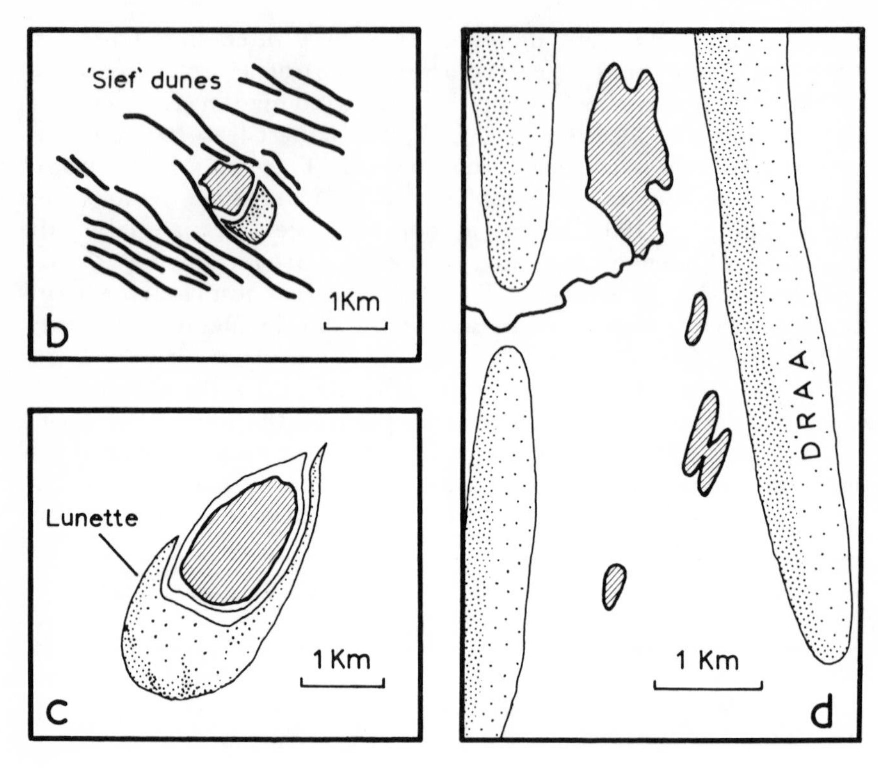

4.8 Pans (wind-eroded basins): (*a*) in the Transvaal (after Wellington, 1955); (*b*) and (*c*) small pans with lunette dunes in southwestern Botswana (after Grove, 1969); (*d*) pans between dunes in the central Sudan (after Warren, 1966).

have been identified (e.g. Daveau *et al.*, 1967; Dresch, 1968; Smith, 1963), but there is seldom any more evidence for their wind-erosion origin than that they are not obviously structural, that they are enclosed and that they occur in deserts. Not even alignment can be invoked since many of the basins have no very obvious trend. If the hollows are wind eroded then the process must be complex and of great age and must have involved several phases of weathering, removal and probably fluvial erosion. Other possible explanations have included solution, downwarping and meteorite impact.

4.3 Bedforms in Loose Granular Material

4.3.1 SAND MOVEMENT BY WIND

(a) Sands. Bedforms in loose granular material are almost invariably built of sand-sized particles.

The sandiness of dunes is primarily due to the range of wind velocities available at the earth's surface, and to the density of air. Winds capable of moving particles bigger than sand-size over rough beds are very rare, and only over temporarily very slippery surfaces, such as ice or wet mud, has the movement of much coarser particles been observed (e.g. Grove and Sparks, 1952; Schumm, 1956*b*; and section 3.5.2.*c*). The coarsest aeolian sands observed in motion over rough beds appear to have been recorded from the Antarctic where Smith (1966) observed ripples built of grains of diameter 5–30 mm. In coastal Peru, Newell and Boyd (1955) observed ripples in which the sand had a modal diameter of about three millimetres, and Sharp (1963) recorded occasional grains over five millimetres in large ripples in the Coachella Valley in southern California. These are all areas with very strong winds.

The lower limit of particle sizes that can be formed into aeolian bedforms is controlled by a number of factors. The primary control is through the settling velocities of particles in air. Particles below a certain size travel in suspension in the wind, whereas above that size the movement is mainly by saltation. The small particles can be lifted by turbulence, and are therefore diffused to great heights, so that they are not concentrated near the ground in the moving saltation curtain as are the heavier grains. The fine particles are thus carried away more quickly and therefore separated from the coarser grains. There is probably no very precise lower size limit to particles that travel in suspension because of variations in wind speed, amongst other things (e.g. Chepil, 1957). However there is a fairly well-defined limit in natural dune sands at about 0·05 mm (Bagnold, 1941; Sharp, 1963), and Kuenen (1960) noted a sharp cutoff below 0·1 mm. Alimen and Fenet (1954) found very few grains below 0·056 mm in Algerian dune sands.

The distinction between dusts and sands is also due to their different susceptibilities to entrainment. Well-sorted fine deposits are not easily entrained by the wind for a number of reasons. Bagnold (1941) was able to subject a pure silty deposit to much higher windspeeds than a sandy deposit before they began to move. This is partly because of the low surface roughness of silts, and partly because of the absence of saltating particles which would initiate movement by bombardment. It is also due to interparticle cohesion by loose chemical bonds and to moisture retention. Fig. 4.5 in section 4.2.1.*c* is an illustration of the combined effects of these processes on particle susceptibility to erosion. In section 4.3.3.*d* we discuss the size characteristics of aeolian sands in more detail.

These properties of the sediments and the wind combine with some rock properties and geological events to ensure that there are large areas of sandy aeolian bedforms in deserts. Most aeolian sands are quartzose, for not only is quartz the most stable of the common rock-forming minerals, but it is also extremely common and very often occurs as primary crystals of sand size. Weathering releases the quartz from the matrix of other more soluble minerals, and the particles are then moved by a number of sedimentary transporting mechanisms. The quartz appears to survive as particles of sand size because, it has been said, only transport in ice amongst common sedimentary mechanisms is capable of grinding it to smaller sizes (Smalley and Vita-Finzi, 1968). Silt-sized quartzose loess is not only associated with glacial environments (for instance in Nebraska the loess comes from eroded Tertiary deposits), but the existence of large amounts of aeolian and fluvially-derived quartzose sand must be an argument in favour of the slow reduction of quartz in these environments.

Most aeolian sands are not derived directly from weathered rock. Lettau and Lettau (1969) have presented evidence that the small barchans in southern Peru were derived in this way, but it is more common for the sand to be presented to the wind as a loose fluvial sediment. Current conversion of fluvial to dune sand has been observed by Harris (1957 and 1958), Melton (1940), Merriam (1969), and Sharp (1966). In other areas sand is derived from the coast (e.g. Broggi, 1952; Clos-Arceduc, 1966; Inman, *et al.*, 1966; Verstappen, 1968). In Western Australia it is said that sand comes from the highly weathered upper part of ancient lateritic profiles (e.g. Mabbutt, 1961; Madigan, 1946).

Quartzose sand is abundant in deserts partly because many sedimentary rocks found in deserts are sandstones (e.g. Brown, 1960; McKee, 1962; Monod and Cailleux, 1945; Sandford, 1937). This in turn can be partly attributed to the continuing continental character of desert areas for long geological periods and the continuous loss of fine soluble material during that time. In Australia dune areas are usually associated with areas of sedimentary rocks (e.g. Mabbutt, 1968).

Most aeolian sand bodies occur in basins in the desert. This, associated sometimes with the inference that the Pleistocene probably saw periods of more intense fluvial erosion than the present, has been used as the basis for the argument that many ergs are fashioned out of deposits of Pleistocene or late-Tertiary alluvium. This is a very widespread belief (e.g. Folk, 1971; Hack, 1941; Monod, 1958; Suslov, 1961). However it is not surprising that ergs occur in basins. The wind is accelerated and its turbulence is increased over mountains and plateaux so that sand will not usually accumulate in such elevated areas but rather in basins. Once there the sand need not be static, as is shown by sandflow diagrams (Wilson, 1971*b*; Fig. 4.3). However slow the move ment, sand is entering at one side of an erg and leaving it at another. Further evidence that suggests that the ergs are not simply wind-moulded alluvium will be discussed in sections 4.3.3.*h*,*i* and 4.3.4.

Non-quartzose aeolian sands are not common in deserts, but in some restricted areas gypsum sand dunes are found. It appears that gypsum crystallizes in shallow lakes, and when these dry out, the sand-sized crystals are strong enough to cohere as they are moved by the wind, and to accumulate as dunes. Gypsum sands have been described from New Mexico (e.g. McKee, 1966), northern Algeria (Tricart, 1967; Trichet, 1963) and Australia (Bettenay, 1962). Small admixtures of non-quartz minerals are more common than pure non-quartz sands. In the Coachella Valley in southern California, biotite and magnetite are sometimes segregated as patches of dark sand locally because of their density (Behiery, 1967; Sharp, 1964) and in many areas magnetite is found sorted into particular zones on ripples (Evans, 1962; Sharp, 1966). Plate 4.7 shows dark magnetite grains on ripple crests in the Great Sand Dunes National Monument, in Colorado.

Aeolian bedforms in material finer than sand are rare but not unknown. In loess rather distinctive bedforms are usually created (e.g. Rozycki, 1968 for south-eastern Europe and China; and Bartowski, 1969 for north-eastern Europe). In Iowa, long boat-shaped ridges, known as *paha*, have been described from the loess areas, although there is some doubt as to whether these are erosional or depositional (Leverett, 1942; section 4.1). Lineations in loess country were also noted by Russell (1929) and Lewis (1960) who both thought the features were dune-like and that they were only found in coarser loess. However in the 'transition zone' in Nebraska where the sandhills merge with the loess there is a very sharp division between the dune areas with distinct bedforms and the loess areas in which the only bedform is the 'loess-lip' that occurs along major river valleys where loess was deposited particularly thickly near to its source (e.g. Simonson and Hutton, 1954).

Clay dunes are, perhaps surprisingly, better authenticated than silt dunes and are more obviously dune-like. They have been reported from the Gulf coast of Texas (Fig. 4.9), northern Algeria, Australia and the Senegal delta. The work on these features has been summarized by Price (e.g. 1963). The occurrence of clay dunes does not contradict the

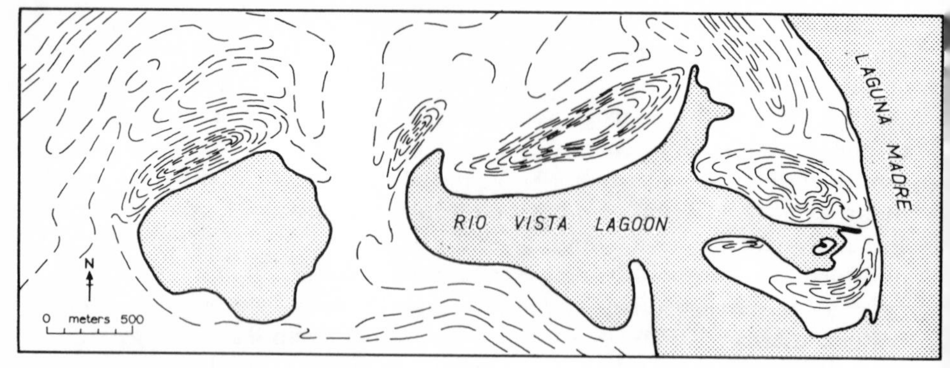

4.9 'Clay dunes' of lunette form on the Gulf coast of Texas (after Price and Kornicker, 1961).

arguments about the settling velocities and cohesion of small particles. There is indeed a large proportion of clay-sized particles in the Texas clay dunes, but the particles that were moved by the wind were actually clay aggregates of sand size. In Texas the clays are deposited on the floors of shallow tidal lagoons and as the lagoons dry, the salts crystallize and the clays crack, curl and blister. The flakes of clay are then blown by the wind to form dunes on the margin of the lagoon. These dunes are usually lunettes (section 4.3.3.*g*). After the dune has formed, percolating rainwater leaches out the salts, the clay particles are dispersed and the dune becomes a solid immovable mass. A very similar process has been described from the Sebkha ben Zïan in northern Algeria and from the Senegal delta (Boulaine, 1954; Tricart, 1954). This process appears to be confined to semi-arid rather than arid areas (Bettenay, 1962). Chepil (1957) also described the movement of clay aggregates from eroding fields to nearby dunes in Kansas.

(b) Sand movement by wind. The understanding of the processes involved in the movement of sand in response to windshear is basic to the understanding of dune forms. This section is a brief treatment of the arguments of various authorities, notably those of Bagnold (1941), to whose book reference should be made for further detail. We feel that the discussion below should be included here for reference since it provides relationships that are essential in understanding dune form and since we have added material published since Bagnold's work.

The discussion of the initiation of sand movement (section 4.2.1) was used by Bagnold (1941) and Kawamura (1953) among others as the start of an argument whose aim was to discover an equation for the amount of sand moved by the wind.

When sand starts to move the pattern of wind movement over the bed is altered. Sand is moved forward in two ways. Most of the sand moves by saltation, but a smaller amount is moved along the surface by creep. It is the coarser grains that move as creep, and their movement depends in large measure on the bombardment by the saltating grains. Although this is probably true in most cases, Lettau and Lettau (in press) have observed creep in motion on barchans in Peru without any saltation in progress.

Bagnold found that creep was usually about one quarter of the total load. Horikawa and Shen (1960) quoted a series of figures for this proportion: Chepil found that with sand between 0·15 and 0·25 mm, the creep load to total load ratio was 0·157, and with sand between 0·25 and 0·83 mm it was 0·249 (very near the Bagnold figure for similar sand); Ishihara and Iwagaki found in the field that the ratio was between 0·065 and 0·166. Horikawa and Shen themselves found that the value was about 0·20 and that it was independent of velocity. Sharp (1964) suggested from his field observations that creep probably became more important as the proportion of coarser grains increased.

As the sand movement starts, the height at which the velocity is constant rises from its former value k to a new value k' (Fig. 4.10). The velocity is no longer zero at this point, but constant at the value of the threshold velocity, U_t. Zingg (1952; quoted by Horikawa and Shen 1960) and Belly (1964) found that the value of k' was about ten times the value of particle diameter or in the order of 0·006 cm. Horikawa and Shen (1960) found it to be about 0·05 cm. Above this point of constant velocity the drag on the wind is increased, changing from τ to τ' and the slopes of the velocity lines also change from lines proportional to U_* to others proportional to U'_*. These changes are illustrated in Fig. 4.10 which is a simplified version of Horikawa and Shen's (1960) experimental results.

The saltating curtain of sand consists of a dense lower layer with a few odd particles travelling above it. Bagnold (1941) found that the sand bounced to maximum heights of two metres above bouldery areas and to maximum heights of about nine centimetres above sand. Chepil (1945) found that on a soil surface, 90 per cent of the load travelled below 31 cm and Sharp (1964) found that on a bouldery surface 90 per cent of the load travelled below 87 cm with a mean height of 63 cm. Maximum observed heights of between six and 19 metres have been quoted.

Bagnold's argument for the derivation of an equation of sand movement begins with the observation that, in moving sand, the wind per-

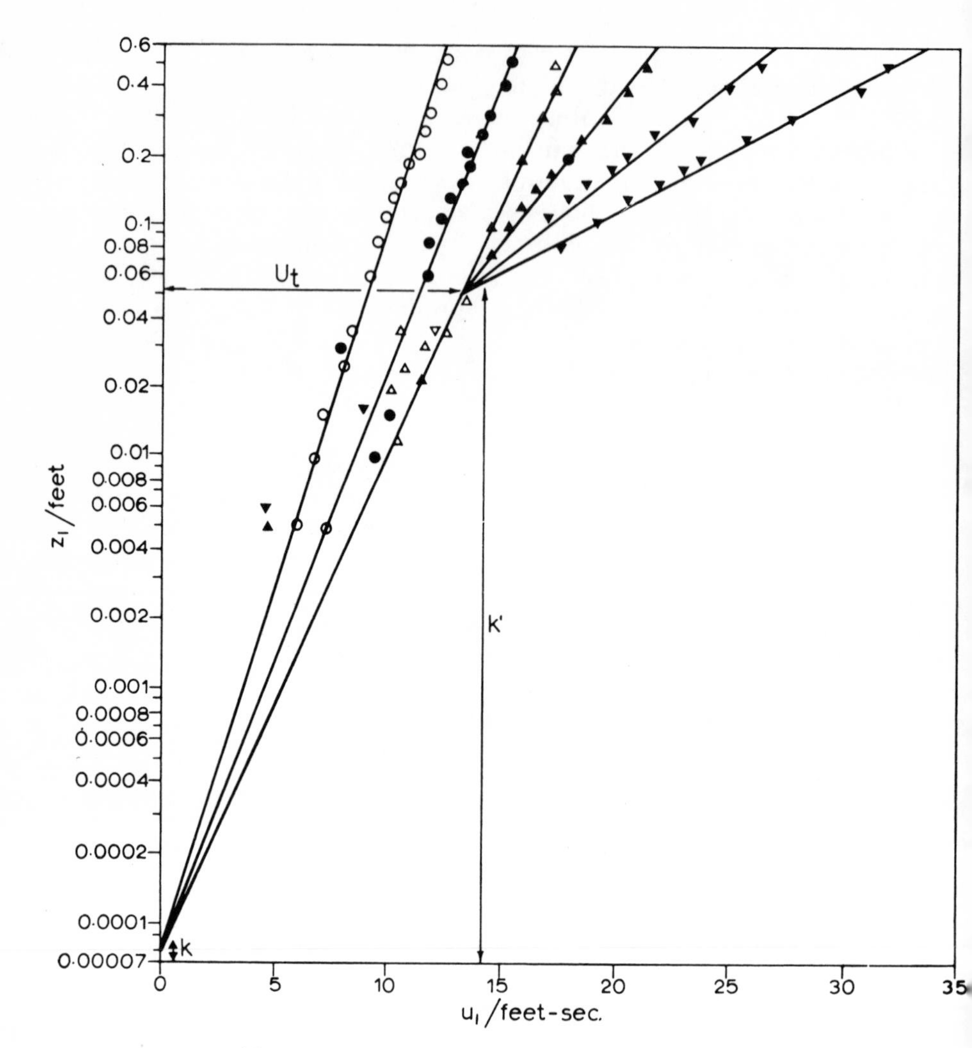

4.10 Changes in wind velocity/height relationships after the initiation of sand movement (after Horikawa and Shen, 1960). (z_1 = height above surface; u_1 = wind velocity)

forms a certain amount of work. This can be expressed mathematically if we first observe the path taken by a single grain (Fig. 4.11). Chepil and Milne (1939) found that most trajectories started with an almost vertical movement. As the grain enters the faster winds above the surface its path is modified, and it is pushed forward; when its initial upward movement is dissipated it is pulled down by gravity, and its path is determined by the balance of this force with the forward force of the wind.

This movement has been powered by the wind, and momentum has been extracted from the wind. If the mass of the grain is m, and the

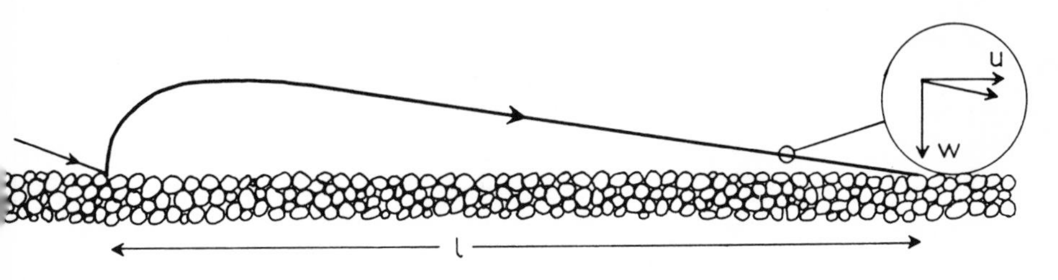

4.11 Path of a single saltating grain.

velocity with which it hits the ground is u (Fig. 4.11) then the momentum extracted from the wind is mu. This is a simplification which is justified since we can ignore the original forward velocity at the surface. If the flight path had had a distance l, then the momentum extracted per unit length would be mu/l. We can convert this momentum extracted by a single grain to that for the whole of the sand in saltation: if q_s is the amount of sand passing by saltation over a unit width of the bed in a second the wind will loose:

$$q_s \frac{\overline{u}}{\overline{l}}$$

units of momentum. This is a great, but very useful, simplification. As an example of the complications which may be involved in the extraction of energy from the wind, Zingg (1953) mentioned that a very large proportion of the energy may be consumed in rotation and De Félice (1956) also noticed that rotation was very important in grain movement.

The expression above is a measure of the drag exerted by the saltating sand on the wind, and therefore:

$$q_s \frac{\overline{u}}{\overline{l}} = \tau' \qquad \ldots (4.10)$$

We know from equation 4.2 (p. 238) that τ is related to U_* (the drag velocity) and to ρ (the density of the air) and Bagnold maintained that the same is approximately true when there is sand in the wind so that:

$$\tau' = \rho U_*'^2 \qquad \ldots (4.11).$$

q_s can now be defined in terms of more measurable quantities than before:

$$q_s = \frac{\tau \rho \, U_*'^2}{\bar{u}} \qquad \ldots (4.12).$$

These are still, however, quantities that are difficult if not impossible to measure in the field or in a wind-tunnel, and to overcome the difficulty Bagnold had to make the reasonable assumption that $\bar{u}/\bar{l}$ was related to the ratio of the gravity constant, g, and the vertical component of the velocity of the grain immediately on impact, $\bar{w}$. He found that $\bar{w}$ was itself proportional to the drag velocity U_*', and he arrived at a proportionality factor for this second relationship of 0·8. The expression that results from these manipulations contains the numerical proportionality factor, and another introduction of the quantity U_*'. It becomes:

$$q_s = 0{\cdot}8 \, \frac{\rho}{g} \, U_*'^3 \qquad \ldots (4.13).$$

Since q_s is only the saltation load, the total load must by Bagnold's calculation be 4/3 of this and the equation for the total load becomes:

$$q = 1{\cdot}1 \, \frac{\rho}{g} \, U_*'^3 \qquad \ldots (4.14).$$

Bagnold was able to confirm this result with some accuracy in the wind-tunnel, as were Chepil (1945) and Belly (1964). We will see how it compares with other measurements when we have discussed the Kawamura derivation.

Bagnold's formula can be confirmed with wind-tunnel measurements, but it includes quantities that are difficult to measure in the field, or which are not available from conventional meteorological measurements. One of the limitations of the formula is that only 0·25 mm diameter sands were used in Bagnold's experiments. To overcome this he evolved a new formula in which the actual grain-sizes could be standardized against his standard size and which included a coefficient C:

$$q = C \sqrt{\frac{d}{D}} \, \frac{\rho}{g} \, U_*'^3 \qquad \ldots (4.15)$$

where d was the actual grain size and D, the standard grain size. C has the following values:

1·5 for nearly uniform sand
1·8 for naturally graded sand
2·8 for poorly sorted sand
3·5 for a pebbly surface.

In other words q increases from a minimum when nearly uniform sand is used, to higher values when poorly sorted sand is used, and to its highest value over a pebble surface. This is because with poorly sorted sand the effects of coarse saltating impacts are probably greater, and because over pebbles the sand saltates more effectively and farther because of its more nearly perfect rebound.

Williams (1964) found q to vary little with initial size distribution over sandy beds, though particle shape made a much more consistent difference. At low velocities q was larger with less spherical particles. At fast velocities q was greater with more spherical particles. In the range $U'_* = 55\text{–}95$ cm/sec shape made little difference.

In equation 4.15 the most difficult value to measure in the field is U'_*. Bagnold achieved it with a bank of manometer tubes, but it is not recorded in conventional meteorological measurements. However U'_* can be conveniently redefined. This is best understood geometrically in Fig. 4.12. U'_* is proportional to the slope of the line of velocity against log-height (line $O'A$ in the Figure). This is the tangent of the angle of the line and therefore U'_* is equal to $CA/CO' \times 0{\cdot}174$ where 0·174 is a

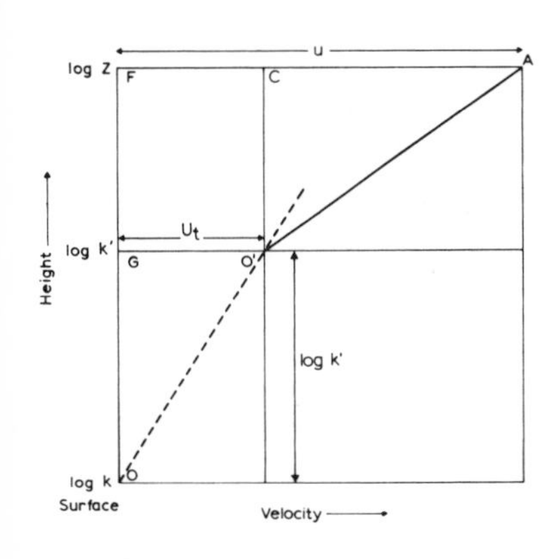

4.12 Geometrical explanation of Bagnold's simplified formula for q (after Bagnold, 1941).

proportionality factor. Now, from the figure and from our definitions earlier in this section, $CO' = \log z/k'$ and k' is a constant for any one grain size and sorting property; z is also a constant if the wind measurements are all taken at one height. Therefore

$$U'_* = \alpha AC \qquad \ldots (4.16)$$

where α is a constant which includes k' and z, and 0·174. Now $AC = (u - U_t)$ where u is the wind velocity measured at height z and U_t is the threshold of sand movement. So we now have:

$$q = C\sqrt{\frac{d}{D}\frac{\rho}{g}}\,\alpha\,(u-U_t)^3 \qquad \ldots (4.17)$$

This expression can be derived from readily available measurements since Bagnold elsewhere defines U_t, in terms of grain size and of z and k', and k' is presumed to be one cm.

Bagnold (1953*b*) later modified the formula even further for general use:

$$Q = \frac{1{\cdot}0\times 10^{-4}}{\log(100\,z)^3}\,t\,(u-16)^3 \qquad \ldots (4.18)$$

where Q = tonnes of sand per m across the wind, t is the number of hours that the wind of u km/hr blows and 16 km/hr is the threshold velocity. Q is summed for winds of all speeds for each direction to give a 'sand movement rose'. Bagnold pointed out that this formula is only really useful in the construction of relative sand movements from different directions, and not for absolute amounts, because of the unreliability of records. Bagnold (1951), Finkel (1959) and Inman *et al.* (1966) have used slight modifications of these formulae in studies of actual dunes.

Kawamura (1951 and 1953) derived a rather different formula in a different way. He set out his detailed argument in his 1953 paper. The following summary is based on an abstract of his Japanese publication (1951) by Horikawa and Shen (1960).

τ, the shear stress at the sandsurface, consists of two components

$$\tau = \tau_s + \tau_w \qquad \ldots (4.19)$$

where τ_s is due to the impact of the saltating sand and τ_w is due directly to the wind. In the equilibrium state, τ_w is found to be equal to τ_t which is the critical shear value ($\tau_t \propto U_{*t}{}^2$); so

$$\tau_s = \tau - \tau_t \qquad \ldots (4.20).$$

τ_s is equal to the loss of momentum from the wind, as in Bagnold's argument. So

$$\tau_s = G_o\,\overline{(u_2 - u_1)} \qquad \ldots (4.21)$$

where G_o is the amount of sand falling on the unit area of sandsurface during a unit period of time and $\overline{(u_2 - u_1)}$ is the mean value of the difference between the final and initial forward velocities of the sand particles. It is now assumed that:

$$G_o \overline{(u_2 - u_1)} = \xi \overline{G_o (w_2 - w_1)} \qquad \ldots (4.22)$$

Where ξ is a coefficient and w_2 and w_1 are the final and initial velocities in the vertical direction. This is generally true for elastic impact. Now since $w_2 - w_1 = -2w_1$ (velocities measured positive upward):

$$\tau_s = 2 \xi G_o \overline{w_1} \qquad \ldots (4.23)$$

(since we are only interested in the absolute value of $(w_2 - w_1)$. We can now substitute this in equation 4.20:

$$2 \xi G_o \overline{w_1} = \tau - \tau_t = \rho (U_*^2 - U_{*t}^2) \qquad \ldots (4.24)$$

It was then found experimentally that:

$$G_o = K \rho (U_* - U_{*t}) \qquad \ldots (4.25).$$

Where K is a constant and ρ is the density of air. From these last two equations therefore:

$$\overline{w_1} = K_1 (U_* + U_{*t}) \qquad \ldots (4.26),$$

where K_1 is another constant. The average particle path length can now be found:

$$\overline{l} = K_2 \frac{(U_* + U_{*t})^2}{g} \qquad \ldots (4.27),$$

where K_2 is yet another constant.
Because the following equation is probably true:

$$q = G_o \overline{l} \qquad \ldots (4.28)$$

we can now write:

$$q = K_4 \frac{\rho}{g} (U_* - U_{*t}) (U_* + U_{*t})^2 \qquad \ldots (4.29)$$

where K_4 is a constant to be determined by experiment.

Two further theoretical determinations should be mentioned. Zingg (1952, quoted by Horikawa and Shen, 1960) arrived at a formula by considering the distribution of the saltating sand above the surface. His formula is:

$$q = C (d/D)^{\frac{3}{4}} \frac{\rho}{g} U_*'^3 \qquad \ldots (4.30)$$

This bears a close resemblance to one of Bagnold's formulae (equation 4.15).

Lettau and Lettau (in press) have attempted to arrive at a formula using more rigorous reasoning. Their formula can be written:

$$q = C' (d/D)^n \rho U_*^2 (U_* - U_{*t}) \qquad \ldots (4.31)$$

where n is an exponent with a value of between $\frac{1}{2}$ and $\frac{3}{4}$. Zakinov (1969) has stated another empirical relationship:

$$q = 0{\cdot}160\,(U_{1{\cdot}0} - 4{\cdot}1)^3 \text{ gm/sec} \qquad \ldots (4.32)$$

where $U_{1{\cdot}0}$ is the wind velocity at one metre above the ground in cm/sec and 4·1 is the value for the threshold velocity of movement.

Some of these formulae are compared on Fig. 4.13. The experimental results of O'Brien and Rindlaub (1936), Williams (1964), Belly (1964) are also included. It can be seen that at fairly high windspeeds the discrepancies between the formular are not too large.

We shall return in later sections to somewhat different methods of measuring sandflow by measuring bulk transport in ripples (Sharp, 1963) and in dunes (Lettau and Lettau, 1969), and to the distinction between actual and potential sandflow rates.

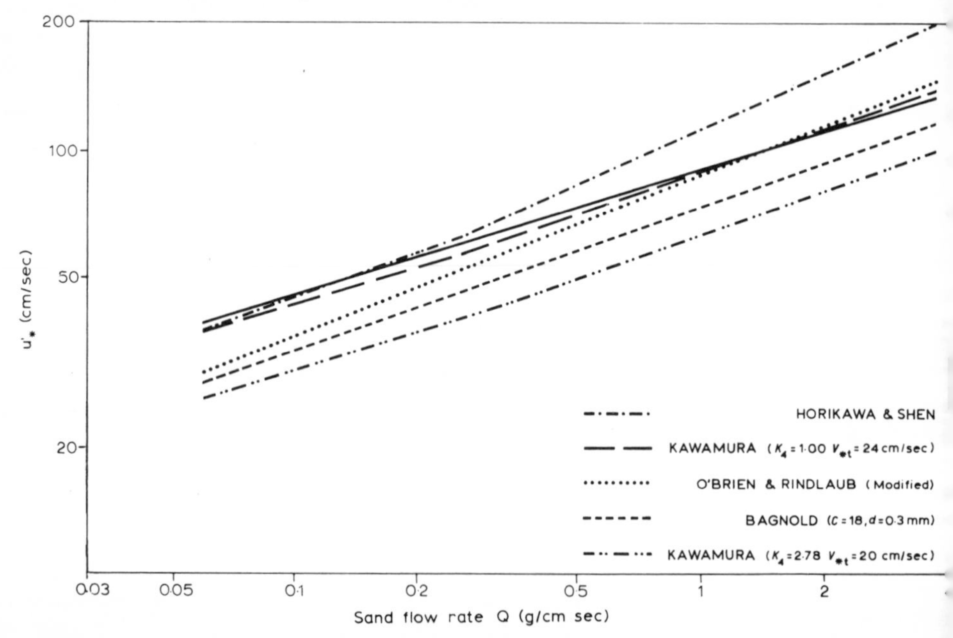

4.13 Comparison of some empirical and theoretical expressions for sandflow rate against drag velocity of the wind (after Williams, 1964).

4.3.2 TWO-DIMENSIONAL CHARACTERISTICS OF SIMPLE AEOLIAN BEDFORMS

(a) Ripples (in cross-section). There are three main groups of theories about aeolian ripples. In the literature of the late nineteenth and early twentieth century ripples were seen as the result of wave-like motions. It was held by some authorities that sand ripples were completely analagous to water waves in that the sand could be regarded as a fluid of high density, and since the air was a lower density fluid, its movement over the sand resulted in the formation of Helmholtz waves. This theory is clearly untenable since there is no fluid motion of the sand beneath the surface. Another idea was that ripples were merely small dunes and, given time, they would grow into dunes (e.g. Cornish, 1914). The evidence that this is not the case has been presented in section 4.1.1. Another view maintained that the saltating curtain could be regarded as a fluid of one density and that the clear air above it was another. Waves were initiated at this discontinuity (e.g. von Kármán, 1947 and 1953).

The second group of theories concerns regular small-scale turbulence in the atmosphere. Bourcart (1928) summarized the views of many European workers when he maintained that ripples were analagous to dunes, ripples being formed by Helmholtz waves in the atmosphere, but that the waves were at a smaller scale. While maintaining that most ripples were ballistic, as we shall see, Bagnold (1941) found a group of ripples in his wind-tunnel that appear to conform to the atmospheric wave-motion theory. These he called 'fluid drag ripples'. They were formed in fine sand at high windspeeds, and were analagous to subaqueous ripples. There is also some evidence in nature that there are small-scale instabilities in the wind. For example, small-scale longitudinal sand streamers have frequently been noted (e.g. Bagnold, 1941; Simons and Eriksen, 1953; Verstappen, 1968; Plate 4.17). It has been maintained that these are the longitudinal equivalents of some transverse system of regular eddies. We have seen that Sutton (1953) quoted evidence for eddies of about ripple size. It is maintained that some ripples are therefore 'aerodynamic' (Ellwood, Evans and Wilson, in press).

The third group of theories can be called the ballistic or impact theories. The original statement seems to come from Bagnold (1941). He found in a wind-tunnel that there was equivalence between the length of the characteristic flight-path of saltating sand grains and ripple wavelengths. The formation of regularly spaced ripples can be described in the following way (Fig. 4.14). When movement starts over a bed of sand, chance irregularities will be emphasized. Thus more grains will land on

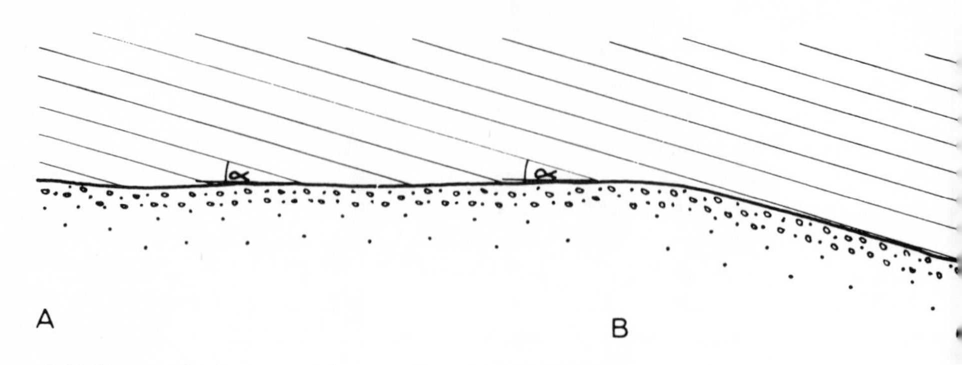

4.14 Ripple formation relationships (after Bagnold, 1941).

section *AB* than on section *BC* which is sheltered by the rise at *B*. This bombardment will release more grains from *AB* than will be released from *BC* and these will land downwind at a distance roughly equal to the characteristic flight-path length. There will be accumulation in zone *BC* and erosion in zone *AB* and the forward movement of the slope. As this happens at one ripple it will happen all along the line and the ripples will move forward conserving their spacing. The ripples have an equilibrium height because any further accentuation of their amplitude would raise the crests into zones of fast flow where the sand would be quickly removed. Von Kàrmàn (1947) also suggested that irregularities would repeat themselves downwind by a characteristic grain-path length.

Bagnold observed that this simple model of ripple growth holds only for sands with almost uniform size distribution, and the ripples formed in these conditions are very low. In the field it is more usual to find sands of mixed grain-size, with the coarser grains travelling along the surface as creep. When these coarse grains reach the summit of the ripple they cannot move on because there are few grain bombardments in the shadow zone (*BC* of Fig. 4.14). The concentration of coarse grains at ripple crests is almost universal. Fig. 4.15 shows ripples that were impregnated with resin and sectioned by Sharp (1963). It is sometimes the case that grains of heavy minerals such as magnetite also accumulate at ripple crests (Plate 4.7).

Sharp (1963) made some observations of wind ripples in the field which give some idea of the controls of ripple geometry. He found that an increase in the wind velocity or in the grain diameter produced an increase in ripple wavelength. Sharp proposed that ripple height might be a more direct control of wavelength than path length. He envisaged a sequence of ripple formation as follows: as the wind blows over a nearly

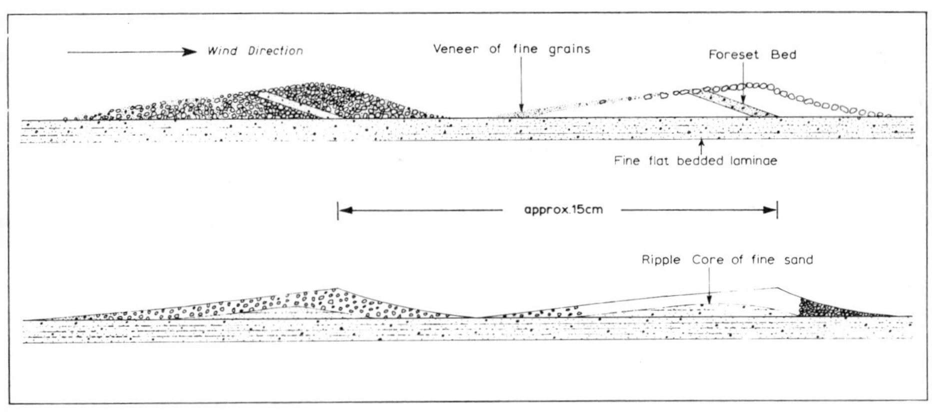

4.15 Sections of ripples produced from resin-impregnated field samples by Sharp (1963).

flat sand bed, saltating sand moves the creep; the creep encounters chance irregularities and piles up; the piles of sand grow until an equilibrium height is reached; this characteristic height and the angle of incidence of the grains (α in Fig. 4.14) control the wavelength. Since α is inversely related to wind velocity, stronger winds produce longer wavelengths.

Many observers have noted a very large kind of sand ripple in the desert. They have been termed 'granule-ripples' by Sharp (1963). They have much greater wavelengths than ordinary ripples; Bagnold (1941) observed some up to 20 m apart. They usually contain a prominent amount of coarse grains: Sharp (1963) noted that 10–20 per cent of the ripple as a whole and 86 per cent of the crest were composed of grains bigger than two millimetres. They are often almost symmetrical in cross-section, especially when they are very large. An example from the Ténéré Desert is illustrated in Plate 4.8. Bagnold (1941) thought that granule-ripples were formed when there had been long-continued deflation of an area so that much of the fine sand had been removed from the surface layers, leaving only coarse grains behind. Continued movement of the creep load was maintained in this situation by a supply of relatively fine saltating grains from upwind, which moved rapidly through the area, keeping the coarse grains in motion. Since the coarse grains could not move beyond the ripple summits, the upward growth of the ripple would continue indefinitely, although it would become very slow and Bagnold accordingly thought that granule-ripples were very old. Sharp (1963) however found that granule-ripples appeared in quite a short time in the Coachella Valley.

He attributed their greater wavelengths to the lower angle of incidence of grain paths at high windspeeds, maintaining that they were basically similar to other ripples, and that they had therefore an equilibrium height. He attributed their symmetry to greater age, since this would allow their modification by winds from different directions.

Ellwood, Evans and Wilson (in press) have agreed there is no basic dissimilarity between granule-ripples and ordinary ballistic ripples. In most cases, they argue, regular turbulence is unlikely to be important in ripples because of the momentum of the sand grains, so that granule-ripples are unlikely to be 'aerodynamic'. They have simulated the growth of ripples on a computer model into which they fed basic data on flight-path lengths. They observed that, in a sand mixture containing a prominent amount of coarse grains, fine grains have much longer flight-paths because of their better bounce off the coarse grains. Because of this, the ripple wavelength is greater. The coarse grains accumulate at granule-ripple crests in the same way as in normal wind ripples.

On further micro-feature of sand surfaces should be mentioned. Plane beds are found in three different situations: *(i)* when the wind velocity is very great, although this seldom occurs in nature; *(ii)* when sand is falling out of the wind on to a calm, gentle slope; and *(iii)* when the secondary coarse mode becomes very prominent and the coarse grains rapidly seal off the surface as a lag deposit preventing any ripple growth (Bagnold, 1941).

We should observe finally that the sand being moved in ripples is the traction load of Udden (1894). Sharp (1963) and Tricart and Mainguet (1965) observed that most of the grains above 0·5 mm move in this way as 'bulk-transport' which we shall meet again in our discussion of dunes. Sharp (1963) made some calculations of the importance of this kind of movement in ripples: under normal wind conditions 545 kg of sand were moved across a line 33 m long in one hour. If surface creep accounted for 20 to 25 per cent of the sand in transport, this produced a figure for total sand transport roughly twice the amount calculated using Bagnold's formula.

(b) Dunes. In explaining the shapes, and particularly the cross-sectional shapes of single dunes, the concepts of sand movement are very useful. The following discussion is a summary of Bagnold's (1941) work on dunes.

(i) The growth of a sand patch. Dunes on the desert surface probably originate in gentle dips, behind small chance irregularities on the surface or where convergent secondary flow occurs. An increase in velocity or divergence in the wind leads to erosion; steady velocity

(which is probably never maintained over a long distance) means neither deposition nor erosion; and decreasing velocity or convergence means deposition. Thus a slight hollow, a change in surface roughness, a forced convergence of a sand-carrying wind, or a sheltered zone in the lee of an obstacle will lower U'_* and so τ' (section 4.3.1.*b*) and lead to deposition. Once sand has been deposited it will itself present quite a different surface to the wind and change the pattern of the drag velocity (U'_*). Since sand particles rebound more easily off a hard surface of rock or a desert pavement than off sand, any wind that can initiate sand movement over the pavement will be able to carry more sand over it (i.e. q will be larger) than over a sandy surface. The value C in equation 4.15 is critical here. Since it is higher over pebbles than over sand, the result of the following expression is deposition:

$$Q_d = q_{\text{pebbles}} - q_{\text{sand}} = (C_{\text{pebbles}} - C_{\text{sand}}) \frac{\rho}{g} U'^3_* \quad \ldots (4.33).$$

Where Q_d is the amount of sand deposited on a unit breadth of the sand patch in a unit time. In this case a sand patch will grow.

There are said to be two important limitations to this simple picture. Firstly, Bagnold (1941) maintained that there is a distinct difference between the effects of strong and gentle winds (see Fig. 4.16). Bagnold's argument is based upon the different rates of transport over a pebble surface and over sand. A wind capable of moving sand over a sandy surface could not move it over a pebble surface, so that a sand patch would be eroded and extended downwind by such a wind. A stronger wind on the other hand might move more sand over pebbles, because of the better rebound from the harder surface (C in equation 4.15 is greater) and this equation 4.33 obtains and the sand patch grows.

A second limitation to the deposition of sand on a patch is that the patch should be of a certain minimum size. This can be explained in the following way. Bagnold found that when a wind passed onto a sand patch there was a lag between the achievement of maximum sandflow and the slowing down of the wind to its final steady speed (x of section 4.2.1.*f*, p.245). This is because the drag exerted by the saltating sand takes time to propogate up through the wind. This actually produces a fluctuating pattern of sand removal and deposition, for the wind is at first able to carry more sand and then, when slowed down, to carry less. The distance before which an equilibrium rate of sandflow was reached, with sand of median size 0·24 mm, was 7 m in Bagnold's tunnel. Chepil and Milne (1939) found figures between 2½ and 10 m and Kawamura (1953) found that over one metre was needed, the actual distance depending on the windspeed. This zone of adjustment means that the size of a patch that will grow by accretion is limited, since it must be at least larger than the distance over which equilibrium can be reached. This limits most dunes to a minimum width of about five or six metres.

(ii) Dune shape. The shape of the upwind slope of the sand patch as it grows will be a gentle curve which meets the desert surface at a feather

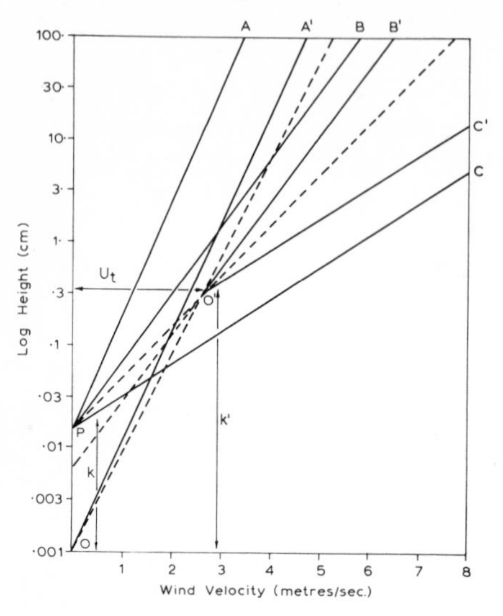

4.16 Bagnold's explanation of the effects of strong and gentle winds. *P* is the point at which velocity lines converge above a pebble surface when no sand is in movement. When a gentle wind, whose velocity is below the threshold needed to move sand on a sandy surface, passes off the pebbles on to a sand patch where the roughness is much less, the point of zero velocity drops from *P* to *O*. A slightly stronger wind, represented by *PB*, has a velocity at height k′ greater than the threshold needed to move sand on the sand patch, but because of the protection afforded by the pebbles no sand is moved there. As it picks up sand over the patch its pattern is changed to *O′B′* (since all velocity lines must converge through *O′*, as in Fig. 4.10). This wind brings no sand to the patch, but removes it to be deposited where the wind passes again over the pebbles and its pattern reverts to *OB*. A yet stronger wind *PC* can move sand over the pebbles and the patch, but on passing over the patch lowers its velocity at all heights because of the drag of the sand in movement. This results in sand being deposited from the wind. Thus, paradoxically, relatively gentle winds appear to erode and elongate the patch, while stronger ones lead to its growth.

edge. This is because of the gradual adjustment of the drag velocity to the new conditions on the patch. Bagnold (1941, p. 164) explained how under similar conditions the equilibrium slope was reached on the back of a ripple under water: 'the curvature of the face of the mound is such that the acceleration it produces in the surface flow is just sufficient to maintain the drag at a constant value from the border onwards and hence prevent further deposition'. It seems likely, however, that the growth of a dune could not take place without some variation in wind-speed and if we are to postulate that sand patches originate at low velocity or convergent nodes in a transverse wave pattern (section 4.3.3.*b*) then we must agree that the upwind slope will be some kind of function of the velocity patterns connected with the wave motion in the wind.

In the absence of such a wave hypothesis, the equilibrium shape of dunes has been the subject of some speculation. Exner (quoted by Leliavsky, 1955 and Graf, 1970) simplified the problem considerably by assuming that the initial mound had a shape that could be defined by a cosine curve (Fig. 4.17, stage 0). He then made the further assumption that the rate of sediment transport was linearly related to the velocity of flow, and that the mound constricted, and therefore accelerated, flow at its crest. The results of his calculations of the resultant change in the shape of the mound are illustrated in Fig. 4.17 stages 1–5).

Bagnold (1941) explained the upwind slope in the following way. As the sand patch grows up into the wind it is growing into zones of higher

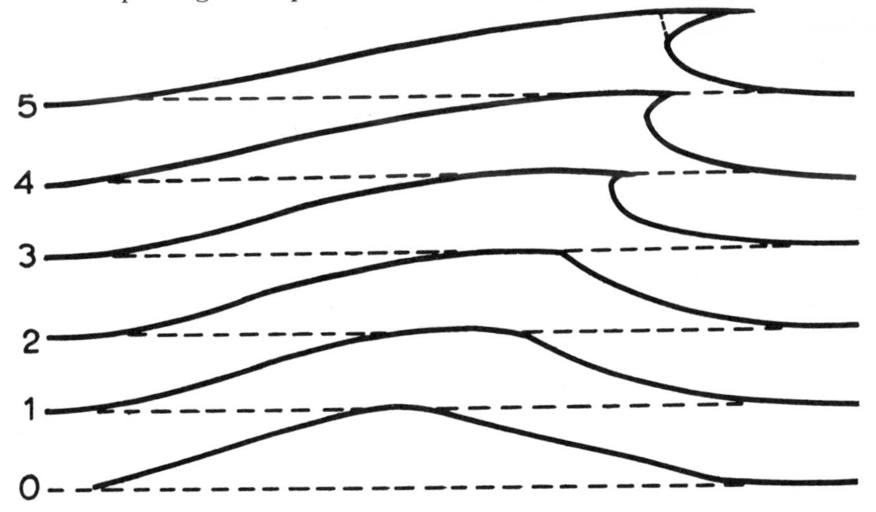

4.17 Growth of a dune from a simple cosine curve (after Exner, 1925, quoted by Graf, 1971).

velocity. On the upwind slope there must be a steady increase in velocity and, as his equations show, this means a steady increase in the sandflow rate, and therefore increased erosion. The rate of erosion is represented as the loss in a unit time from a unit area. In Fig. 4.18 this is represented by the area of the parallelogram, *ABDE*. The amount of sand in this area can be represented also as dq/γ where γ is a packing function, and dq is the amount of sand released from the area to the wind. *ABDE* is equal in area to the rectangle *ABEF*; *AF* is the distance in the horizontal direction through which the slope retreats, and can be called c; and *FE* is the height through which the level at *A* is reduced and can be called h. Therefore:

$$dq/\gamma = c \times h \qquad \ldots (4.34)$$

Now:

$$\tan \theta = h/dx \qquad \ldots (4.35)$$

where θ is the angle of the surface at *A*, and x is measured along the horizontal. Therefore

$$h = dx \tan \theta \qquad \ldots (4.36)$$

and therefore, using equation 4.34:

$$dq/\gamma = c\, dx \tan \theta \qquad \ldots (4.37)$$

or:

$$dq/dx \text{ (the rate of loss of sand per unit area)} = \gamma\, c \tan \theta \qquad \ldots (4.38)$$

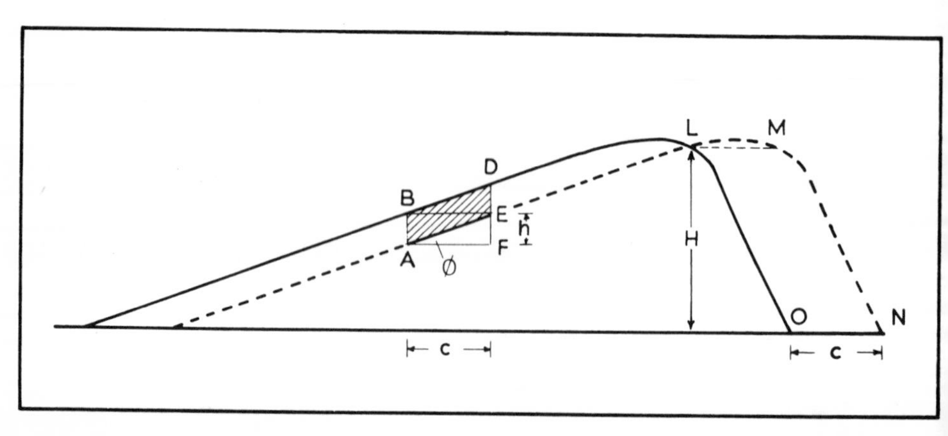

4.18 Diagram to explain the geometry of dune shape and movement (after Bagnold, 1941)

In other words, the greater the angle θ the greater the rate of removal. There will be an equilibrium angle (adjusted to the wind velocities at any one time) which will mean that there will be neither erosion nor deposition.

The actual slopes that have been measured on backslopes are usually in the region of 10° to 15° (e.g. Inman *et al.*, 1966; McKee, 1966; Sharp, 1966).

Bagnold (1941) and King (1916) assumed that the dune would have an equilibrium height controlled by faster windspeeds above the ground. It is more probably controlled by some property of wave motion in the atmosphere. It has of course been observed that there is an equilibrium height for dunes in any one area. Fig. 4.19 is a plot of dune height from Peru which illustrates this point.

At the crest of the dune the surface angle is zero, since there is neither erosion nor deposition. On the far side of a growing sand patch the slope angle declines, tan θ becomes negative and there is deposition.

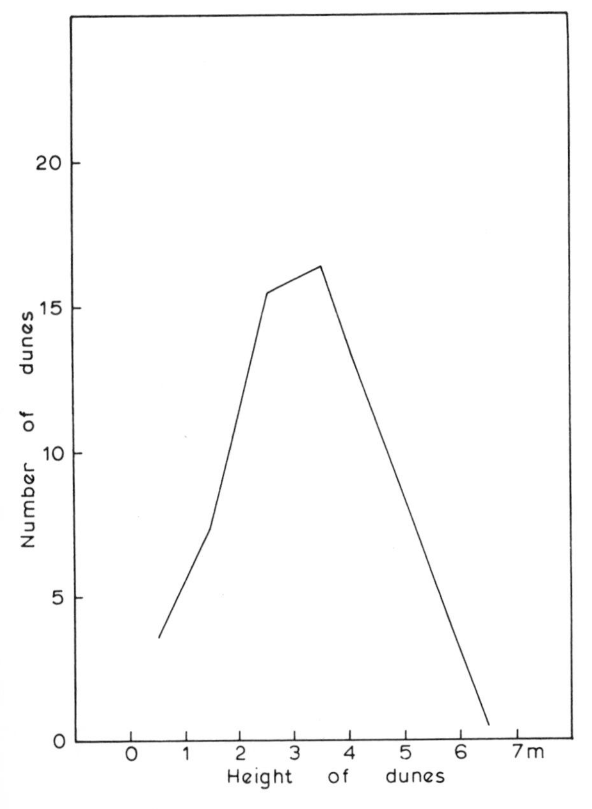

4.19 The distribution of dune height in a Peruvian barchan field (after Hastenrath, 1967).

There will be a lag between the actual change in slope and the effect this has on the wind (*cf*. section 4.3.2.*b.i*), so that theoretically there should be a distinct distance between the crest and the point of maximum deposition and the crest will be convex, as it is indeed usually found to be. Since the lag remains the same for any one sand and any one wind velocity, this separation should be relatively less important in the case of large dunes. Both Sharp (1966) and Inman *et al*. (1966) observed convex crests but Hastenrath (1967) noticed that while there was a distinct convex crest on small dunes, on large ones it disappeared. Verlaque (1958) also found that on a large barchan near In Salah, in Algeria, there was no convex crest. In both cases, however, a reversing summit was recorded for short periods, and this may explain the anomaly.

As the dune grows, deposition on its lee side becomes relatively nearer to the summit. The lee slope becomes oversteepened, and when the angle of repose for dry sand is reached downward slipping results and a slip face is formed. McKee *et al.* (1971) found that the main processes on slip faces were either flowage or slumping of whole blocks. Bagnold (1954*b*) also examined these processes. Inman *et al.* (1966) observed that each individual slip resulted in an advance of the slip face of some five centimetres. The angles of slip faces reported in the literature vary little from $33° \pm 1$. Sharp (1966) described slight over-steepening at the top (usually relieved by slipping) and slightly lower angles towards the base of the slope, so that slip faces were actually gently convex.

Bagnold (1941) observed that there was a minimum size for a slip face. If the slope can be overstepped by saltating grains (if its horizontal projection is less than the characteristic flight-path length) it will not accumulate enough sand for slipping. Only when the slip face has a horizontal projection greater than this will a slip face develop.

This discussion illustrates some simple properties of the transverse ridge. There clearly remains much more to be defined, especially of the relationships between secondary flow and dune shape.

(iii) Dune movement. Bagnold was also able to derive a formula for the rate of dune advance. The quantity c in equation 4.8 is the rate of retreat of the windward face. If the dune is an equilibrium form then it will move forward as a whole, conserving its shape. This means that the forward movement of the windward slope will be matched by a forward movement of the lee, the same amount of sand being deposited there as was removed from the windward side. Therefore c is the rate of advance of the dune.

If we now look at the lee slope, the amount of advance c takes place with an addition of a parallelogram $ABCD$ to the whole of the slip face

(Fig. 4.18). This contains all the sand that has passed over the dune and is now trapped in the slip face, and is equivalent to q/γ in a unit time. By simple geometry we know that the area of $LMNO$ is equal to H, the height of the dune, times c its advance in a unit time, so that:

$$c \times H = q/\gamma \qquad \ldots (4.39)$$

where γ is, as before, a packing factor. This defines the rate of advance in terms of the rate of sandflow (q) and the height of the dune (H):

$$c = \frac{q}{yH} \qquad (4.40).$$

These theoretical considerations are amply supported by observation. Dunes certainly appear to advance while conserving their shape. In observations over several years of barchan movements in Peru, Finkel (1959) and Hastenrath (1967) were able to identify barchans quite easily despite considerable movement. They, Norris (1966), and Long and Sharp (1964) observed small changes in dune shape, but that these were readily explained by changes in the wind regime. Inman *et al.* (1966), McKee (1966), and Kerr and Nigra (1952; Fig. 4.31) observed that barchans moved forward conserving their plan shape and by implication also their cross-sectional shape. Lettau and Lettau (in press) also made observations of sand transport which clearly indicated conservation of shape.

Several measurements have been made on the rates of barchan advance. Bagnold was able to quote Beadnell's evidence that barchan rate of movement was related to height. This has been confirmed by Finkel (1959), Hastenrath (1967), Long and Sharp (1964), and Norris (1966). Fig. 4.20 shows a plot of some of these measurements. Long and Sharp (1964) noted that barchan shape might significantly influence rates of advance: a growing dune that is wider than normal moves more slowly than one that is narrower, because a greater volume of sand is needed to produce the same amount of growth. The opposite is true for dunes that have reached their equilibrium size and have stopped growing, because a wider dune is then turning over more quickly.

Actual rates of advance will, of course, vary with local conditions. Observed rates vary from 17 to 47 m p.a. in Peru (Hastenrath, 1967). Beadnell (1910) measured rates of 15 m p.a. in the Abu Moharic dunes in Egypt.

Bagnold (1941) compared his formula with Beadnell's measurements. He calculated, using approximations for windspeed, the amount

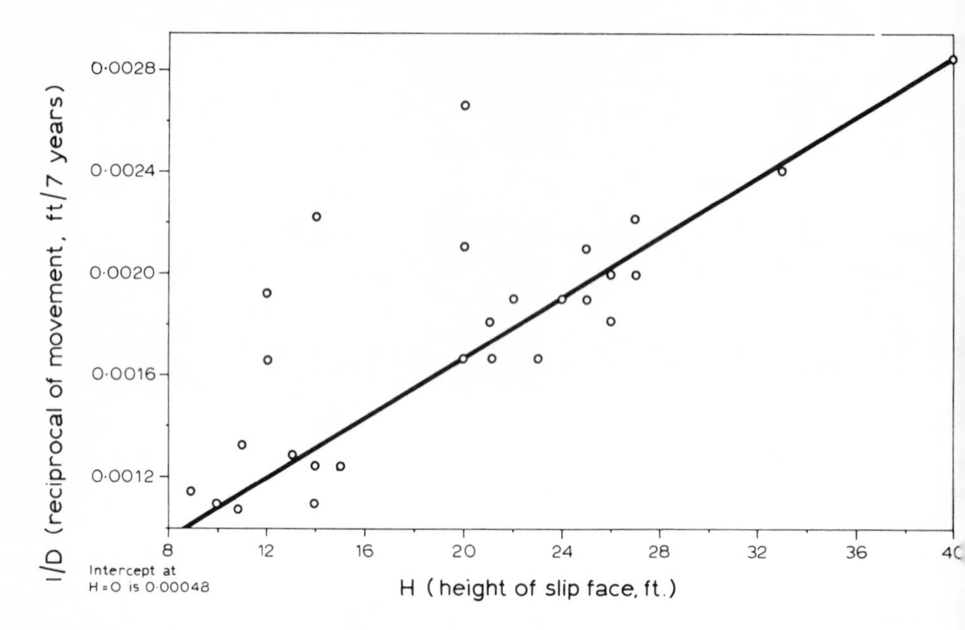

4.20 Relation between height of slip face and rate of advance of dunes in Peru (after Hastenrath, 1967).

of sand in movement, and fitting this with dune height into equation 4.40, he compared the results with the actual measurements. The two were in very close agreement. Finkel (1959) made a further comparison using measurements of dune movements in Peru. He found that Bagnold's formula gave good results for dunes with a height range between two and seven metres, but for smaller dunes there was a wide discrepancy. This may have something to do with the influence of dune shape as we have discussed it, since smaller dunes below equilibrium height would probably have more varied shapes. Inman *et al.* (1966), however, found a much greater discrepancy in their measurements of transverse dune ridge movements in Baja California. They compared measured amounts of dune advance with calculated ones and found that the measurements were always greater, and commonly by a factor of about two. They were unable to give an adequate explanation of the discrepancy. Lettau and Lettau (in press) have closely observed barchan movement in Peru. Fig. 4.21, taken from their report, illustrates the close relationships between movement and meteorological variables.

Bagnold's formula for the amount of sand blown by the wind can be

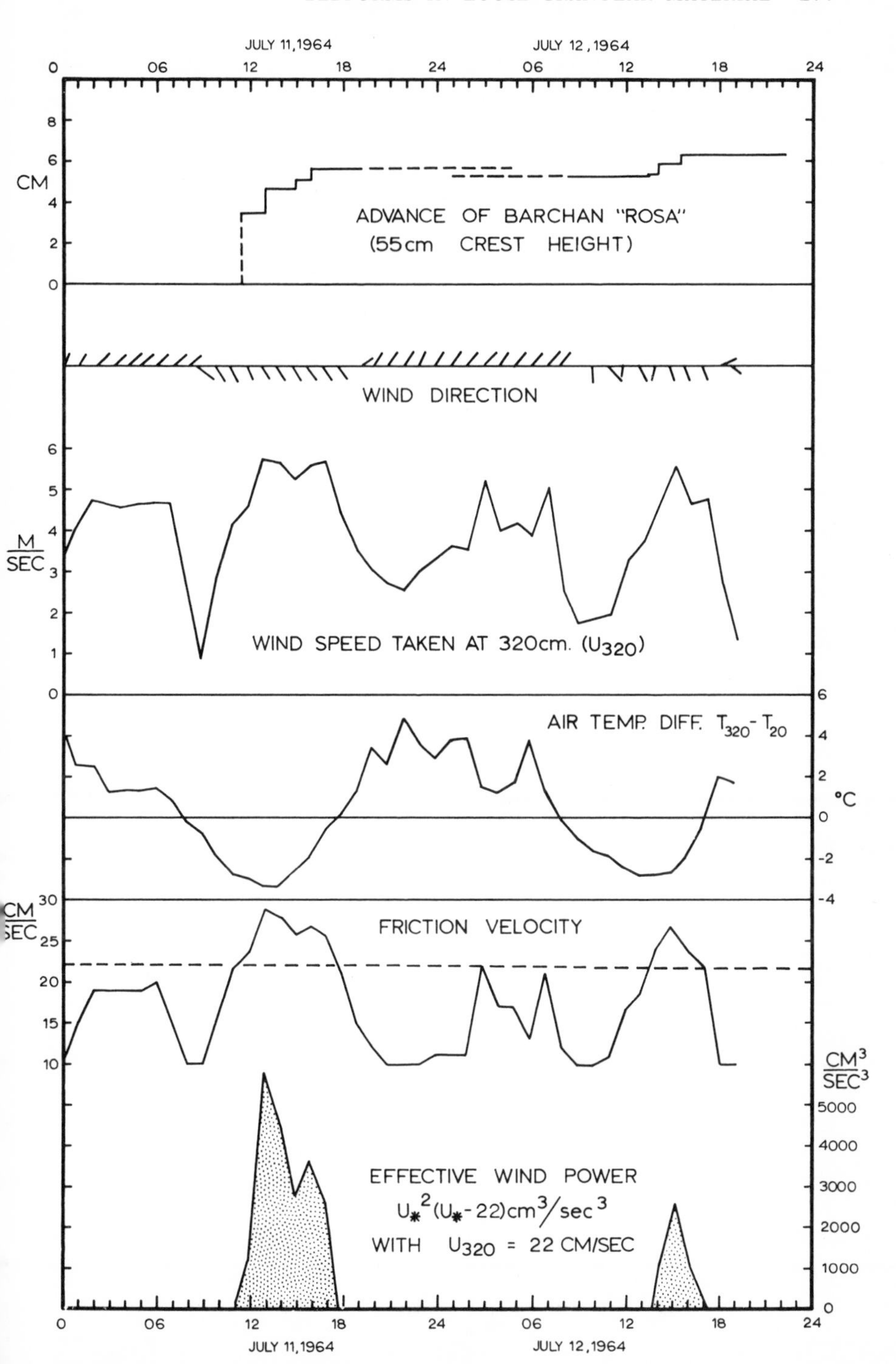

4.21 Detailed observations of meteorological conditions and dune movement in Peru (after Lettau and Lettau, in press).

used with a knowledge of dune geometry to find the time taken for a dune to reach its present size (Wilson, 1972*a*). This is, of course, only the growth time to equilibrium, not the time after this; it is assumed, too, that all the sand passing over the dune is trapped in the lee face. Using sandflow rates of 3·6 m^3/m breadth/p.a. (typical of the Great Eastern Erg in Algeria) we find that a dune with a height of three metres and wavelength 100 m would need about 40 years to grow.

One further point should be made about barchan movement. The movement of the dune involves a new mode of transport. Lettau and Lettau (1969) have drawn the distinction between bulk transport of sand in the dune and streamer transport which is the movement by saltation. They calculated that a three metres high barchan in Peru transported 20,550 m^4 a year past a line at right angles to the wind. Using data for the proportions of the width of the desert floor occupied by barchans, as against open desert pavement, they calculated that bulk transport varied between 0·4 and 0·6 of the total sand in transport. They were able to use these figures to calculate total rate of sand transport. This kind of calculation might be a useful check in barchan areas on the accuracy of sand transport figures arrived at in other ways. Fig. 4.22 shows the bulk transport across a dune belt in Peru.

(iv) Regularity of spacing. Transverse dune ridges are commonly found in groups in which dune spacing is very regular. This is clear from an examination of any of the accompanying aerial photographs (e.g. Plates 4.1, 2, 3 and 4), but it is a property that has not often been measured. Cornish (1914) measured some small dunes near the Nile at Helwan. He found that his 24 measurements gave him an average wavelength of 9·3 m and a height of 0·5 m. He does not mention the dispersion around these mean values. Matschinski (1952) published the results of 2,000 measurements of dune wavelengths in the vicinity of Beni-Abbes, Algeria. He found two distinct peaks on the frequency diagram, one between 100 and 300 m, and another between 10 and 100 m. He observed that dunes with similar wavelengths were found in the same fields. Working in the Sudan, Warren (1966) made a number of measurements of the wavelengths of transverse dunes. He too found that dunes of similar wavelength were found together, and that spacings were very regular.

Regularity is a property of dune forms that has been remarked upon for very many years. We have seen in section 4.1.1 how some early authorities saw ripples and dunes as essentially the same, and attributed them to wave motions in sand initiated by shear with the wind. However, by the early twentieth century it was generally agreed that regularity in dune forms must be attributed to some regularity in the patterns of

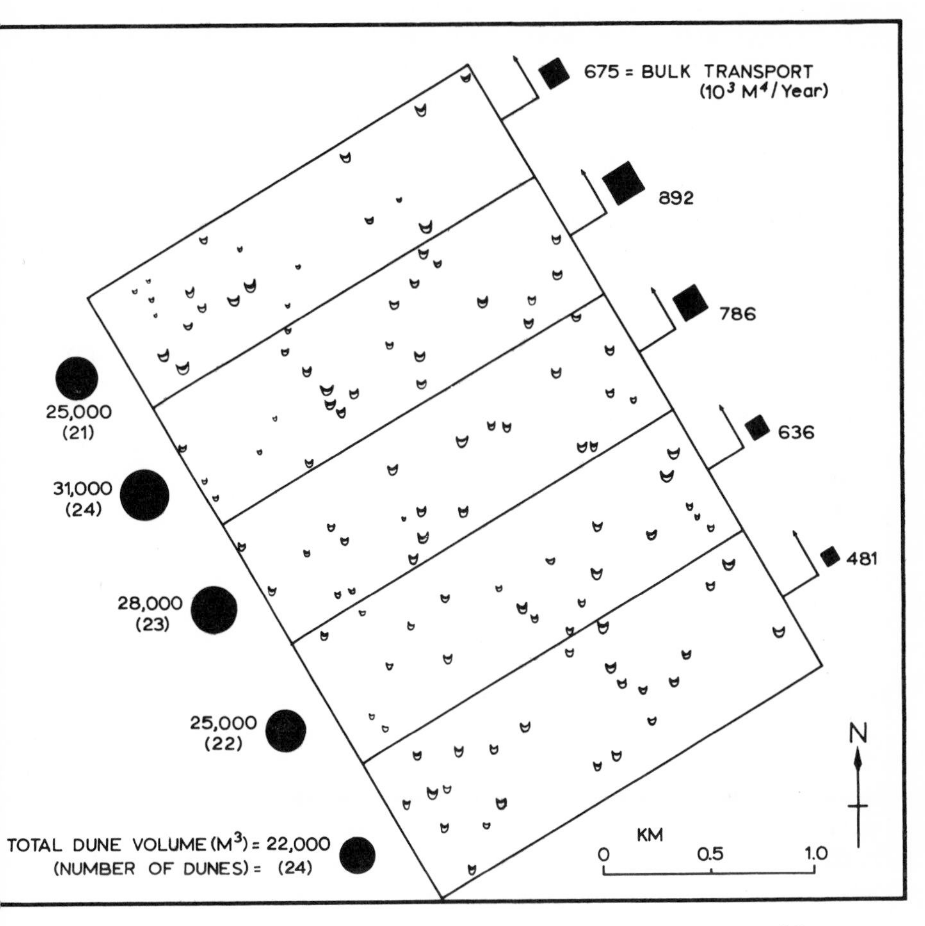

-.22 The bulk transport of sand by barchans in Peru (after Lettau and Lettau, 1969).

turbulence in the wind (e.g. Bourcart, 1928; Dobrolowski, 1924; Högbom, 1923). This conforms to some ideas of 'two-dimensional turbulence' that have been evolved for subaqueous bedforms (e.g. Allen, 1968*a*; Fig. 4.29b), although the mechanisms that give rise to regular eddy patterns may be rather different. It is believed by some authorities that beneath water a bedform arises because of some variation of the current speed on the bed, and as the bedform grows an eddy is fixed from the current. It has sometimes been supposed in dune studies that the first dune to grow may either fix a whole eddy pattern and so give rise to a regular repetition of forms downwind, or may actually initiate eddy motion in the wind (e.g. Cooper, 1958).

The fixation of an eddy form means that there is a permanent eddy in the lee of the bedform (Fig. 4.29b). The exact form of this eddy may well be modified by other kinds of secondary flow (as outlined in the next section), but there can be little doubt that it exists. It has been observed by numerous field workers using smoke, pieces of paper, threads and the like (e.g. Cooper, 1958; Coursin, 1964; Hoyt, 1966; Melton, 1940; Sidwell and Tanner, 1938; Verlaque, 1958; Volkov, 1955). What is more in dispute is the efficacy of this current. Cornish (1914) maintained that the lee eddy actually eroded the lee slope, but this has often been disputed. For example, Inman *et al.* (1966) found that even in high winds the lee currents never exceeded five metres per second, and Sharp (1966) observed that they merely brought light vegetation debris to the foot of the slope or lightly rippled the surface. However, Sevenet (1943) working in northern Mali, Hoyt (1966) in south-west Africa, and Glennie (1970) in Arabia, have cited evidence of erosion of the bed in the lee of transverse ridges. Similar evidence can be cited from the central Ténéré Desert in Niger (section 4.3.3.*d*). In Peru, Lettau and Lettau (in press) observed reverse transport of sand towards the slip face in the lee hollows of barchans and consider that this is important in maintaining barchan shapes.

(c) Draa. At their simplest, the cross-sectional forms of draa are analagous to those of dunes. This is illustrated in Fig. 4.23 which shows the Pur-Pur dune in Peru. This draa-sized barchan is covered with dune-sized barchans and has a slip face like them. Its movement and geometry are probably controlled in very much the same way as those of dunes. Simons (1956) noted that its movement was of the order of 0·45 metres p.a., whereas the movement of associated barchans was about nine metres p.a. Wilson (1970) cited evidence from the Great Eastern Erg in Algeria of slip faces on draa and Glennie (1970) mentioned 140 m high slip faces.

However, it is more common for draa to be covered both on their windward and leeward faces by dunes (see Plate 4.3, and Smith, 1956). The dunes on the lee tend to have large slip faces, and must therefore be contributing to the advance of the draa, but the dune forms are nonetheless migrating over the back of the draa, carrying some sand with them. The whole has a complex motion, since ripples are passing quickly over the dunes, dunes move rather less quickly over the draa, and draa are moving forward at a few centimetres a year or even less.

Draa, like dunes, are regularly spaced. Draa spacing is perhaps easier to attribute to regular atmospheric motion than is dune spacing since, as we have seen, regularly spaced convective plumes and their combina-

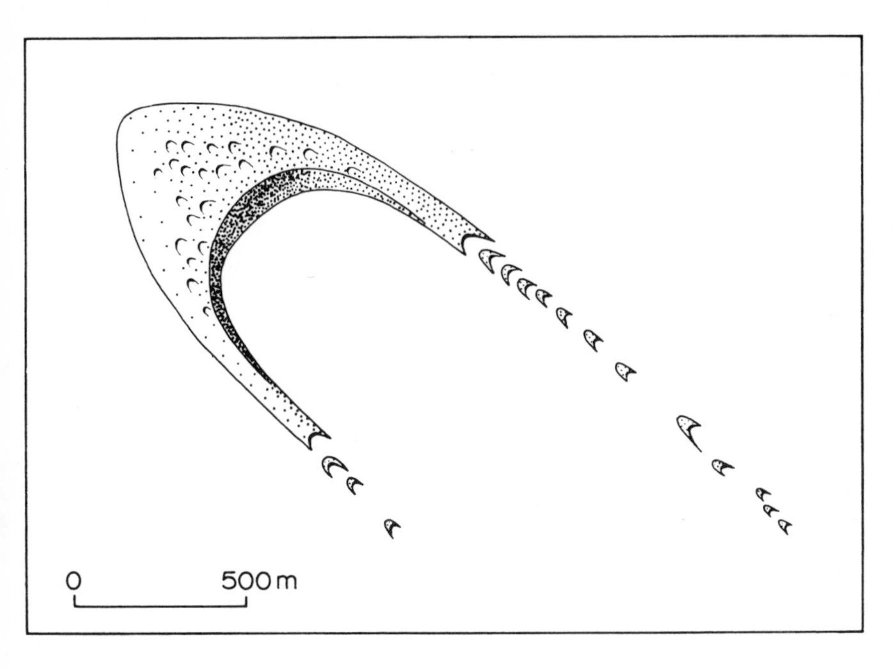

4.23 Pur-Pur dune in Peru: a *draa*-sized barchan (after Simons, 1956).

tion into regularly spaced long vortices is one of the better understood atmospheric motions. Lee-waves behind large hills too may also be important in the spacing of some draa, but lee-waves behind draa themselves seem less likely since some draa-spaced features are found that have very little height. Some of these points are further discussed by Wilson (1972*a*).

The idea of reconstitution time can be applied to draa as well as to dunes. Wilson (1970) has made some estimates for the massive draa of parts of the Great Eastern Erg using sandflow rates calculated from wind data. He estimated that if all the sand passing over a large draa 25 m high and about 1,000 m wide in the direction of sandflow were trapped in the slip face, its minimum age would be 1,400 years. Since some of the sand probably moves over the draa without being incorporated, its age to equilibrium is more likely to be of the order of 4,000 years.

4.3.3 DUNE PATTERNS

(a) Introduction. The preceding discussion has been simply of the cross-section of transverse ridges. We now look at the patterns of aeolian bedforms in plan. This is a much more complex matter, and one that has

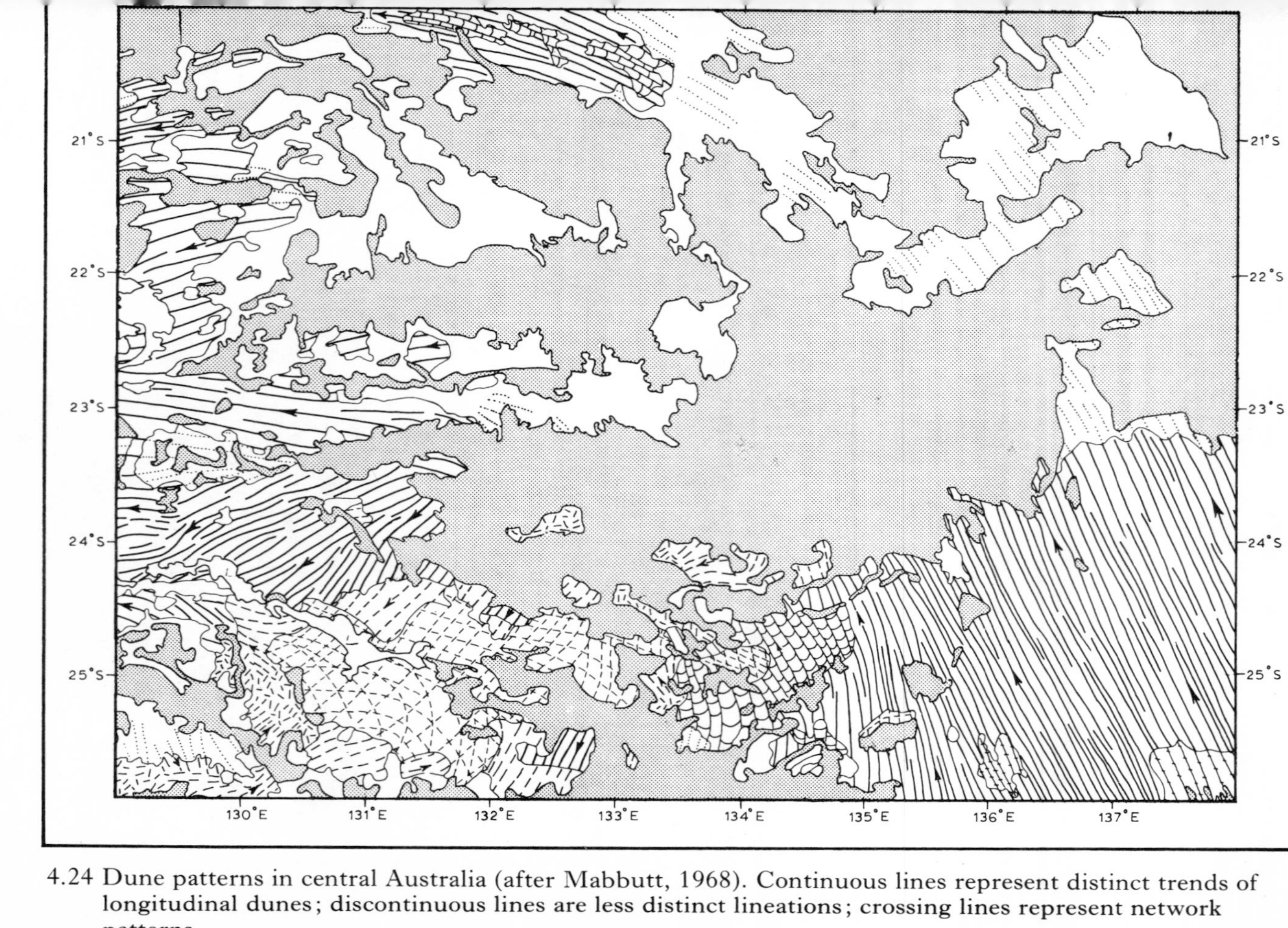

4.24 Dune patterns in central Australia (after Mabbutt, 1968). Continuous lines represent distinct trends of longitudinal dunes; discontinuous lines are less distinct lineations; crossing lines represent network patterns.

yet to yield to the kinds of analysis we have cited for cross-sections. Its complexity arises not only because of the patterns of secondary flow in the wind, but also because of complicating factors such as vegetation, sand grain-size, topography and palaeoforms.

Dune patterns are not simple in nature. The huge number of aerial photographs of arid areas that have recently become available have revealed that complicated rather than simple patterns are the rule (e.g. Plates 4.1, 2, 3, 9, 12, 17, 18 and 19). It has been common to describe dunes as either transverse or longitudinal, but one of the most obvious properties of dune patterns is that they fail to conform to these simple classifications: oblique forms are not only common, they are virtually universal. In almost every description of dune patterns, oblique forms are mentioned. Examples from central Australia, from the western Sahara and from Algeria are shown in Figs. 4.24, 25 and 26.

These patterns have been explained in a number of ways. They may be the result of crossing winds (section 4.3.3.*c*); where this explanation seems inadequate, one can invoke either underlying topography (section 4.3.3.*f*), or palaeoforms (section 4.3.3.*h*). However, many of them can be explained more simply and fully by considerating the basic flow patterns in the wind and the ways in which they interact with the bedform; it is patterns generated in this way that we discuss first.

(b) Patterns in unidirectional winds. The evidence for complex secondary flow patterns in the atmosphere which was cited in section 4.1.2 can be supplemented by the evidence contained in the aeolian bedforms themselves.

If wave-like motions in the atmosphere were the only flow pattern responsible for dune patterns, the latter would consist simply of long, straight equidistant ridges transverse to the wind (Fig. 4.29b). This pattern can be found in ripples, although not commonly, or over great areas (Plate 4.10), but it is seldom, if ever, found among dune or draa patterns, although it may be found in some *zibar* (section 4.3.3.*d*).

In dunes the simplest pattern is a network, as shown in Fig. 4.27f and h and Plate 4.11. In the western Sahara and in the French literature this pattern is known as *aklé* (e.g. Monod, 1958). The pattern is also found in draa-sized features (Fig. 4.28 and Plate 4.2). Aklé patterns seem to require relatively unidirectional winds and a considerably quantity of sand (Bourcart, 1928; Brosset, 1939; Cooper, 1958; Cornish, 1914; Dobrolowski, 1924; Doubiansky, 1928; Dufour, 1936; Hack, 1941; Hefley and Sidwell, 1945; Inman *et al.*, 1966; Melton, 1940; Monod, 1958; Sevenet, 1943). Detailed illustrations and descriptions can be found in Cooper (1958) and Inman *et al.* (1966). We can start by describing the

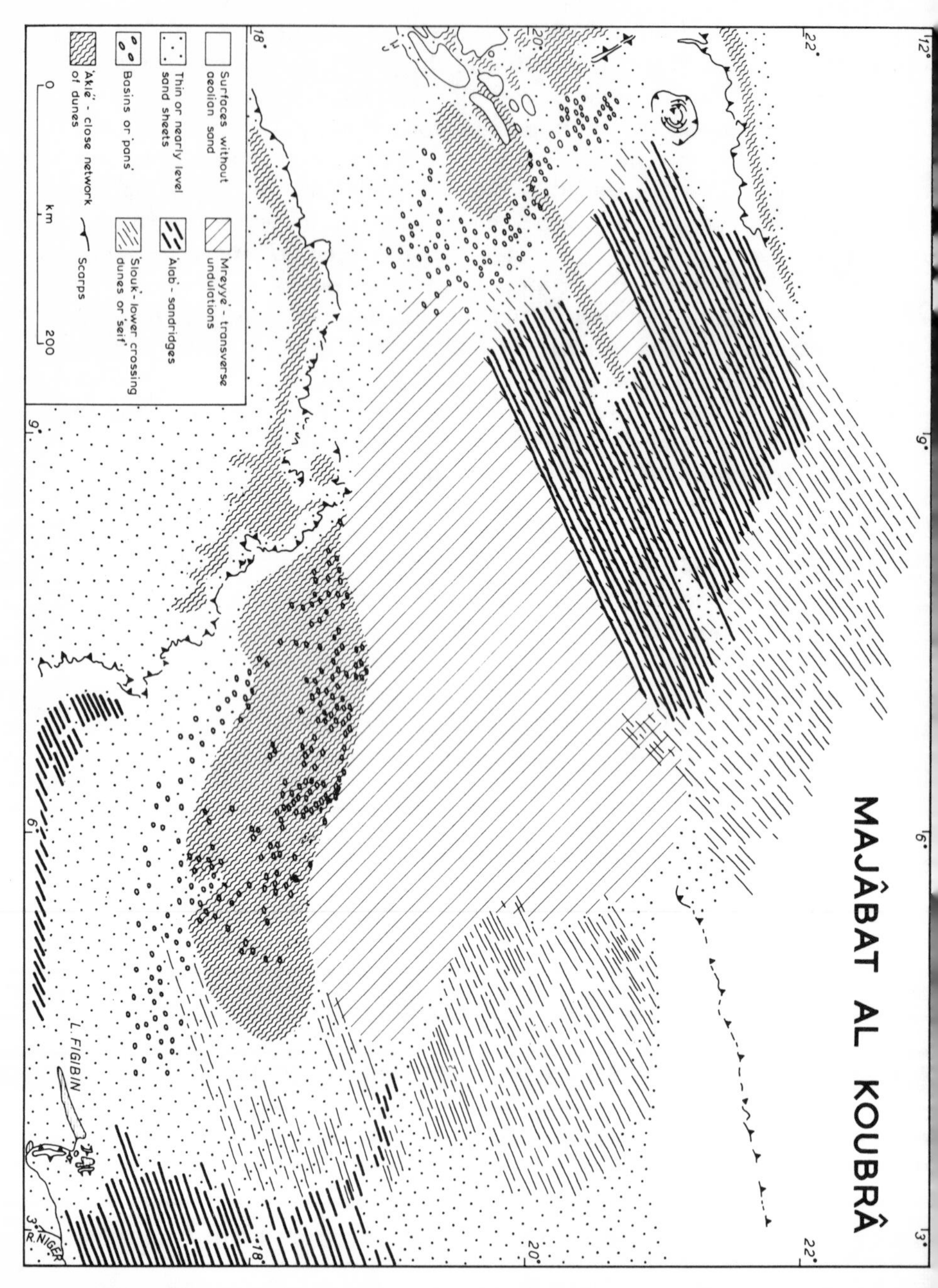

4.25 Dune patterns in the western Sahara (after Monod, 1958).

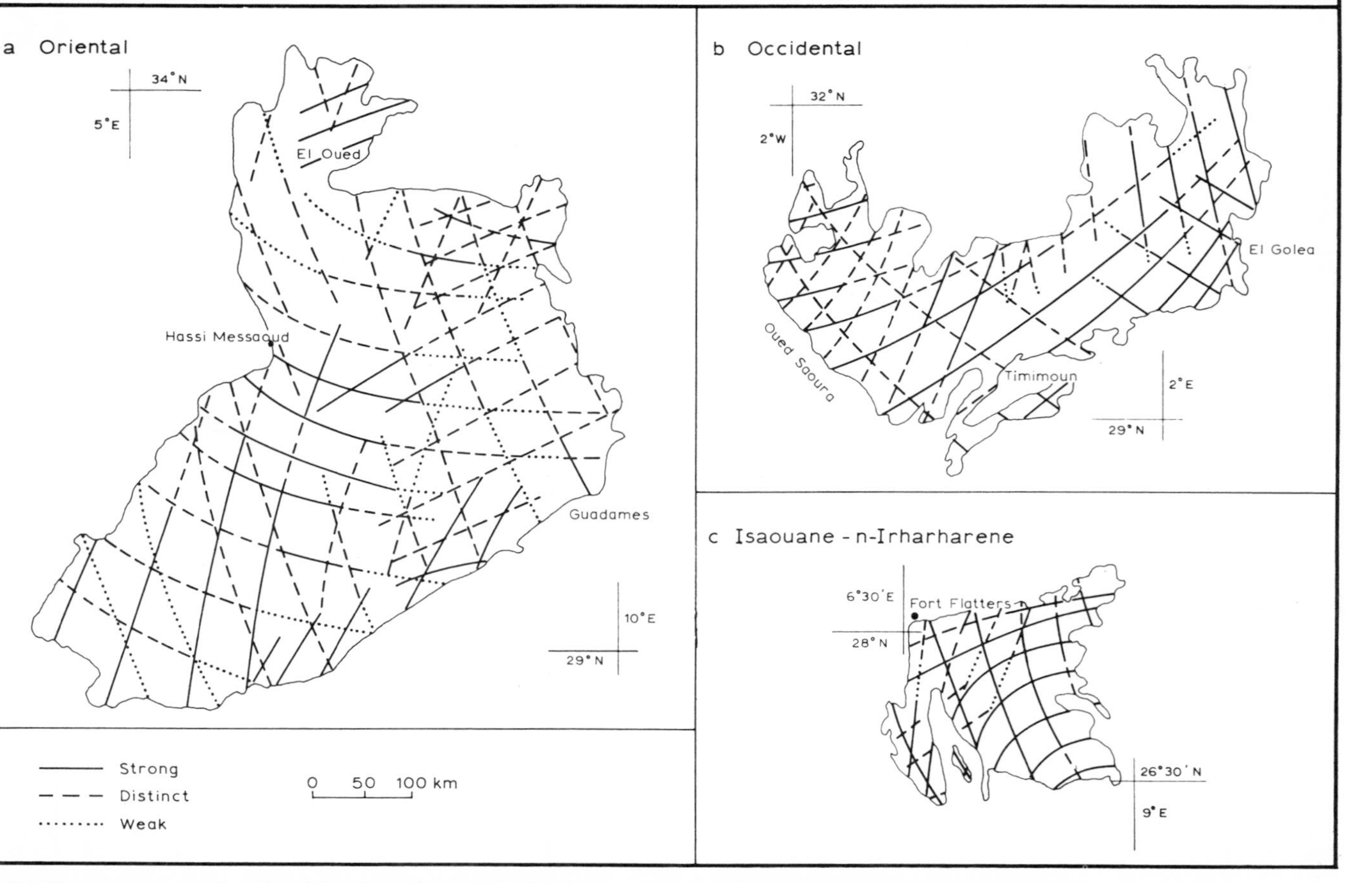

4.26 Dune patterns in the Algerian Grand Ergs. Only the main trends, not individual alignments, are shown (after Wilson, 1970).

simple unit of the pattern. This is a sinuous ridge (Plate 4.12) transverse to the wind, made up of crescentic sections alternately facing into and away from the wind. We call those facing the wind *linguoid* (*cf*. Bucher, 1919; Dobrolowski, 1924) and those facing away from the wind, *barchanoid*. These units alternate in a regular way along the ridge. The pattern is closely analagous to similar patterns found in ripples

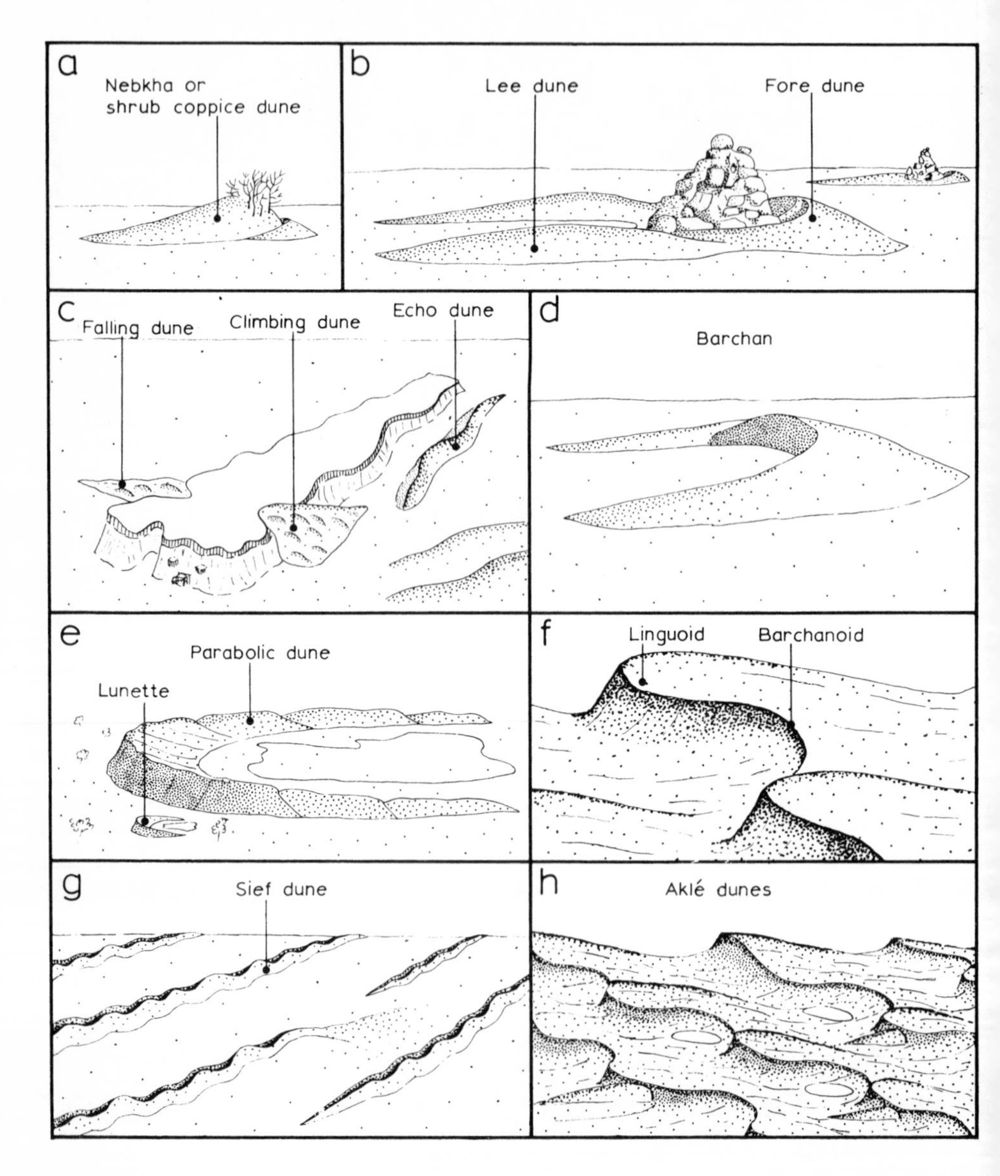

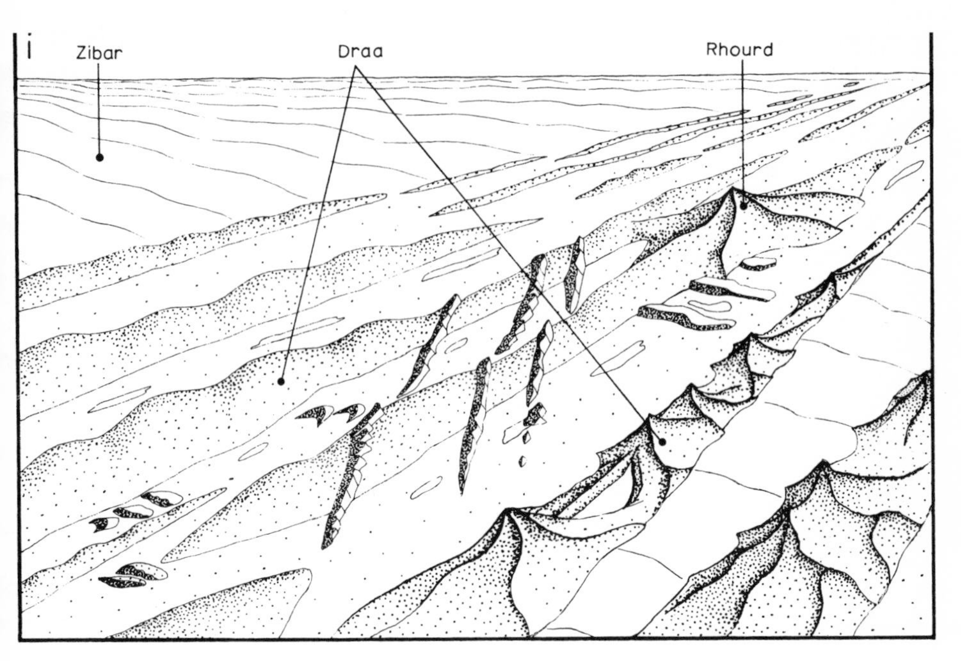

4.27 Dune types.

(e.g. Allen 1969*b*; Bucher, 1919). In dunes, as in subaqueous bedforms, it has been assumed that the pattern could be explained by postulating fast and slow lanes of secondary flow parallel to the mean flow. The fast lanes push the linguoid sections forward, leaving the barchanoid sections in the slow lanes. The early assumption summarized by Bourcart (1928) was that these fast and slow lanes were simply parallel filaments of the wind, but it is unlikely that these could travel side by side without secondary flow from the fast to the slow lanes, and if continuity is to be maintained, this would require return flows at some level above the ground. This is in essence a description of Taylor-Görtler vortices (or three-dimensional flow patterns) as shown in Fig. 4.29*a*. Simple patterns of two- and three-dimensional flow do not, however, explain a further common property of aklé patterns. It is observed that the barchanoid element of one ridge is followed downwind by a linguiform element in the next ridge (e.g. Inman *et al.*, 1966 and Fig. 4.35). The barchanoid and linguoid elements enclose a hollow which is often bounded by a small longitudinal ridge (e.g. Bourcart, 1928; Cooper, 1958; Cornish, 1914). An explanation of these will be discussed after we have considered further evidence of secondary flow patterns in dunes.

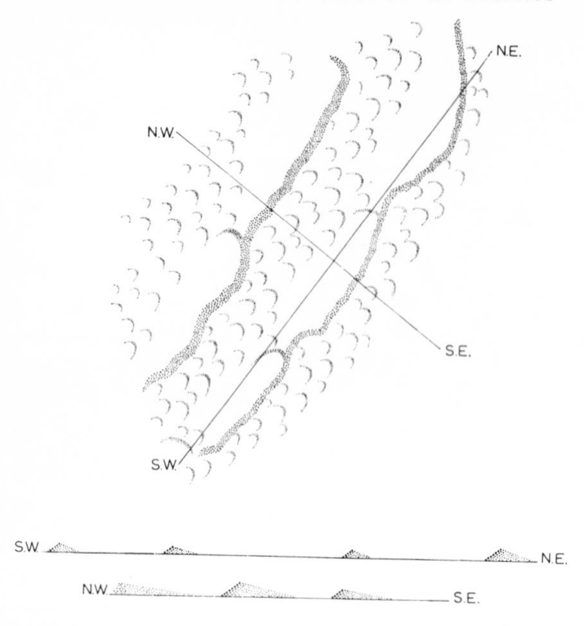

4.28 *Draa*-sized aklé in the Tarim Basin (after Hedin, 1904).

Evidence for both two-dimensional and three-dimensional flow patterns can also be found in barchans. Barchans, like aklé, are found in unidirectional winds, but they occur only on hard desert surfaces with sparse supplies of sand (as stated by every authority on this kind of dune).

Bagnold's (1941) description of barchans started with the assumption of an arbitrary width of sand patch across the wind, and of no lateral variation in sand supply. Because the flow over the desert pavement at the sides of this patch is faster (C in equation 4.17 is greater over pebbles than over sand), the patch will grow in the middle and taper down at the sides. Since the slip face at the sides is lower, but still traps the same amount of sand as the slip face at the summit, it will move forward more quickly and the dune will become crescentic. Bagnold explained the eventual slowing down of the arms by suggesting that they moved into the zone protected by the main mass of the dune. One might explain it in another way by saying that, since the arms are too low to have slip faces to trap any sand, they cannot advance any further than just beyond the end of the slip face.

This explanation is, of course, only partial. It does not explain the reasons for the regularly repeated width of barchans (Fig. 4.30). Nor does

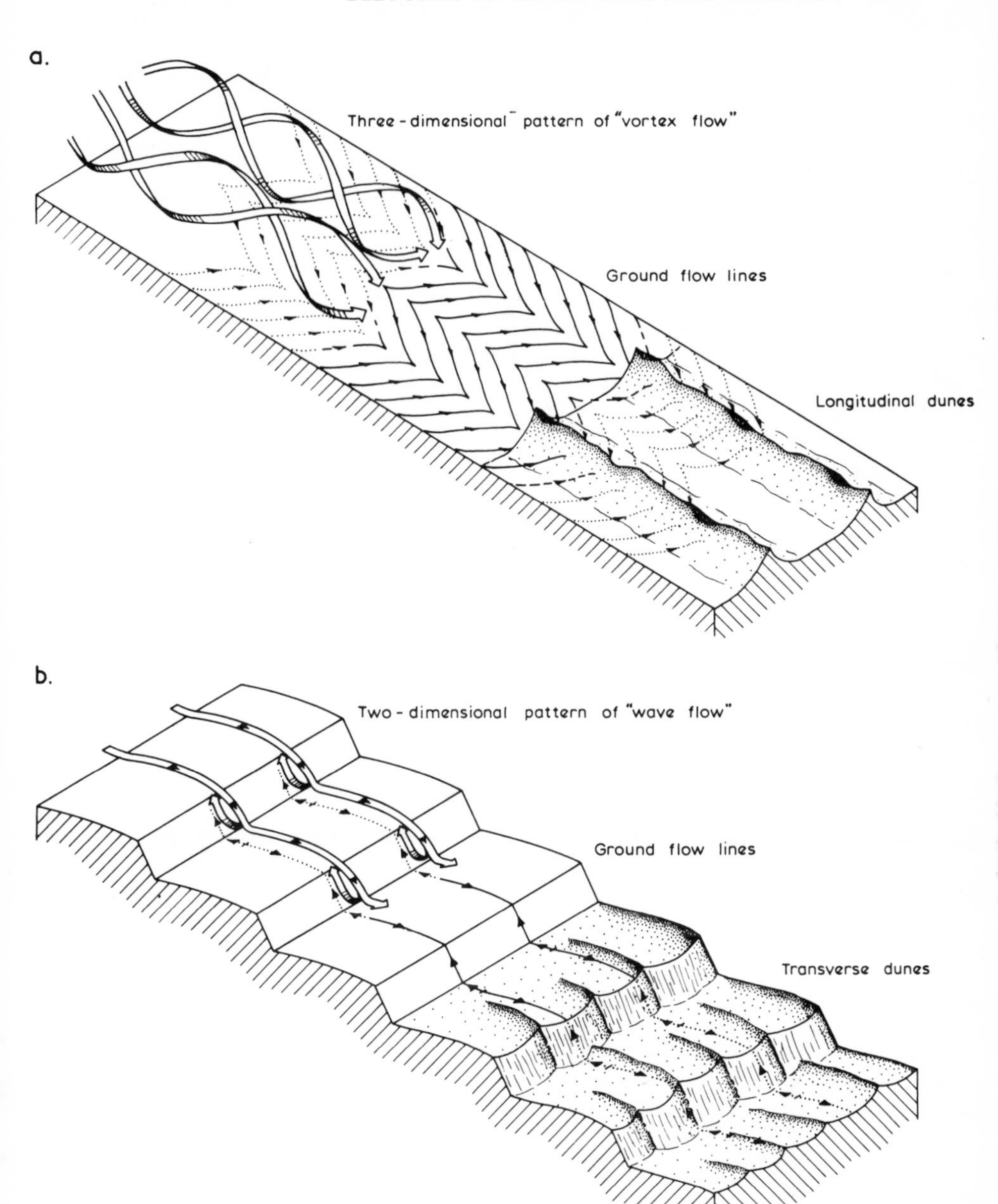

4.29 Patterns of three-dimensional and two-dimensional flow: (*a*) Vortex flow known as Taylor-Görtler flow; the relationship of the flow to ground flow lines and to longitudinal dunes is shown; (*b*) wave-like motion and its relationship to transverse dunes.

it explain some observations on the elongation of the arms. In many parts of the world, barchan arms are elongated more in one direction than another, and this can often be easily attributed to an assymmetry in the wind pattern (e.g. Holm, 1960; King, 1918; McKee, 1966; Melton, 1940; Norris, 1966), to an asymmetry in the sand supply (Rim, 1958), or to a slope in the desert surface (Lettau and Lettau, 1969; Long and Sharp, 1964). But in some areas both arms are distinctly elongated (Fig. 4.31, c, d, e and f). This is the case at In Salah in Algeria, on the coast of the Persian Gulf (Kerr and Nigra, 1952), and near Faya in Tchad (Capot-Rey, 1957). Another difficult problem to explain with an hypothesis involving seasonal cross winds is the situation illustrated in Fig. 4.31b, c, and f, and in Fig. 4.32e where two barchans have their

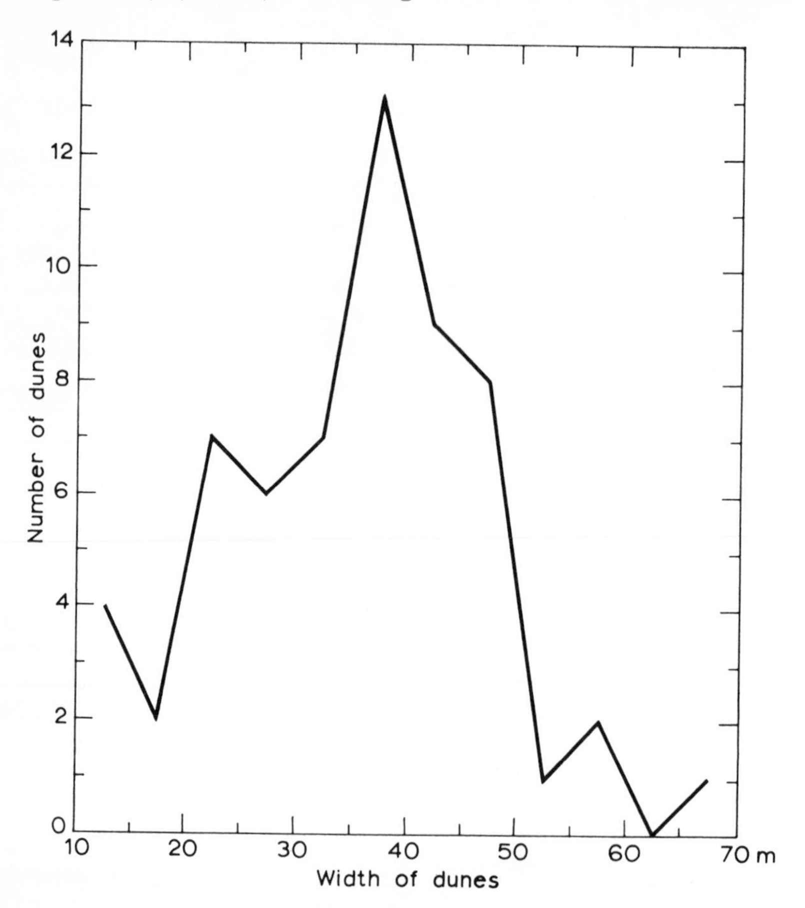

4.30 Widths of barchans in Peru (data from Hastenrath, 1967).

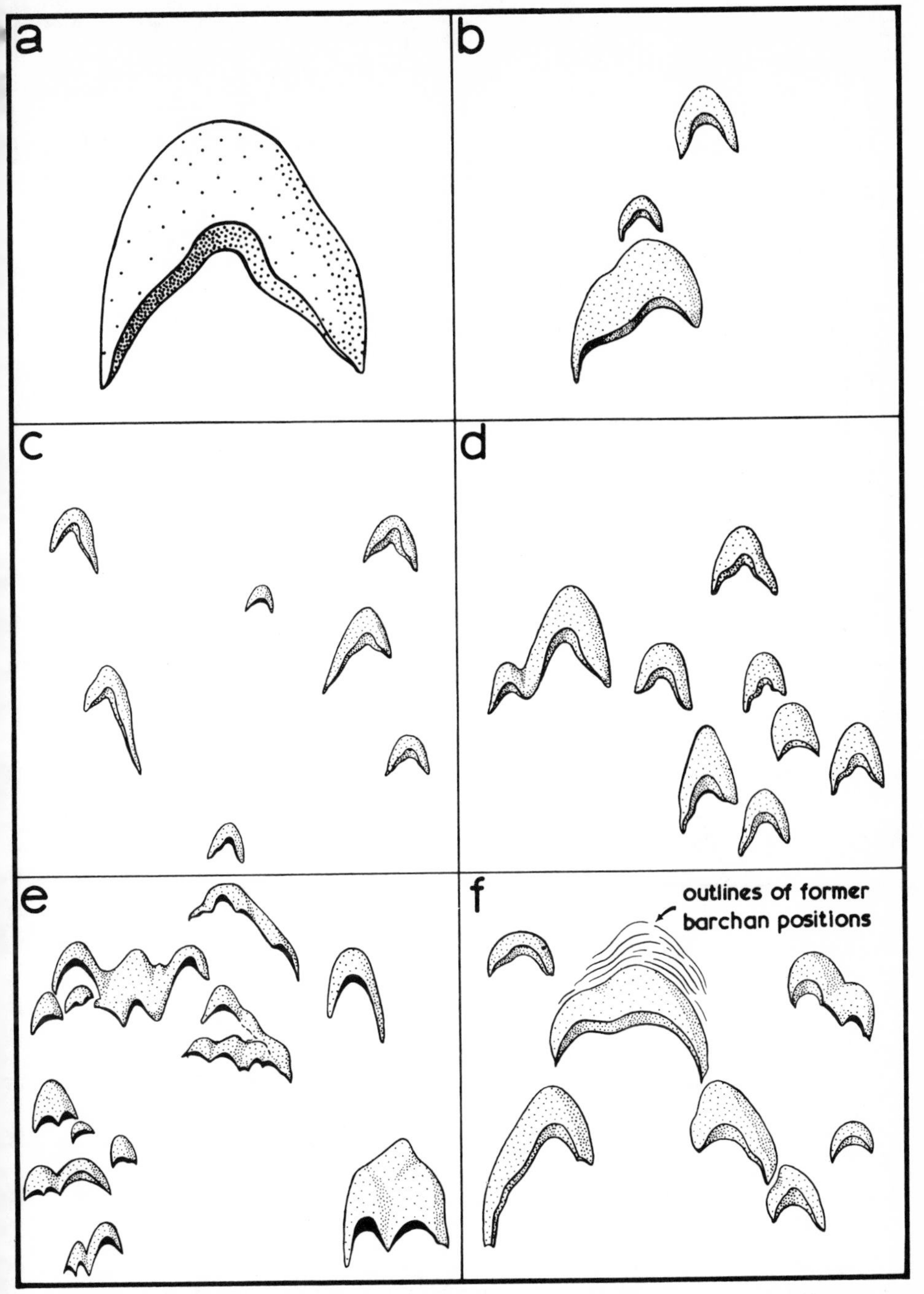

4.31 Barchan types: (*a*) barchan with incipient development of both oblique elements; (*b*) barchans in the same area as (*a*) in which there is development of the left oblique element; (*c, d, e,* and *f*) the development of different oblique elements. (*a, b* and *e*) after Capot-Rey (1957*a*); (*c* and *d*) after Clos-Arceduc (1967*b*), and (*f*) after Kerr and Nigra (1952).

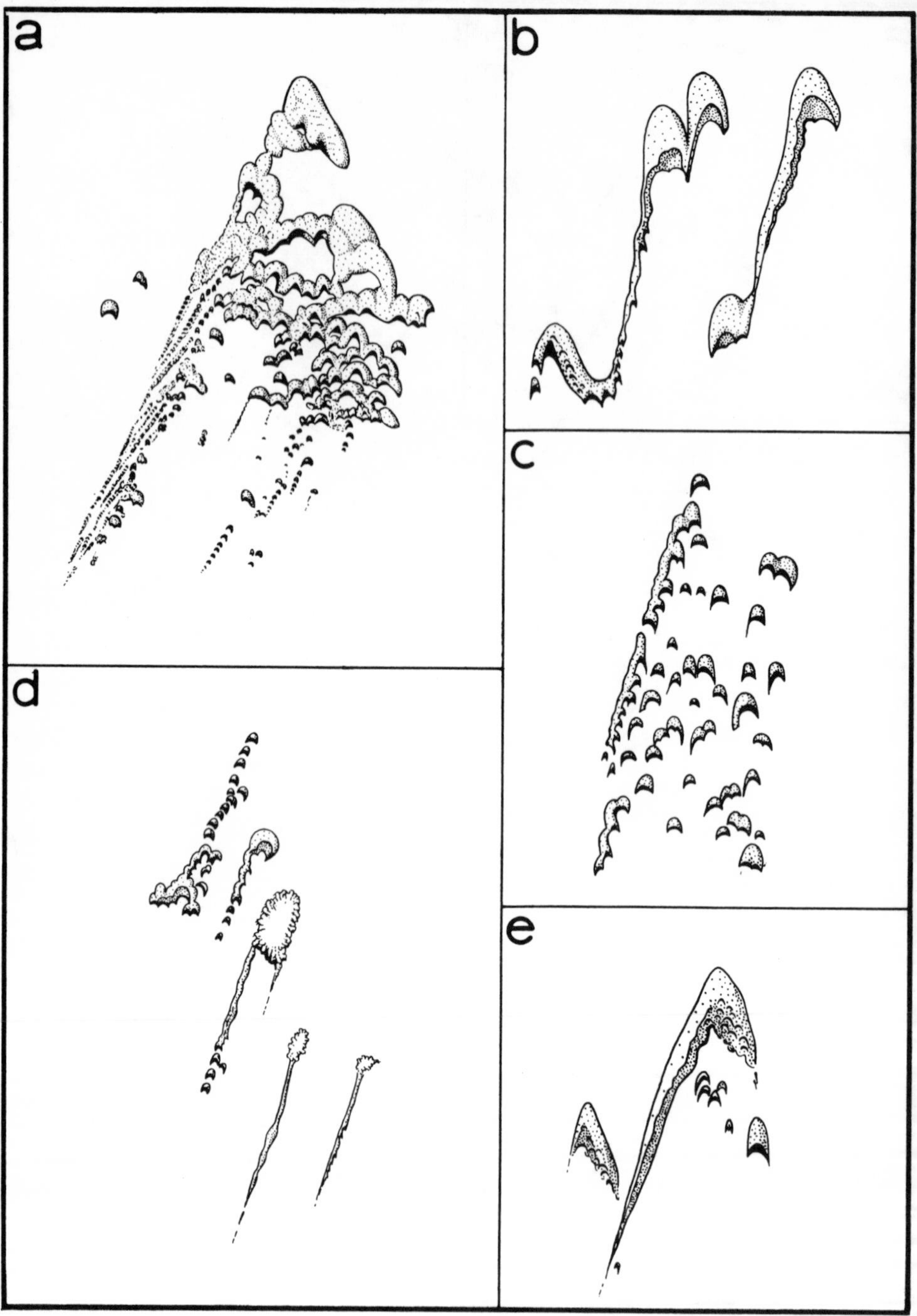

4.32 Barchan and lee-dune patterns: (*a*) From near Bilma, Niger; the oblique element on the left of the diagram is emphasized; (*b*, *c* and *e*) the development of different oblique elements in barchans; (*d*) barchan oblique elements aligned with lee dunes, (*b*, *c*, *d* and *e*) after Clos-Arceduc, 1967b).

adjacent arms elongated. These patterns are more easily explained with a knowledge of common patterns of secondary flow in the wind which will be discussed below.

Further evidence for regular 'three-dimensional' secondary flows comes from the long almost straight dunes that are common in many deserts. These dunes occur in regular patterns, and they can be distinguished from transverse dunes by their straightness; often it is only the summits of the ridges, and not their bodies, which exhibit regular sinuosity. Since they often appear to be nearly parallel to the dominant wind they are known as longitudinal dunes (Plate 4.4, Fig. 4.33). The early literature on the great sand seas was confused in its interpretation of these dunes. Madigan (1946) cited the accounts of the first travellers in the Simpson Desert in Australia which describe the ridges as transverse, and some of the travellers in the Western Desert of Egypt were of the same opinion (de Lancey Forth, 1930). In opposition to this a later group of authorities saw the ridges as longitudinal, and moreover thought that only such ridges could survive at equilibrium with the wind (Aufrère, 1928; Bagnold, 1941; Beadnell, 1934). We now know that longitudinal, transverse and oblique dunes are all found widely in nature. However longitudinal dunes are very common and cover large areas of some deserts such as the Rub al Khali (Glennie, 1970), the western Sahara, the Libyan Desert (Bagnold, 1933), and the Australian deserts (Folk, 1971).

Longitudinal dunes are more obviously connected with regular flow-parallel variations in windspeed or with three-dimensional vortex flow patterns than are transverse patterns, and many authorities have postulated just such a connection (e.g. Bagnold, 1953*a*; Clarke and Priestley, 1970; Folk, 1971*a*; Madigan *et al.*, 1969). The hypothesis is illustrated in Fig. 4.35a. There is much circumstantial evidence from the dune patterns themselves to support the proposal: in Saudi Arabia where the longitudinal features are of draa size smaller siefs are aligned diagonally across the sides of the major ridges at an angle which would conform to the secondary flow lines illustrated in Fig. 4.29 (Glennie, 1970); and in the Simpson Desert not only these minor features but also the Y-junctions exhibit the 30–50° angle to the main flow that one might expect of the secondary flow (Folk, 1971*b*). Closely analagous forms are found in subaqueous bedforms at many scales (Allen, 1968; Dzulynski and Walton, 1965; Fig. 4.33), and Folk (1971*b*) produced very acceptable reproductions of longitudinal patterns, complete with Y-junctions, by passing rollers across grease and acrylic paint. The evidence of the dunes themselves has been augmented with further meteorological evidence by Hanna (1969).

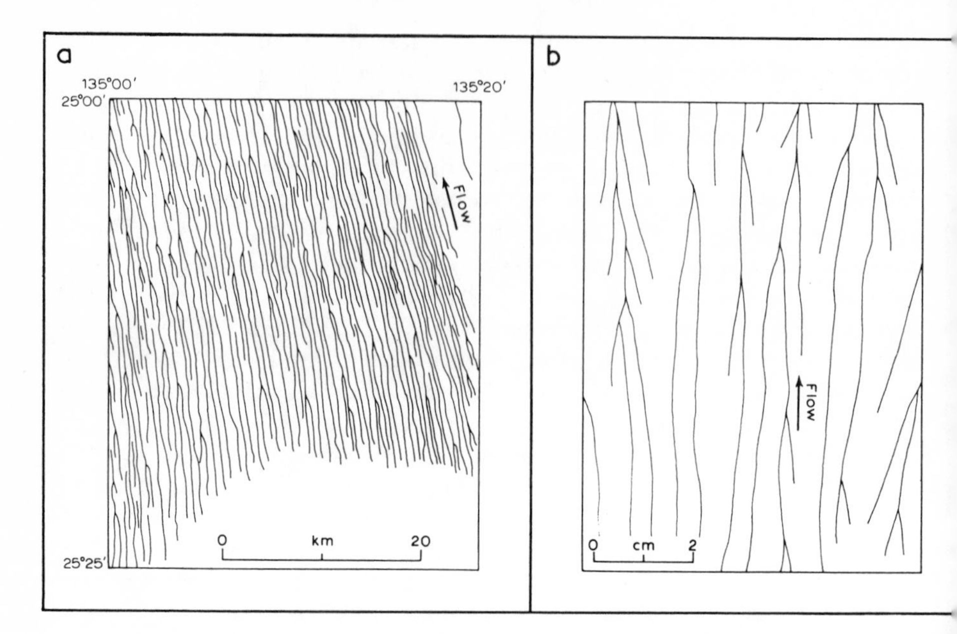

4.33 Longitudinal bedforms: (*a*) Dunes in the Simpson Desert, Australia (afte Folk, 1971); (*b*) marks in the Oligocene Krosno beds, Poland (after Dzulynski and Walton, 1965).

It is important to note that the longitudinal 'three-dimensional' flow conforms to the same hierarchical pattern as two-dimensional flow (Wilson, 1972*a*). There are, in other words, vortical flows of aerodynamic-ripple, dune and draa scale. With impact ripples it has been suggested that longitudinal vortices are generated in concave ripple troughs during saltation (Ellwood, Evans and Wilson, 1972).

If dune patterns were made up of simple combinations of transverse and longitudinal elements, they would be simple rectangular reticules as in Plates 4.14, 15, and 16. However we have seen that complex patterns are more common. Plate 4.17 is an illustration of a complex pattern of crossing dune ridges from the Great Eastern Erg in Algeria. It can be seen that there is an apparently transverse system of low dunes with steeper slopes facing to the bottom left of the picture, upon which are superimposed two sief systems at an oblique angle to each other and to the draa ridges. Most of the forms appear fresh and active, so that it is difficult to attribute this kind of pattern to ancient wind patterns as did Capot-Rey (1947) in his discussion of a similar pattern in the Idehan Marzuq, in Libya. Nor can it be explained by ascribing the siefs to resultants, as did McKee and Tibbitts (1964) in their studies of siefs in an area not far to the east of this area in Libya, since one cannot have more than one resultant in the same wind pattern. Plate 4.10 shows that crossing patterns also occur in ripple-sized features.

Similar contradictions between the idea of simple transverse and longitudinal forms, actual patterns on the ground, and the wind regime have been noted in several areas. In the Simpson Desert in Australia, Madigan (1946) noted that the ridges often swing away from the wind. Mabbutt (1968) and more precisely Brookfield (1970) have made recent and more accurate observations of deviations in central Australia; in the Algodones dunes in southern California, Norris and Norris (1961) observed that the mega-barchans appeared to be oblique to the wind whereas the smaller barchans were more transverse to it; Hedin (1904) observed that some of his draa ridges appeared to be oblique to the prevailing wind (Fig. 4.34).

An explanation of the displacement of barchanoid and linguoid elements in aklé, of the odd elongation of barchan arms, and of crossing patterns in sand seas, can be made by considering the actual patterns of combined transverse and longitudinal secondary flow, and the ways in which these interact with the bedform itself. Wilson (1970) has produced a number of diagrams which show this kind of interaction, some of which are illustrated in Fig. 4.35. Allen (1969*b*) produced similar diagrams to explain some patterns in subaqueous bedforms.

Simple combinations of transverse and longitudinal flow elements

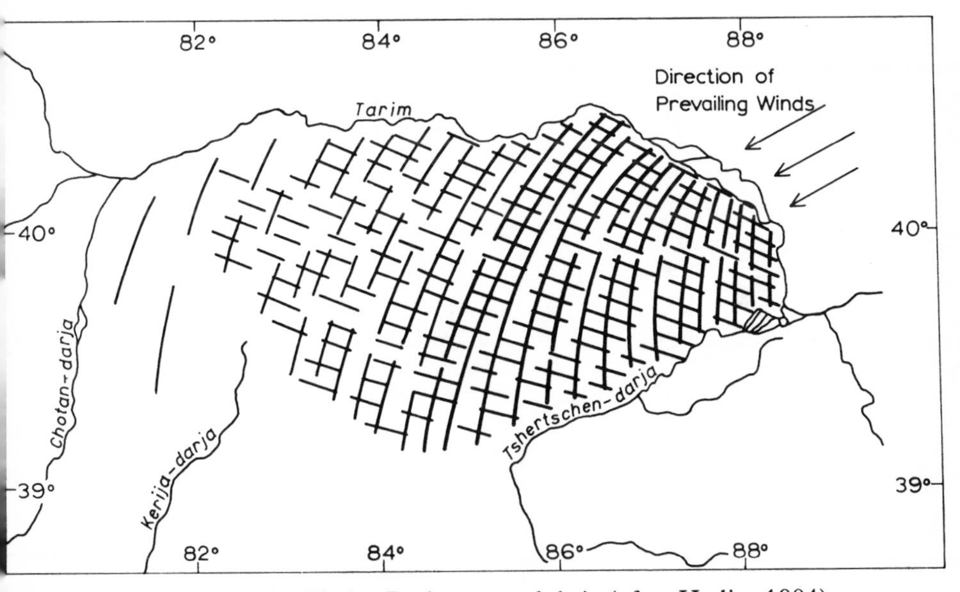

4.34 *Draa* patterns in the Tarim Basin, central Asia (after Hedin, 1904).

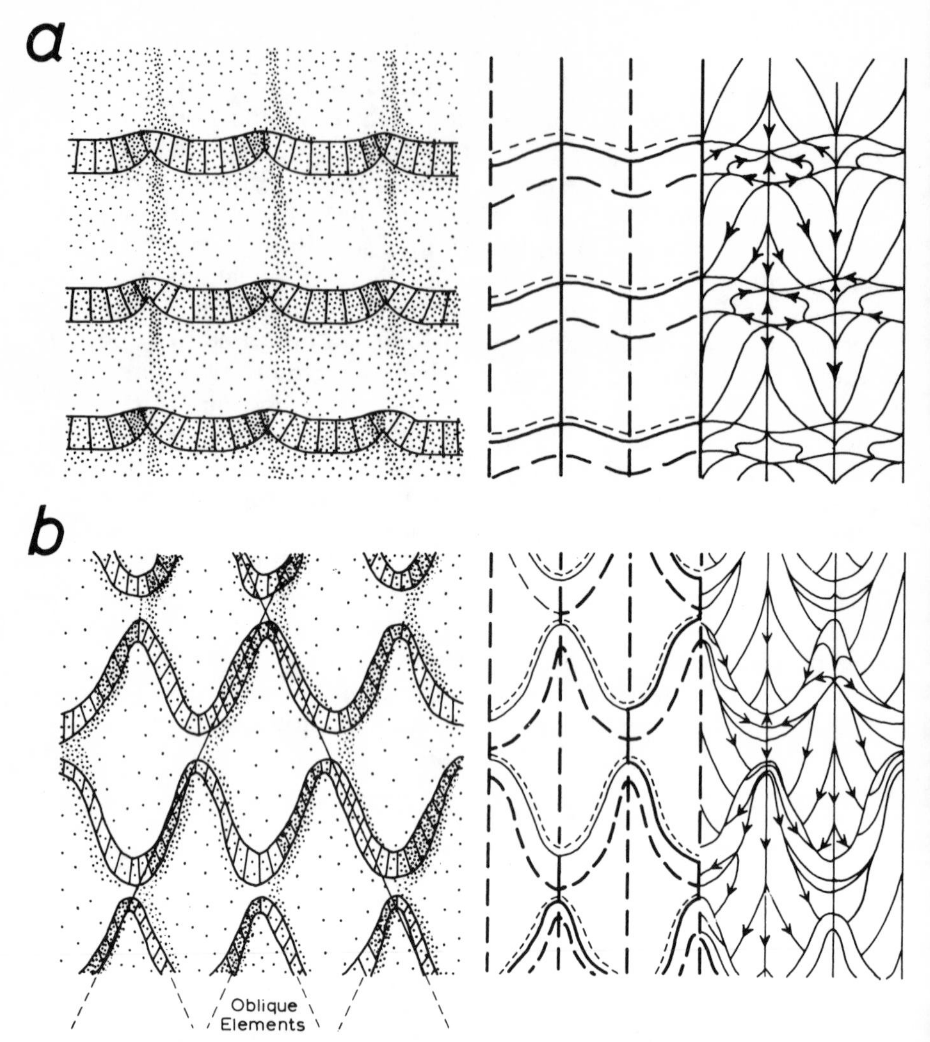

4.35 Dune patterns (aklé) resulting from combined 'wave' (two-dimensional) and vortex (three-dimensional) flow patterns: (*a*) with 'in-phase' vortices; (*b*) with laterally displaced vortices (after Wilson, 1970).

including return flow of the lee eddy type can be seen in Fig. 4.35a. This gives a regular rectangular reticule (Plates 4.14, 15 and 16). However, Allen (1968*b*) pointed out that this form is unusual in subaqueous ripple patterns (as it is among dunes) because of secondary patterns of return flow in the dune corridors. It is more usual to find that the longitudinal elements are displaced sideways by half a wavelength.

This type of flow was noticed by Bucher (1919) over water ripples. It produces patterns as in Fig. 4.35b. When this kind of flow pattern is well developed, the linguiform elements are pushed well forward. This conforms to our description of typical aklé. The details of the flow pattern can be seen in Fig. 4.36.

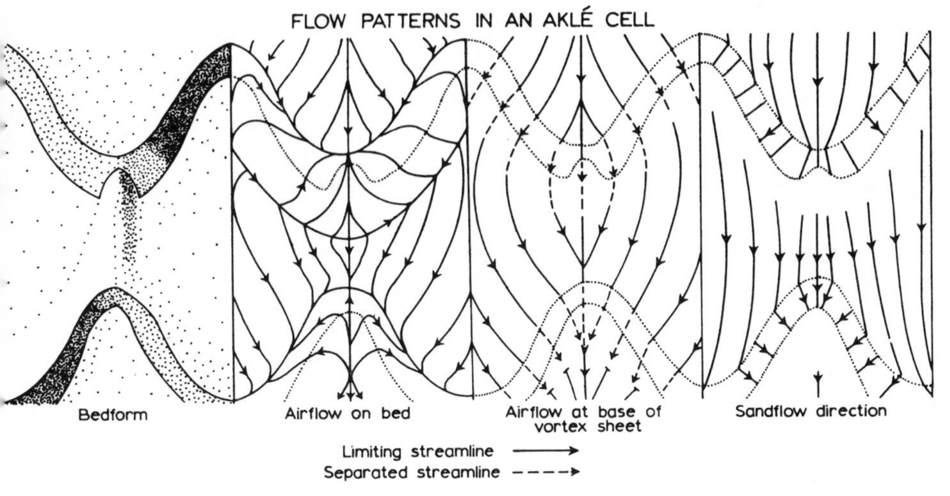

4.36 Details of flow patterns over aklé ridges (after Wilson, 1970).

The result of this kind of flow is to produce dune-ridge elements that are oblique to the flow. Wilson (1970) has called these *en echelon* elements because of their origin in the *en echelon* offsetting of barchanoid and linguiform elements. (They have been emphasized in Fig. 4.35b). Their angle to the main flow can be defined by $\tan \theta = {}^{l}/_{2}\, \lambda$ where l is the separation distance of the longitudinal flow elements and λ is the transverse wavelength. If these were equal, the value should be in the order of 26 degrees. Actual oblique elements observed from the field have angles of 10° (Sevenet, 1943); 20° (Glennie, 1970) and an often repeated figure of about 18° to the prevailing wind direction (Warren, 1970). Experiments with rolled grease and observations in the Simpson Desert have however produced junctions angles of 30 to 50° (Folk, 1971*b*). These data may point to slightly different longitudinal and transverse wavelengths. Some convincing examples of oblique patterns can be seen in Fig. 4.32 b, c and d. The oblique alignment in Fig. 4.32 are continued downwind. Oblique elements tend to be preserved in bedform patterns because of the development of a spiral lee eddy at the base of the slip face, as has been explained by Allen (e.g. 1970). These kinds of flow can be used to explain dune meanders and dentintric dune patterns which are not uncommon in nature (Folk, 1971*b*; Wilson, 1972*a*). Goudie (1969) illustrated dune junctions and meanders in the Kalahari.

The questions posed by barchan geometry can also be explained with these flow patterns. In the first place, the barchan is probably initiated in one of the convergent nodes in the flow pattern. Since these are regularly spaced it will tend to have a regular size. Barchans with one arm longer than the other could develop almost by chance. Once one arm had tended to grow it might extend downwind oblique to the flow, as in Fig. 4.32 b and c. Its growth would be maintained by the spiral eddy in the lee and by the fact that it would trap most of the sand coming from upwind. In the western Sahara, Clos-Arceduc (1967*b*) found that sometimes one *en echelon* direction and sometimes another was emphasized. Some of his examples are also shown on Fig. 4.32. Barchans in which both horns are elongated, as in Fig. 4.31, can be explained as having developed both of the oblique elements, and probably occur in very regular unidirectional flow (Wilson, 1970).

Longitudinal dunes are more obviously due to vortex flow. It is interesting that in subaqueous longitudinal patterns, Y-junctions very similar to the Y-junctions in the Simpson Desert are quite usual (*cf.* Allen, 1969 and Fig. 4.33).

Oblique draa patterns in ergs can be seen as the intersection of two or three of the elements of a dune pattern—longitudinal, transverse or left- or right-oblique. The emphasis of one or two elements at the expense of others is probably, however, the result of the selection and emphasis of one of the elements by a cross wind. It is to the more common case of multidirectional winds that we now turn.

(c) Influence of multidirectional winds. Most deserts experience wind regimes that have some degree of variation in direction. The degree of variation from unidirectionality that effects distinct changes in dune patterns is not easy to decide and observations of the degree of variation from one direction that can effect a change in aeolian bedform patterns have been very few. Sharp (1963) observed that winds which varied no more than 20 degrees about a mean direction could be regarded as unidirectional in regard to ripple patterns. This might be a reasonable first approximation in discussions of dune patterns.

It has often been very reasonably assumed that the main components of a pattern in any area are aligned to a 'prevailing' or 'predominant' wind while minor winds lead only to asymmetry of the form. This appears to be the case in the Simpson Desert in Australia where, in any one part, steeper slopes always face in the same direction (e.g.

Folk, 1971). Bagnold (1941) made use of his findings about strong and gentle winds and these ideas about dominant and cross winds to evolve an explanation of *sief* dunes. Sief (meaning 'sword' in Arabic) is a term which is used by Arabs simply to describe the long, sharp, sinuous crestlines that are common in the desert. It is now used in dune literature to mean a dune that is generally straight and very elongated which has a sharp crest (Plate 4.20). The upwind end of the sief is always a low, rounded mound whereas the downwind end is sharp. The sinuosities along the summit are regularly spaced. The slopes are of the order of 20 degrees on either side but they may be topped by an active slip face.

Bagnold's line of reasoning showed that only strong winds could build up a sand mound on a desert-pavement surface (*cf.* section 4.3.2. *b.i*). These winds would build up a barchan-like dune. When, in a regular seasonal cycle, the wind changed to become a gentle but persistent wind, the dune would be elongated in the direction of this second wind. Repeated cycles would create a long dune aligned with the persistant wind. Bagnold later (1951 and 1953) proposed that siefs, and perhaps the draa chains in the Arabian deserts, would be aligned, not with the persistent wind, but with the vector sum or resultant of all sand-moving winds in an area.

There is a large body of evidence to suggest that sief chains are very similar in trend to the resultant wind. Madigan (1946) found that although his wind data were sparse, dune trends in the Simpson Desert corresponded to the resultant of the winds. Wopfner and Twidale (1967) illustrated the effects of cross winds on the upwind end of a sief dune in the same area; the result of two cross winds appeared to be the extension of the sief in a resultant-wind direction. Cooper's (1958) detailed study of free dunes on the coasts of Oregon and Washington showed that some large 'oblique' ridges inland from the coast were aligned with a resultant of the summer and winter winds (Fig. 4.37). In the Tibesti area, Durand de Corbiac (1958) observed that 'longitudinal' dunes appeared to have a trend that was independent of barchan, lee-dune and yardang trends; the trends of the longitudinal dunes followed one pattern of deflection around the massif, and the other dune trends followed another deflection pattern. He proposed that strong winds alone were responsible for the second group, and resultants could account for the trends of the 'longitudinal' dunes. In the same area, Wilson (1970) has also shown that the winds are deflected around the massif, and that where the two winds reunite on the far side of the mountains, sief dunes appear, evidently in a direction which is the resultant of two winds. In Libya, McKee and Tibbetts

(1964) noted that the sief dunes which they examined could be seen as being aligned parallel to the resultants of morning and evening winds. In Israel, Striem (1954) found that he could explain sief trends by calculating the directions and amounts of movements of dunes as induced by summer and winter winds; the sief trend was aligned with the resultant. In Australia, Brookfield (1970), and Clarke and Priestley (1970) have suggested asymmetry in wind regimes as the cause of sief-like dunes with marked regional asymmetry. Using resultant trend patterns of modern winds and the trends of sief-like dunes among others, Warren (1970) was able to present a coherent picture of climatic changes in a fixed dune field in the Sudan. In this area, as around Tibesti, the independent deflection of two trends by a topographic feature (Jebel el 'Ain) can be used as an argument that two winds are involved in forming the dune pattern (Fig. 4.38).

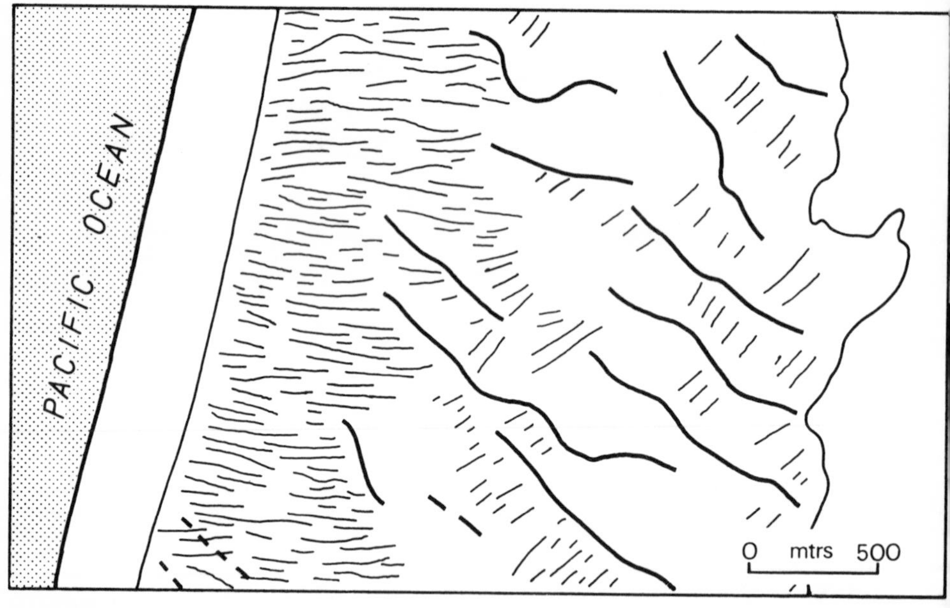

4.37 Patterns of transverse and oblique dunes on the Oregon Coast (after Cooper, 1958).

We have already pointed out some of the objections to the universal use of the resultant in explaining sief-dune trends. For example, if two sief trends are found in the same area, the hypothesis is clearly untenable. To these objections we can add some evidence that sief trends are not always aligned with the resultant. For example Warren (1971)

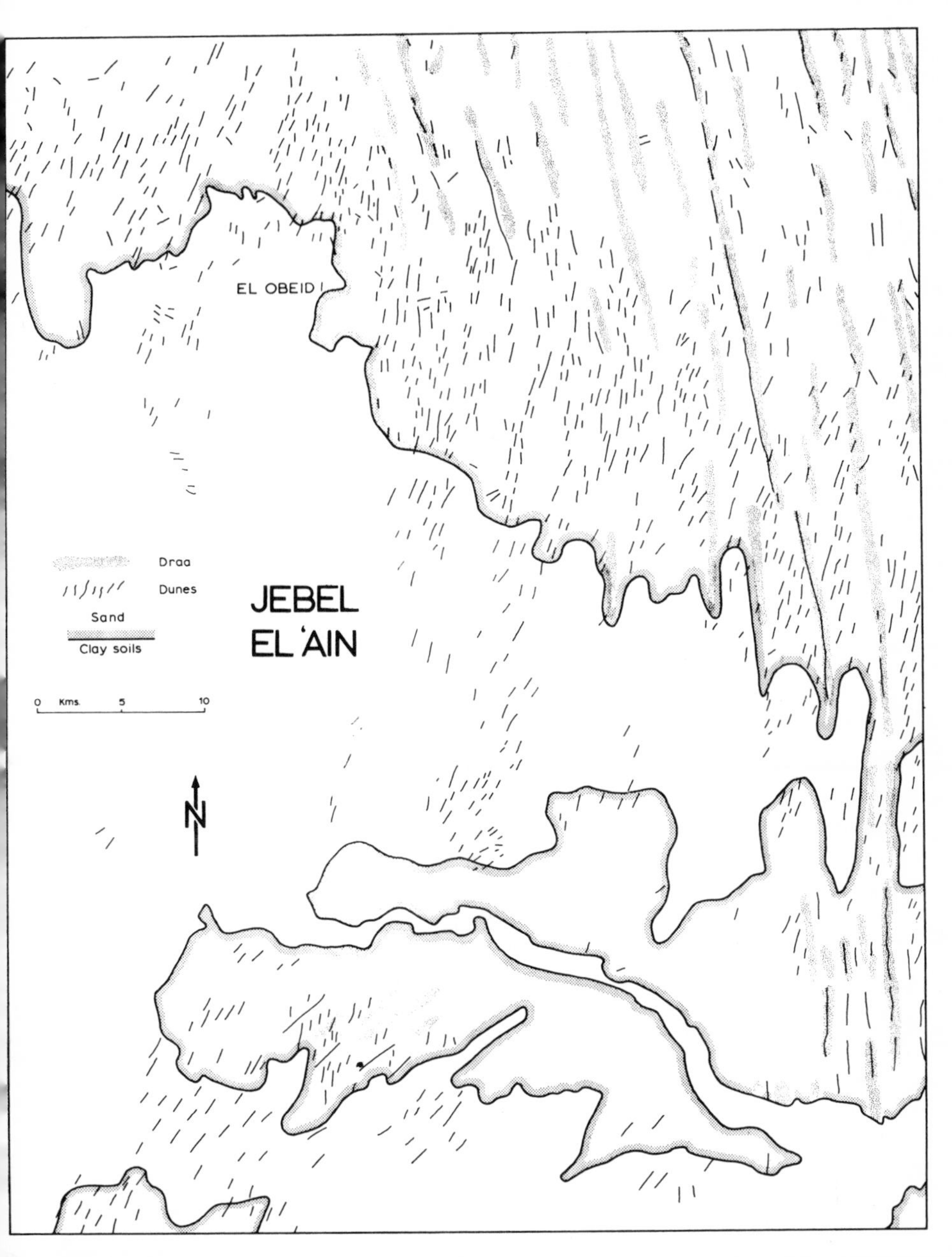

4.38 Deflection of *draa* and dunes in different patterns around the Jebel el 'Ain Dome, central Sudan (after Warren, 1966).

working in the Ténéré Desert in Niger, found that sief trends were distinctly at variance with the resultant trend of the appropriate winds; and Mabbutt (1968) and Brookfield (1970) have pointed to cases where there are distinctly oblique alignments in the Australian deserts.

Although some siefs may be explicable using the 'resultant' wind it seems more likely that many siefs, particularly in areas where there is complete sand cover, are oblique elements of a pattern created by a dominant wind that have been emphasized by a cross wind. Fig. 4.39, taken from Wilson (1970), shows some ways in which this might occur, both in incomplete sand cover and in complete sand cover. These explanations apply to both dune-sized and draa-sized features. The diagrams can be used to explain some of the crossing patterns that have been noticed in the main sand seas. Plate 4.19 shows the selection of one barchan arm by cross winds in Niger.

In parts of the open Simpson Desert (Fig. 4.24) there are zones where crossing patterns suddenly appear, sometimes facing one direction and sometimes another. Wilson has suggested that these might be in the nature of a 'phase change' i.e. a sudden change in the pattern in which one of the obliques is selected, in response to a rather gradual change in grain size or in the wind pattern (Wilson, 1970). In the Saharan patterns illustrated in Fig. 4.26 there can be seen the more gradual extinction of one pattern and its replacement with another, still composed of some of the same elements.

This explanation involves the selective filtering of the basic elements of a pattern of cross winds but crossing elements have been explained by other workers in many different ways. One explanation has been that each set of dune trends was adjusted to one wind in a complex wind regime. One wind would build up one set of transverse or longitudinal dunes, whereas another would build up a completely distinct set. This has been proposed for the central Sahara by Aufrère (1935), for central Australia by Mabbutt (1968) and for some parts of the Kelso dune pattern in southern California by Sharp (1966). We examine other explanations in the following sections.

Crossing patterns involve the creation of nodes where two trends come together. In the Great Eastern Erg of the Sahara, and in parts of the Arabian sand seas, these nodes in draa patterns have been developed into very large and very spectacular 'sand mountains' or *rhourds* (Plate 4.3). It can be shown that rhourds are usually part of a pattern of crossing trends since they merge with more distinctly aligned elements in other parts of the same sand sea (Wilson, 1972*a*). However, other explanations have been offered for these features. Cornish (1914) and Clos-Arceduc (1966) proposed that rhourds originated at nodes between heated convection cells. As the wind swept in towards the centre of a rising plume of warm air, it would bring in sand from the surrounding area and this would accumulate. Clos-Arceduc simulated the process with standing waves in a tank. He was also able to find what appeared to be aeolian erosion marks on air photographs at the same scale, and these he attributed to the same process. It is doubtful, however, if the secondary flows generated in convection cells could be

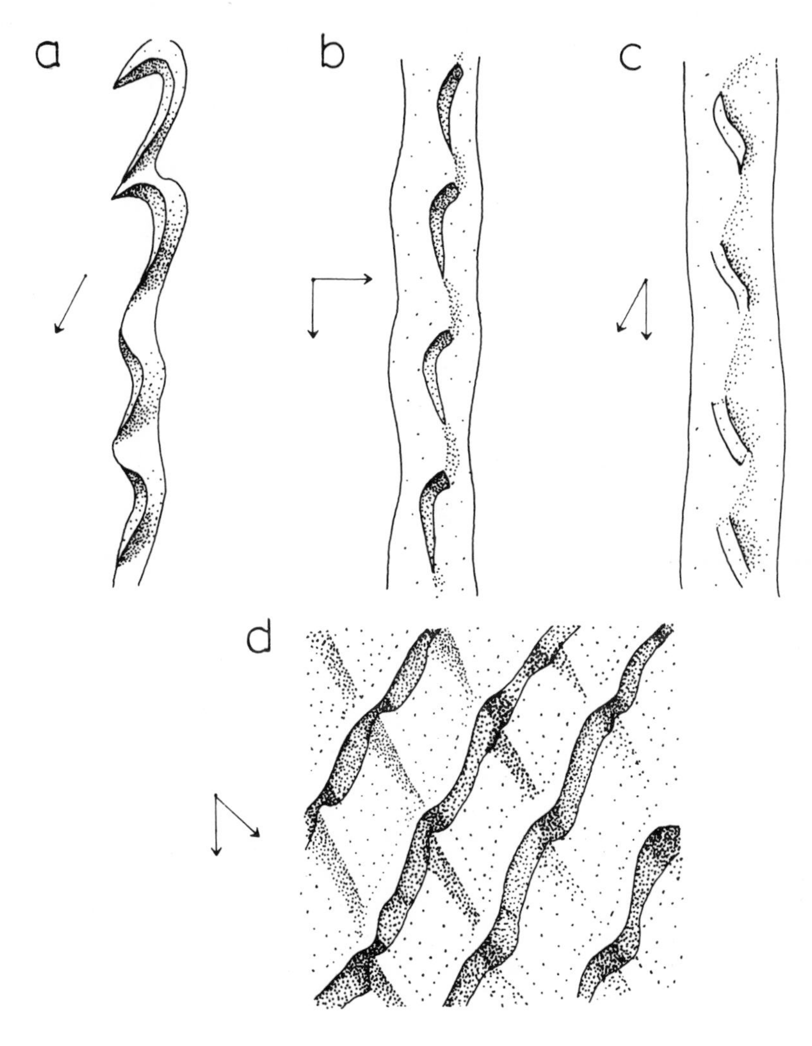

4.39 Sief and aklé patterns resulting from variable wind patterns (after Wilson, 1970).

strong enough to effect this kind of movement. It has been more common to propose that the sand mountains are the result of winds from several directions, which blow the sand back and forth (e.g. Aufrère, 1935; Bagnold, 1953*a*; Brown 1960; Capot-Rey, 1943; Holm, 1957; Smith, 1963; Suslov, 1961). McKee (1966) found some evidence to support this theory in bedding patterns on top of dunes in Arabia. Apart from the fact that evidence such as presented for the Sahara by Wilson (1971*b*; Fig. 4.3) shows that there are continuous sand flows across all the ergs in which these forms occur, it seems unlikely that large uniformly-thick areas of sand would have existed over extensive areas ready for moulding by the wind into these features.

Cross winds undoubtedly effect minor modifications on many dune forms. Several workers in the Simpson Desert, for example, have pointed out how cross winds create asymmetry in the cross sections of sief dunes. Complete reversal of winds is quite a common occurrence and its effects have been noticed by many observers. Usually the effects are minor (e.g. Hastenrath, 1967). Sometimes reversing winds mean that there is very little net movement of a dune in any direction , in spite of the movement back and forth of enormous amounts of sand. Sharp (1966) has observed this kind of process in detail in the Kelso dunes in southern California, as have Lettau and Lettau (in press) in Peru. In the Sahara many minor dune forms are completely reversed from season to season. This is very obvious in the El Golea area in Algeria, where small barchans are completely reversed between winter and summer.

(d) Grain-size: processes and effects. Sands are sorted in aeolian transport. The wind selects certain sizes of sand from pre-existing mixtures, carries them in a number of ways, and lays them down in distinctly different types of deposit. Whether aeolian sorting produces sediments that are easily distinguishable from sediments laid down by other processes does not concern us here, since we are concerned merely with the interacting dune-sediment system.

A variety of sand-size characteristics have been reported from dunes, but most aeolian sediments can be seen as part of a continuum between two distinctive types. The best known of these is the very well sorted, positively skewed mixture: one in which there is a very limited range of sizes with a particularly sharp cut-off towards the coarse grains, and a tail of finer material (Fig. 4.40). Another very common sand found in dune areas is bimodal: a coarse mode (often about 0·6 mm) is well separated from a much finer mode at about 0·02 mm (e.g. Fig. 4.41, curve 1; Bagnold, 1941; Folk, 1971*a*; Warren, 1971). In areas in which

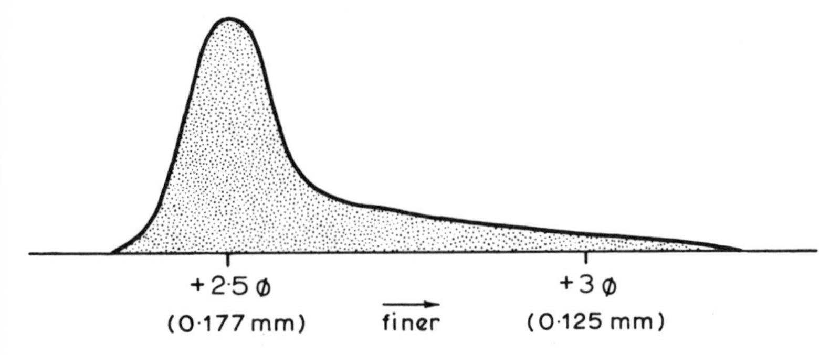

4.40 Typical size characteristics of common aeolian sands.

these two types of sand are found close together the mode of the well sorted sand fills the gap between the two modes of the bimodal sand (e.g. Folk, 1971*a*; Warren, 1972). This is illustrated for some sands from the Ténéré Desert in Niger in Fig. 4.41. It is reasonable to suppose that these two sands are the result of an aeolian sorting process, and were therefore once combined in a unimodal pre-existing sand.

The sorting processes that produce these two sands can be explained if we examine the ways in which sand is transported by the wind. As a pre-existing sediment comes into aeolian transport, fine material is moved away quickly in suspension although some may settle back in low areas between the dunes, as has been observed in some areas (e.g. Madigan, 1946; Simonett, 1951). In the remaining sand the finer material will be moved as saltation load, and the coarser as creep load. The one will move faster than the other so that they will separate in transport. In this process, however, sands of the finer grades will be able to penetrate in between the coarser grains, but slightly bigger grains will be too large to fit into the voids between the coarse grains and will therefore move on more quickly. The creep load becomes a bimodal mixture of the coarse and fine sands. This process has been used to explain bimodal sands in the Ténéré Desert by Warren (1972). A similar process was suggested by Verlaque (1958) to explain bimodal sands on barchan arms and Sharp suggested interparticle penetration by fine sands to explain bimodal sands in ripples. Folk (1968 and 1971*a*) has suggested that the finer sands are less mobile because of their cohesion and smaller surface roughness, but this would not be the case where they are found in an original mixture of several grain sizes. Only where particles are fine enough to cohere as structure-units (when they are of clay size) could we expect an increased resistance to wind erosion. Indeed Chepil (1957) found that clay was often more common than

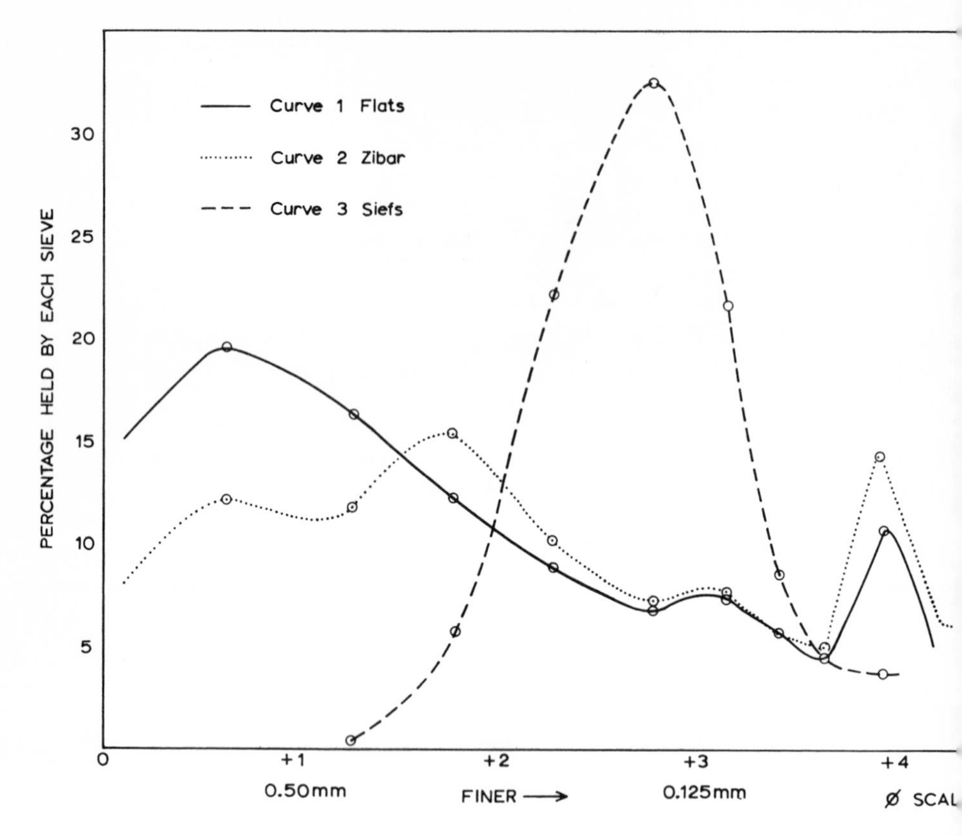

4.41 Sands from the three different environments shown in Plate 4.20. It appears that the intermediate grade sand has been sorted out to leave bimodal sands in the flat and zibar areas.

silt in dunes and drifts downwind of eroded fields because the clay moved as aggregates. The clay aggregates being less dense than sand grains were larger than the latter in the dunes. Wood (1970) has suggested a fractionation process to explain bimodal sands, by analogy with differentiation processes in chemical columns.

In the Ténéré Desert in Niger, this sorting process is associated with distinct dune patterns (Warren, 1971 and 1972). Long-continued removal of the intermediate grain-sizes leaves a sand in which there is a prominent coarse mode (Fig. 4.41). Bagnold (1941) observed that when the coarse mode becomes very prominent it seals off the surface and no dunes can form. This is indeed the case in the Ténéré (Plate 4.18) where flat areas with very distinctly bimodal sands are found downwind of most sief dunes. It is also true in many other parts of the world (e.g. Folk, 1968 and 1971*a*; Lewis, 1936; Madigan, 1945).

Downwind of the flat areas in the Ténéré low dunes without slip faces appear where the sands have a slightly less prominent coarse mode (Fig. 4.41, curve 2). These low dunes have often been described in the corridors between higher dunes (e.g. Alimen, 1958; Bagnold, 1931 and 1933; Capot-Rey, 1947; Capot-Rey and Grémion, 1965; McKee and Tibbetts, 1964; Monod, 1958; Plate 4.14). They are known in Saudi Arabia as *zibar* (Holm, 1953). Most observers have noted that they are associated with sands having many coarse grains. The zibar are lower and have longer wavelengths than dunes of unimodal intermediate-sized sand. This is because the coarse grains seal off the surface so that only strong winds can build dunes, and these winds will produce longer wavelengths than gentler winds. Upward growth may be inhibited because the sand supply from upwind will be moved very rapidly over these dunes (Bagnold, 1941), but it may also be the case that the sharper velocity gradient of strong winds will put an effective limit on upward growth. Since the zibar are low they are never able to induce flow separation and no slip face can form.

In the Ténéré the sands of intermediate size are sorted out from the bimodal sands and are carried downwind to where they accumulate on high sief dunes (Fig. 4.41, curve 3 and Plate 4.18).

It is a general rule that grain-size affects dune height and spacing. Fig. 4.2 is a plot of grain-size against the wavelengths of ripple, dune and draa features. It is probably true that quite a small percentage of large grains can seal off a dune surface and act as an inhibiter to the effects of gentle winds, and so lead to greater bedform spacings (Wilson, 1972*a*).

It is often the case that the stronger winds in an area come from one particular direction, so that one would expect the coarser sand to be built into dunes aligned with these winds. Finer sands are moved by gentler winds as well as the strong winds, and since the directions of this whole group of winds is usually more varied than that of the strong winds alone, the finer sands will be blown first in one direction and then another, and one might expect dunes formed of fine sand to be aligned with the resultant trend. The coarser sands would probably form an aklé pattern since, for them, the winds are effectively unidirectional, whereas the finer sands would form siefs which might well be at an oblique angle to the trend of the transverse dunes. This appears to have occurred in the central Sudan (Warren, 1970). On the other hand the siefs in the Ténéré (described above) are oblique both to the direction of the stronger winds and to the direction of the resultant of all the winds capable of moving sief sand (Warren, 1971).

Apart from the association of coarse sands with zibar, the evidence for there being distinct sand types in different dune types is meagre. Bellair (1953) and Alimen *et al.* (1958) found quite different pictures in their investigations in the northern Sahara, and Capot-Rey and Grémion (1965) found no distinct patterns. On a larger scale it has been suggested that the rather distinct morphology of the Australian sand seas may be due to their fine sands (Madigan, 1946).

Once on a dune, the sand is subject to a further group of sorting mechanisms. On barchans it has often been observed that coarse grains accumulate at the foot of the windward slope and are led round to the arms rather than up on to the back (e.g. Amstutz and Chico, 1958; Finkel, 1959; Hastenrath, 1967; Sidwell and Tanner, 1938; Simonett, 1960; Tricart and Mainguet, 1965; Verlaque, 1958). Coarse grains also accumulate around the base of other kinds of dune. This can be explained in the following way. There is divergence of sandflow from windflow when a sand-carrying wind impinges on a slope which is oblique to the wind. The formula for the angle of this divergence has been given by Wilson (1972*a*):

$$\tan \psi = \frac{R \sin 2\theta \sin \phi}{1 - R \sin 2\theta \sin \phi} \quad \ldots \quad (4.41),$$

where θ is the slope angle, ϕ is the angle between the slope and the wind direction, and R is a coefficient depending on the sorting of the material (this affects the perfection of grain bounce). With more poorly sorted material in which the range of sand sizes is greater, R increases and there is greater deflection. Coarse grains are therefore more readily deflected round the base of the slope than fine ones. This applies only to the saltating grains. Coarse grains travelling as creep are difficult to move up the slope because of the influence of gravity and because of the lower bombarding power of the saltating grains on a slope arising from shortened flight-paths. Coarse-particle creep is therefore also deflected around the base of the slope. Hastenrath (1967) noted that because of these sorting mechanisms, small dunes were generally composed of coarser sands, and this may be one reason for the discrepancy between the calculated and actual rates of movement of small barchans, as described by Finkel (1959). Incidentally we can note that the coarseness of the sand on barchan arms which is a corollary of this process will be yet another feature that will slow down their movement. These kinds of sorting mechanism at the foot of a dune apply to most draa as well as dunes. It is almost always the case that large draa are found to rest on a base of coarse grains (e.g. Alimen, 1953; Bagnold, 1941).

Once the sands have passed through these processes at the base of the slope they are subject to even further sorting on the dune itself. However the sorting mechanisms are probably different on different kinds of dune. Some workers have found crests that are coarser than flanks and some have found finer crests (the literature is reviewed by Folk, 1971*a*). Whatever the relative median size of crest sands they are almost always the best sorted. This is sometimes helped by the reversal or reworking of the crests by many different winds. In aklé it may be the case that the funnelling of coarse grains along the foot of the slope of barchanoid elements leads them on to the linguiform elements, so that barchan crests are relatively fine and linguiform crests are relatively coarse. This process may slow down the advance of the linguiform crests and help to maintain the aklé pattern.

Further sorting processes take place on the slip face. There are two opposing tendencies here. The projection of the saltation and creep load over the brink of the dune (Plate 4.10) means that coarse grains accumulate at the top of the slope and finer ones lower down. However, in the slipping process coarse grains are brought to the surface (Bagnold, 1954), and will then tend to roll to the bottom more readily. These two processes may cancel themselves out.

Hastenrath (1967) has described a process that may oppose these sorting processes. Barchan slip faces face inwards, so that the coarse grains channelled round the sides are moved in towards the middle of the dune on the slip face, and are mixed with the finer grains from above. Thus the barchan is remixing its constituents and may therefore be expected to retain its grain-size mix as a whole. One would imagine that the opposite might be the case with linguiform elements in an aklé pattern, so that these might be engaged in a further sorting process.

These intra-dune sorting processes probably lead to an overall sorting pattern in a dune field. It has sometimes been possible to follow the development of an aeolian sand bed from its source to the full development of a dune pattern. It is usually found that there is a zone in which the sands still bear some resemblance to their pre-existing distributions, but there is no agreement on the actual dimensions or characteristics of this zone. Chepil (1945) followed a sand downwind after one storm. He found a completely aeolian (very well sorted) sand had been formed in only 0·34–0·68 km. Sharp (1966) followed sands from a source area on the Mojave River in southern California, towards the Kelso dunes. He found that the fluvial character was lost by between eight and 20 kms. Simonett (1951) followed these patterns in much more detail in

New South Wales and found that there were still distinct signs of the original grain-size patterns 1·37 km from the sand source. In Kansas he found (1960) that there was actually a very narrow zone in which grain-size characteristics changed from fluvial to aeolian.

As the sand is transported downwind by bulk transport, it appears to become slightly finer and better sorted in most dune fields (Chepil, 1957; Finkel, 1959; Hastenrath, 1967; McKee, 1966; Warren, 1966). Aufrère (1931*a*) and Lewis (1936) suggested that, because sands became finer and better sorted to the south-west, there had been south-westerly sand movement in the Great Western Erg and in the Kalahari. It is probably true in these two large ergs that both fluvial and aeolian sorting patterns are involved in the grain-size distributions (e.g. Bond, 1948; Poldervaart, 1957). Mabbutt (1968) quoted heavy-mineral evidence in support of this last point in his discussion of the sands of the Simpson Desert.

(e) Effects of differential sand supply. We noted in section 4.3.3.*b* that where there is abundant sand in an area with unidirectional winds, aklé patterns develop, whereas if there is only a limited supply, only barchans develop. On the surface of a stone pavement over which sand is transmitted very quickly (section 4.3.3.*b*) dunes can only form at the convergent nodes of the flow pattern where barchanoid dunes are found; in the divergent nodes the flow is too rapid for the development of linguoid forms.

The availability of sand has played an important part in several theories about dune patterns. In some instances sand supply has clearly been critical to the form of a dune area: in Egypt the Abu Moharic dune belt stretches downwind as a long narrow zone of dunes from a narrow sand supply (Bagnold, 1941; Beadnell, 1910). Similar narrow dune zones stretching downwind of coastal sand supplies have been observed in Peru where they are known as *chiflones* (Broggi, 1952). Some authorities have made an analogy between these long narrow dune belts and longitudinal dunes (e.g. Hack, 1941). Melton (1940) suggested that longitudinal dunes might stretch downwind from regularly spaced sand supplies on point-bar deposits of rivers. If longitudinal dunes are partly erosional and partly depositional forms (section 4.3.3.(1)) then we might suppose that they would develop where there was an initial relative shortage of sand (e.g. Bourcart, 1928). Shortage of sand could of course be a function of faster windspeeds, so that faster windsppeds might also be linked to lonitudinal dunes (Högbom, 1923). We have, of course, seen that these explanations are not general enough to explain most dune patterns (section 4.3.3.*b*).

Shortage of sand probably does have an effect on the development of draa. In the Australian deserts it is rather remarkable that there are no draa, and this has been explained by invoking a shortage of sand (Madigan, 1945). Wilson (1970) has calculated that if the sands in dunes were to be spread out flat the Simpson Desert in Australia would be covered by a layer only one metre thick, whereas the Great Eastern Erg in Algeria where draa are well developed would have a cover 26 m. thick. It should be added that not only must the sand in the Simpson Desert have been in short supply, but it must also have been spread fairly evenly over the whole of the erg, since otherwise draa would surely have developed in some areas.

In a series of experiments in which he compared actual sand dune patterns in Israel with patterns of ripples in a flume, Rim (1958) proposed that the initial distribution of the sand might be important. When he arranged the experiment so that sand was available only on one side of the flume he found that a diffusion process made the dunes fan out downcurrent obliquely to the flow. He used this to explain some dune patterns near the Israeli coast: winds blew onshore at an angle to the sand supply along the beach and the resulting dunes then diffused inland obliquely to the wind. He showed with another experiment that once initial oblique ridges had been formed they channelled subsequent sand movement in the same direction.

Most of these explanations involving sand supply as a factor in dune formation are concerned with dune morphology in relatively humid areas. In the very arid desert, sand supply becomes less of a factor because of the shepherding effect of the ergs: within the ergs one finds an abundance of sand; outside them there is usually an extreme shortage partly because of the nature of the stone pavement surface, and elsewhere because of the rugged terrain. Only occasionally does one find dunes such as barchans whose form is due in part to a shortage of sand.

(f) Effects of underlying topography. Apart from the unresolved problem of whether an initial obstacle is needed to fix or to initiate regular eddy motion in the wind (section 4.2.1.*b*) the effects of topography on dune forms are not difficult to understand.

If a large obstacle such as an inselberg or bornhardt is found in the path of a sand-carrying wind, a fixed eddy is formed on the upwind side of the hill, the wind is channelled round the flanks in a couple of fixed roller vortices and a relatively calm zone is found in the lee. Dunes form at the convergent nodes in this sytem of secondary flows (Richardson, 1968). Narrow hills form single lee dunes whereas broader ones form two parallel lee dunes (Fig. 4.27*b*). This kind of dune form has very frequently been observed in the desert (e.g. Bagnold, 1941; Bosworth, 1922; Cornish, 1914).

Experimental studies in wind-tunnels, aimed often at the solution of design problems for buildings in arid areas, have shown the effects of different heights, widths, arrangements and shapes of obstacles on the lee and fore dunes (Dobrolowski, 1924; Duchemin, 1958; Queney and Dubief, 1943). Fairly steep slopes presented to the wind usually mean the accumulation of a fore-dune which is separated from the slope by a long roller vortex. Some large dunes formed in this way run parallel to scarps for many kilometres in the Sahara. Some of these have been illustrated by Clos-Arceduc who has called them 'echo dunes' (1967). Duchemin (1958) showed that buildings which had a sloping face to the wind lead the sand up and over them so that no fore-dune collects. On broad hills facing the wind such gentle slopes are often found in small gullies. Sand is led up these, collecting in irregularities and between the stones to form a 'climbing dune'. Then sand passes over the watershed to the other side of the hill where it forms a 'falling dune' (Plate 4.20; Bagnold, 1941; Evans, 1962; Hack, 1941; Kádár, 1934; Smith, 1954).

Dunes, of course, act as obstacles in ripple development, and draa act as obstacles in dune development. Trailing dunes behind draa-sized features have been illustrated by Norris and Norris (1961) from the Algodones dunes in southern California.

Some lee dunes can reach considerable size. In the western Sahara, Clos-Arceduc (1967*b*) illustrated some lee dunes that were over three kilometres long (Fig. 4.32). Similar lengths have been noted in the Thar (Cornish, 1914) and in the Simpson Desert (Madigan, 1946). In fixed-dune areas, lee hollows and trailing dunes may attain even greater lengths (Warren, 1966). In the Katsina area of northern Nigeria, Grove (1958) noticed some trailing sand patches and lee sand-free areas that were over 120 km long.

It is usually supposed that lee dunes are a good indication of the direction of the resultant wind. However Clos-Arceduc's (1967) illustrations suggest that they may sometimes be aligned with an oblique trend (Fig. 4.32).

Observations on simple dunes formed in the lee of obstacles led many early geomorphologists to suggest that all dunes were initiated in the lee of hills (e.g. Sokolow, 1894, and others quoted by Högbom, 1923). This idea has persisted in many explanations of longitudinal dunes (e.g. Durand de Corbiac, 1958; Edmonds, 1942; Enquist, 1932; Lewis, 1960; Melton, 1940; Stokes, 1964). A variety of obstacles was favoured including small buttes, sphynxes, dead donkeys and camels. We have seen that many of the early Saharan travellers thought that rhourds were simply too big to be dunes and suggested they were sand-covered hills (e.g. Chudeau, 1920). These explanations of course do not explain the regularity of the spacing and size of the features and they ignore the fact that the wind and therefore the sandflow is accelerated over plateaux (Wilson, 1971*b*). Further evidence against the 'obstacle' theory will be cited below.

When a dune field piles up against a hill or scarp a confused pattern is usually the result. Monod (1958*d*) has called this *une maladie de l'erg*. Downwind of the hill there is usually a wide sand-free zone which closes up only after several kilometres. Urvoy (1933) described some of these sand-free zones in the fixed sands of southern Niger. Another of these zones is filled with Lake Faguebine in Mali (Fig. 4.25; Urvoy, 1942).

Despite these observations, many small hills within a dune field appear to have hardly any effect on its morphology. Capot-Rey (1953) and Jutson (1934) noticed that small hills hardly affect the morphology of longitudinal dunes, and Mabbutt (1968) noted in addition that although orientations were altered as sand ridges passed through a gap between two hills, there was no change in spacing. Plate 4.21 illustrates a dune pattern near Alice Springs in which topography has remarkably little effect. A similar aloofness of the dune pattern can be seen in Fig. 4.42a, taken from a dune pattern in central Sudan. The transverse pattern passes over the hill without a change. Elsewhere in the same area the dunes formed in the lee of small hills are actually quite distinct from the main draa pattern (Fig. 4.42b). This can be cited as good evidence that many dune patterns are independent of hill features.

Most sand seas, however, occur in desert basins where there are few hills. Maps of the desert floor beneath the dunes usually show very gentle slopes over large areas.

In parts of deserts where there are large hill massifs, winds are more distinctly deflected. We have noted this is the case of the Tibesti Massif (section 4.3.3.*c*), where patterns of dunes and of deflation furrows clearly show the deflection of the winds. There is some evidence in the central Sudan that even quite low domes in the underlying topography can deflect winds in a significant way (Fig. 4.38). In valleys in the larger hill massifs, winds can be funnelled between the hills so that they achieve very constant directions and strong windspeeds. Perhaps the best-known example of this is the Coachella Valley in southern California (Russell, 1932).

(g) Effects of vegetation. When moving sand comes into contact with plants, or when a sandy surface is itself colonized by plants, distinctive

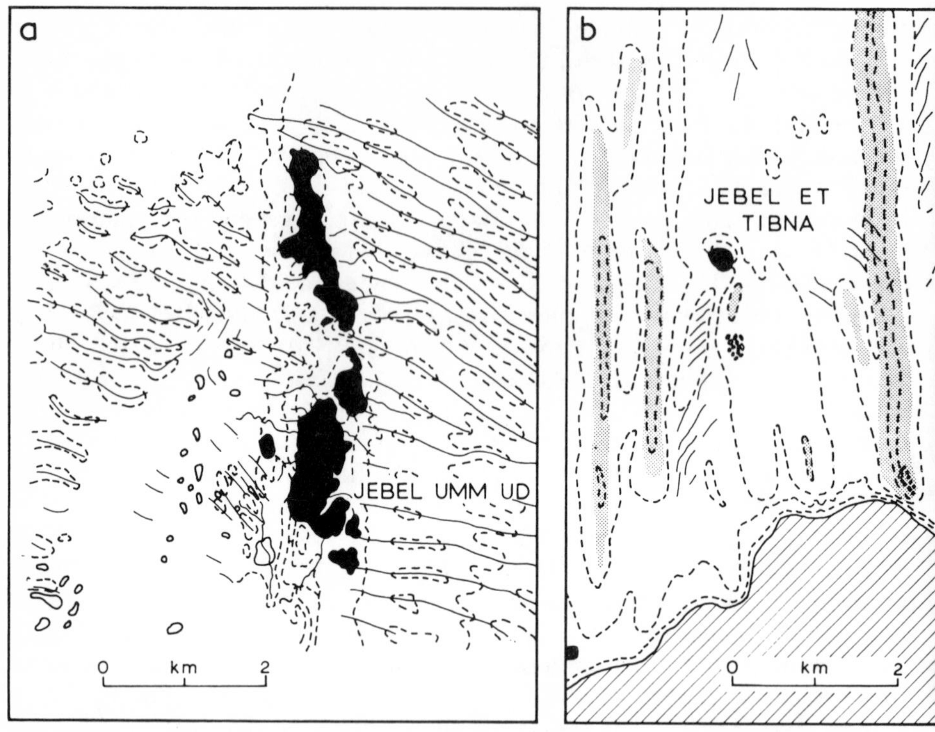

4.42 The negligible influence of topographic features on *draa* and dune patterns. The hills (small inselbergs) are shown in solid black; the *draa* are stippled and the dune alignments are shown by black lines (Warren, 1966).

dune forms develop. The form of these dunes is well known, since they occur not only in the more accessible margins of the desert, but also on coasts, and detailed descriptions and explanations of them have existed since the mid-nineteenth century.

In semi-arid areas there are a number of ecological controls on vegetation. Rainfall is, of course, a basic source of moisture but there are a number of edaphic factors that are also important. In the first place, soil texture controls the availability of moisture, and, as we have seen (section 2.3.2), sandy soils tend to hold water and are therefore frequently more densely vegetated than surrounding heavier soils. Their greater permeability also tends to mean that sandy soils are less saline, and this, too, favours colonization. Counteracting these attractions is the mobility of sand, and this mobility is itself subject to a number of sedimentary and aeolian controls. Dune crests are both drier and more mobile than flanks, so that flanks are more likely colonization sites (e.g. Capot-Rey, 1943; Högbom, 1923; Madigan, 1936;

Rempel, 1936). Aklé dunes, siefs and barchans are more mobile than rounded longitudinal dunes (e.g. Monod, 1958), and coarse-textured rolling zibar are in turn more stable than these (e.g. Monod, 1958). In the Ténéré Desert, for example, where angular sief dunes have formed on a base of zibar, it is the zibar that bear the very few clumps of grass, whereas the sief are completely bare. In Western Australia it has been maintained that the active downwind end of longitudinal dunes is not colonized by plants, whereas the more stable upwind is colonized (Jutson, 1934).

In the southern Sahara there is the additional complication that the alternating wet and dry periods of the Quaternary have left ancient series of dunes whose surfaces have been stabilized by soil development in wet periods, and are thus firm and inactive, and are sometimes colonized by plants; these surfaces transgress the desert-savanna boundary (e.g. Monod, 1958; Warren, 1966).

Beyond the acknowledged desert boundary, where most of the sand appears to be fixed, there are zones of active sand movement, often on the dune crests. Small fields of active dunes have been reported in many areas with quite high rainfall, and although they have sometimes caused speculation as to whether the ancient sands were perhaps deposited in a humid climate (Prescott and White, 1960; Urvoy, 1933), they are usually ascribed to overgrazing by domestic stock (e.g. Dresch and Rougerie, 1960; Flint and Bond, 1968; Grandet, 1957; Grove, 1960*a*; Suslov, 1961; Tricart, 1959; Tricart and Brochu, 1955; Vaché-Grandet, 1959; Warren, 1966). This complex ecological picture has meant that it is usually difficult to decide where on the desert borders fixed sands begin and active ones end (e.g. Brosset, 1939; Monod, 1958).

On the ground the most obvious dune form in lightly vegetated areas is the *nebkha* or 'shrub-coppice dune' (Fig. 4.27c). These are also known simply as 'coppice dunes' or when large as *rebdou*. These have been widely reported (e.g. Bourcart, 1928; Chudeau, 1920; Coque, 1962; Frere, 1870; Hefley and Sidwell, 1945; Melton, 1940; Petrov, 1962). As the wind reaches a bush the value of k (section 4.2.4) rises up to 30 times its original height (Olson, 1958). In other words the depth of the zone near the surface in which the wind velocity is low is very greatly enlarged. This means that U'_* changes drastically and sand is deposited. The sand tends to build up in the lee of the bush where there is a wind shadow. The form of the new dune depends on the plant species involved and its growth form (e.g. Capot-Rey, 1957*b*; Coque, 1962; Gimmingham, 1955; Melton, 1940).

Nebkha are found in quite arid as well as in semi-arid areas. Wherever the water-table comes near enough to the surface, as around oases and wells, small fields of nebkha appear (Bourcart, 1928; Capot-Rey, 1957*b*; Petrov, 1962). In the case of the larger nebkha, wind erosion may well play a more important part than wind deposition. The large *tamarix* or *Salvadora*-topped mounds around some of the Saharan wells are built of fine material held together by numerous fossil and living plant roots, and their steep sides and the absence of free sand undoubtedly indicate that they are largely erosional (e.g. Bourcart, 1928; Capot-Rey, 1957*b*; Coque, 1962).

Where sands have been colonized by vegetation, and the vegetation cover has later been removed in small areas (for instance by overgrazing or trampling) winds tend to channel into the breach because of the lower value of k over bare sand (Olson, 1958). The sand is dried and eroded by the wind, and the process is accelerated when the upper cemented soil mantle has been removed. Initially a small accumulation of sand downwind of the hollow forms a dune and the assemblage is known as a blowout; if this dune migrates downwind as the hollow is enlarged, trailing dunes are left on either flank of the hollow and a parabolic dune is formed (Fig. 4.27). Migration of several dunes down the same path gives a nested parabolic system. If the small transverse dune at the advancing head is blown away, only the parallel longitudinal dunes may survive. The feature is thus formed by both erosion and deposition. This kind of dune is well known in coastal areas (e.g. Cooper, 1967; Landsberg, 1956; Steensrup, 1894), and from the inland dunes in northern Europe (e.g. Högbom, 1923). It has also been reported from semi-arid areas, especially where there are pre-existing fixed sands (e.g. Hack, 1941; Hefley and Sidwell, 1945; Hurault, 1966; Melton, 1940; Verstappen, 1968). McKee (1966) made measurements of the movement of a parabolic dune in New Mexico. We shall return to some of the implications of parabolics in section 4.3.3.*h* and *i*.

The lunette is a form that is related to the parabolic dune, but it is usually built of finer material (Figs. 4.8 and 4.27d). The processes in which pseudo-sands are created in saline ephemeral lakes was explained in section 4.3.1.*a*. These materials are often blown off the lake-bed to form a crescentic dune downwind, the crescent opening to the wind leaving a lake-hollow from which the material is removed and which is often wholly wind-eroded (see section 4.2.2.*c* and *d*). The lunette is very common in semi-arid Australia where it was first named and described by Hills (1939) and where the optimum climate for its formation has caused some debate since clearly there must be enough water to bring in salt and clay to the lake, and enough wind activity to blow it out. Campbell (1968) suggested that wave action in the lake was an important effect in lunette formation although wind action was dominant. Bettenay (1962) claimed that the optimum rainfall was about 38 cm p.a. for lunettes in Australia. Most lunettes in Australia are indeed now thought to be palaeofeatures. Lunettes have also been described from northern Algeria (Boulaine, 1954), the Gulf coast of Texas (Price, 1963), New Mexico (Everard, 1964), the Senegal Delta (e.g. Tricart, 1954), coastal Tunisia (Jauzein, 1958; Trichet, 1963) and in the Kalahari (Grove, 1969).

(h) Dune palaeoforms in deserts. There is undeniable evidence in most deserts and in almost all semi-arid areas that there have been repeated changes in climate during the Quaternary. Some of the best evidence is contained in ancient lake deposits from between dunes in the desert and from the existence in areas now densely covered with vegetation of aeolian sands and dune forms (e.g. Grove and Warren, 1968). One of the methods of interpreting dune patterns in areas of fixed sands is to search for analogues in dune fields that are active at the present day (e.g. Smith, 1968; Warren, 1970), since a more arid climate undoubtedly existed in the now fixed dunes at the time of their formation. In the same way, faced with the great complexity of patterns in arid areas, it has sometimes been tempting to attribute them to periods of even greater or long-continued aridity in the past. It has also been tempting to compare desert dune forms with forms now covered by vegetation, and thus to make the original argument circular (e.g. Monod, 1958).

The argument that desert dunes were formed in more humid conditions seems to have originated in the preconceptions of European geomorphologists who had first worked with dunes in humid climates. It was often suggested, for example, that longitudinal (or any elongated) dunes in the desert had originated as the trailing arms of parabolic dunes, and therefore that the longitudinal dunes had been formed in a more humid climate (e.g. Aufrère, 1931*b*; Blandford, 1877; Capot-Rey, 1943; Dubief, 1952; King, 1960; Melton, 1940; Raychaudri and Sen, 1952). Often the evidence is convincing (e.g. Verstappen, 1969*a*) but sometimes the mere appearance of a linguiform dune, sometimes a zibar formed of coarse sand, has been enough to suggest a change in climate (Capot-Rey, 1957*a*). The parabolic-longitudinal hypothesis is closely connected with the idea of erosion and deflation as the cause of dune patterns, which will be discussed in the next section.

Whether or not a change in rainfall is suggested, it has often seemed unavoidable to postulate a change in wind patterns to explain complex patterns of crossing dunes and draa. This has been suggested in the western Sahara (Grandet, 1957; Monod, 1958; Sevenet, 1943), in the central Sahara (Capot-Rey, 1947; Smith, 1963), in Arabia (Bagnold,

1951; Glennie, 1970); for the Pur-Pur dune in Peru (Simons, 1956), for the Kelso dunes in the Mojave Desert (Sharp, 1966), and for central Australia (Mabbutt, 1968; Mabbutt *et al.*, 1969). The evidence of dune trends has been examined in an attempt to understand past climates in fixed dune fields of the Sudan (Warren, 1971), of Tchad (Aufrère, 1928), of the Kalahari (Grove, 1969; Lewis, 1936), and of the American Mid-West (Cooper, 1938; Melton, 1940; Smith, 1965; Warren, 1968). The literature on the possible wind shifts connected with northern and eastern European dune fields is enormous (e.g. Dylikowa, 1969; Högbom, 1923).

The existence together of major and minor forms (draa and dunes) has often been attributed to changed conditions. Major forms for example have been attributed to stronger winds in the past (Alimen, 1954; Aufrère, 1931*a*; Bagnold, 1953; Glennie, 1970; Melton, 1940; Lelubre, 1950; Prenant, 1951).

The acceptance or rejection of these interpretations leaves untouched the evidence that in most dune fields there have been several phases of dune building separated by wetter periods. The evidence is in the form of ancient calcreted sands beneath the modern ones (e.g. Hack, 1941; Huffington and Albritton, 1941), archaeological remains on dunes (Bagnold, 1931; Kennedy-Shaw, 1936; Schoeller, 1945), terraces and calcified crusts in the dune hollows (Capot-Rey, 1943; De Villers, 1948; Prenant, 1951; Schoeller, 1945), lake deposits between the dunes (Faure, *et al.*, 1963; Monod, 1958) and palaeosols within the dunes (Hills, 1940). Most ergs, in fact, seem to include evidence of repeated phases of dune formation (e.g. Aufrère, 1931; Capot-Rey, 1945 and 1947; Mabbutt, 1957 and 1967; Norris and Norris, 1961; Smith, 1965; Suslov, 1961; Wopfner and Twidale, 1967).

(i) Hypotheses of erosion and deposition. The proposal that many of the Saharan dunes originated as parabolic forms (outlined in the last section) involved the idea that both erosion and deposition have been involved. An erosion/deposition hypothesis was evolved on this basis to explain many draa patterns: they had been formed by the erosion of hollows and the piling up of the sand on to the intervening ridges. The idea is very old and has passed on from decade to decade with very little substantial evidence. It has been proposed for the Saharan ergs by many authors (e.g. Aufrère, 1928 and 1931*a*; Bourcart, 1928; Capot-Rey, 1943, 1945 and 1948; Enquist, 1932; Monod, 1958 and 1962) and for some large elongated features in Arabia (e.g. Holm, 1953), in the Kara Kum (Heller and Doubiansky, quoted by Jorré, 1935) and in the Thar (Blandford, 1877) and it has also won support as an explanation for some smaller features in the Australian ergs (Folk, 1971*a*; King, 1956 and 1960).

There is a little evidence in favour of erosion. In the first place it is widely accepted that erg sands are derived from pre-existing alluvial sands (section 4.3.1.*a*), so that it seems reasonable, in the absence of actual bedded remnants of these sands, to propose that they were eroded by the wind and the resulting sands piled on to the intervening ridges. This idea also helped to explain the existence of large aeolian bedforms alongside smaller ones—the first were erosional, the second were depositional (Aufrère, 1931*a*; Bourcart, 1928; Enquist, 1932; Monod, 1958). The existence of alluvial cores has, however, never been proved in the Algerian ergs, and there appears to be only one area in which they have been found: King (1956) drilled into some ridges just north of Lake Eyre in Australia and found such cores which evidence he later used to extend the idea of erosion to all the Australian sand ridges, claiming the forms to have originated as parabolic dunes (King, 1960). More recent sections of sand ridges elsewhere in the Simpson Desert have shown some of them to be made up only of sand and others of sand over clayey (possibly elluvial) cores (Mabbutt and Sullivan, 1968). It is certainly true that many dune fields do not transgress beyond the edges of deep alluvial basins, a fact which has been cited in support of the deflation hypothesis (King, 1960; Grove, 1959; Tricart and Cailleux, 1961; Warren, 1966), but this may simply mean that sands are moved more quickly over the harder floors away from the basins. Another piece of evidence for the erosional hypothesis is that many of the sands of the ergs do not appear to have been moved far in aeolian transport, although in the state of our knowledge about the characteristics of sands subject to aeolian transport, this would be hard to substantiate.

It is true that there are signs of small-scale and large-scale wind erosion of rocks and clays in the desert and in semi-arid areas (section 4.2.1) and direct analogies can be drawn between the scale of these features and the scale of certain dune features, particularly those of draa size (Clos-Arceduc, 1967; Wilson, 1970), but this analogy can be opposed by the shortage of direct evidence for wind erosion between dunes, and to the directly contradictory evidence of stream patterns amongst dunes. On the western side of the Indus Delta in Pakistan the ancient, quite shallow patterns of the ancient Nara can be traced in the corridors between large dune ridges; in the Simpson Desert, similar patterns can be traced. The Saoura River has left traces within the Great Western Erg, and very clear channel patterns can be found in Mali leading northward from the present Niger River in the region of Timbuctu into a zone, first of transverse dunes, and then of longitudinal dune ridges (Palausi, 1955). Similar ancient river patterns exist amongst the dunes in the Kalahari (Grove, 1969), and amongst the active

Algodones dunes in California (Norris and Norris, 1961). Some of these patterns, it is true, post-date some dune phases, but they quite clearly do not post-date the major phase in most cases.

One last interesting point has sometimes been made about the protective effects of dunes. It is said that dunes, because of their high infiltration capacity, may inhibit water erosion, whereas this would continue on nearby dune-free areas so that interdune corridors might be lowered. This was suggested for the Great Western Erg by Capot-Rey (1943). Similar suggestions have been made for the Australian deserts and the Colorado Plateau (Stokes, 1964; Webb and Wopfner, 1961).

4.3.4 ERGS

Isolated dunes incorporate only a small fraction of the total amount of aeolian sand: it has been calculated that 99·8 per cent of aeolian sand is found in ergs that are greater than 125 km^2 in area and 85 per cent is in ergs greater than 32,000 km^2. The distribution of erg sizes is fairly peaked (*i.e.* with a small standard deviation) and has a modal size of about 188,000 km^2. The largest erg seems to be the Rub al Khali in Arabia whose area is over 560,000 km^2 (Wilson, 1972*a*).

The smaller ergs include dunes of only one pattern, but most ergs show patterns of different kinds in different parts (Figs. 4.24–26). The study of the ways in which these patterns vary can give some answers about the individual patterns themselves, and the study of the ergs as units can give some answers to wider problems of aeolian geomorphology (e.g. section 4.2.1.*f*).

The importance of the context of a dune pattern within a dune field has been appreciated for many years (e.g. Aufrère, 1931*a* and *b*; Beadnell, 1934; Brosset, 1939; Capot-Rey, 1945; Doubiansky, 1928; Högbom, 1923; Urvoy, 1933). It was once widely held that changes in pattern could be linked in a time-sequence. For example, many workers maintained that patterns with a distinct transverse element, or ones that were very complex, were 'young', and patterns of simple, evidently longitudinal, ridges were 'old' (e.g. Aufrère, 1931*a* and *b*). Bagnold (1941) also maintained that transverse ridges were unstable and would be replaced eventually by longitudinal dunes. The evolutionary model was sometimes linked to size of dunes (section 4.3.3.*h*) and sometimes to sedimentary changes; for example Beadnell (1934) believed that the whalebacks (draa?) in the Libyan Desert were old, since they contained many coarse grains that appeared to have accumulated from the passage of many smaller dunes over long periods of time.

There is of course, plenty of evidence that there are changes in dune form from one part of a dune field to another, and that the changes occur both in space and in time. In the Abu Moharic dune field in Egypt, in Peru, and in California, dunes in barchan fields appear to grow until they reach an equilibrium size (Bagnold, 1941; Hastenrath, 1967; King, 1918; Norris, 1966). The same pattern of growth has been observed in coastal aklé dunes by Cooper (1958) and Inman *et al.* (1966). In Peru and in Egypt there appears to be a decline in dune-size towards the outer edges of the field, but this is not a function of bedform development but of sand supply, since the dunes disperse downwind and are therefore less able to trap streamers of sand from their upwind neighbours. In the Algodones dune field in California, sand supply controls dune patterns in a rather different way, since the sand in the field seems to have come from the beaches of an extinct Pleistocene lake. The main mass of sand and the larger dunes have moved downwind since the disappearance of the lake, leaving a trail of smaller and more dispersed dunes in the zone behind (McCoy *et al.*, 1967). Similar patterns of sand thickness have been described from Nebraska by Warren (1968). If it were not for such changes in the controls of dune-size, it is unlikely (by analogy with other bedforms) that subaerial dunes would change their form once they had reached a size and pattern that was in equilibrium with prevailing conditions. This statement can be further supported by reference to draa patterns many of which must be old (section 4.3.2.*c*) and many must therefore be regarded as being at equilibrium with conditions prevailing for a long time. These facts lead us to reject the earlier ideas about sequences or cycles of dune development.

The significance of different dune patterns and the role of ergs are better appreciated if we study the context of ergs in the regional patterns of aeolian sediment movement. Wilson (1971*b*) has produced inductive models of erg growth covering a number of possible situations; the following discussion is a summary of his paper and a discussion of one of his simpler and more widely applicable models.

The general situation of aeolian sediment movement in deserts is of movement of sand along pathways from 'sandflow peaks'. The situation is illustrated for the Sahara by Fig. 4.3. Most ergs are 'flow-crossed', which is to say that the sandflow lines pass over the erg in more or less parallel lines. If we consider one of these flow-crossed ergs we can first examine the situation of erg growth without bedforms (Fig. 4.43). Deposition of sand in such a situation can only take place if the flow is decelerated or if there is convergence. Deposition is negative erosion, so that, using the terms explained in section 4.2.1.*f*:

$$\text{deposition rate} = -\overline{E}_s = dQ/dx \text{ kg per m}^2 \text{ p.a.} \quad \ldots (4.42)$$

where $\overline{E}_s$ is the mean sand deposition rate (if negative) and Q is the potential sandflow rate as in section 4.2.1.*f*. Away from the sandflow peaks and divides, as shown in Fig. 4.3, ergs may form in any position where there are local decelerating conditions (this will be chiefly in basins). We should note here that convergence of the flow may also initiate deposition, and that the nature of the deposition will be rather different from the conditions where there is only deceleration, but that marked convergence does not seem to occur frequently in the Sahara (Fig. 4.3). The plots of the changes in potential and actual sandflow rates and in the deposition rate shown in Fig. 4.43 indicate that the erg without bedforms would be a convex lens with edges tapering down to meet the desert floor at both ends.

In the Sahara most large bedforms are separated by wide areas of stone pavement, so that it is clear that the sandflow is not saturated throughout its passage across an erg: there are local alternations between saturated and unsaturated flow which depend on the pattern of secondary flows at the scale of the bedform. Bedforms can therefore develop in areas where the flow is not saturated; Wilson (1971*b*) defined such a condition as 'metasaturated'. Bedforms can develop wherever the flow is above the metasaturation point.

We can now proceed to an erg model which incorporates bedforms. The bedforms will start to grow when metasaturation is reached. As these bedforms grow they will tend to trap much of the sand, but when they reach equilibrium all the sand is passed on; in this way an equilibrium front passes downwind (Fig. 4.44). When the erg has grown downwind to its fullest extent, flow near the centre may pass a saturation point and the erg will start to grow upward over a lens of sand although this situation seems to be rare in the Sahara. The model can be varied to accommodate a number of different environmental conditions, such as when deflation exists within the erg (e.g. in Peru, as discussed by Lettau and Lettau, 1969), or when the bedforms trap all the sand or very little of the sand as they grow (Wilson, 1971*b*). The Australian ergs for example have dunes which are efficient 'sand-passers', whereas the Algodones dunes in California are mega-barchan dunes which are probably efficient 'sand-trappers'.

Since the wind is decelerated on reaching a sand patch (section 4.3.1.*b*) and since the bedforms too must decelerate the overall flow, the zone between the metasaturation point and the saturation point is usually narrowed and the erg margins are sharpened. Sharpening will also occur if sand supply upwind is decreased, as for instance when a stone pavement seals off the upwind surfaces (Wilson, 1971*b*). The sharp contrast between sand-free and *ensablées* zones has often been remarked upon in the Algerian ergs (e.g. Aufrère, 1931*a*).

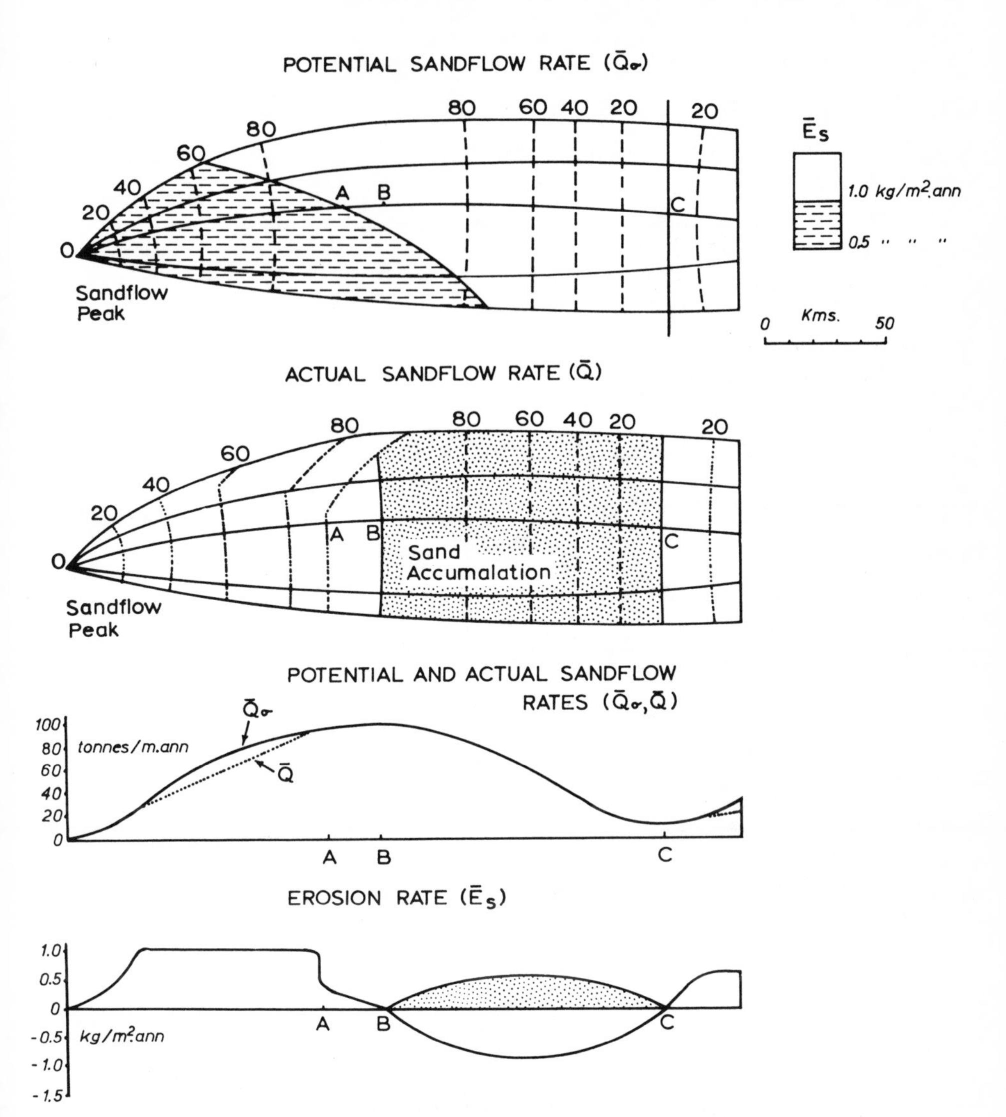

4.43 A simple model of erg development without bedforms in a 'sandflow sector' downwind of sandflow peak, cf. Figure 4.3 (after Wilson, 1971*b*).

4.3.5 DUNE PATTERNS: CONCLUDING STATEMENT

The preceding sections have shown that the major elements of dune patterns can be explained with fairly simple models. We may conclude by summarizing the argument and assigning relative roles to the various mechanisms at work.

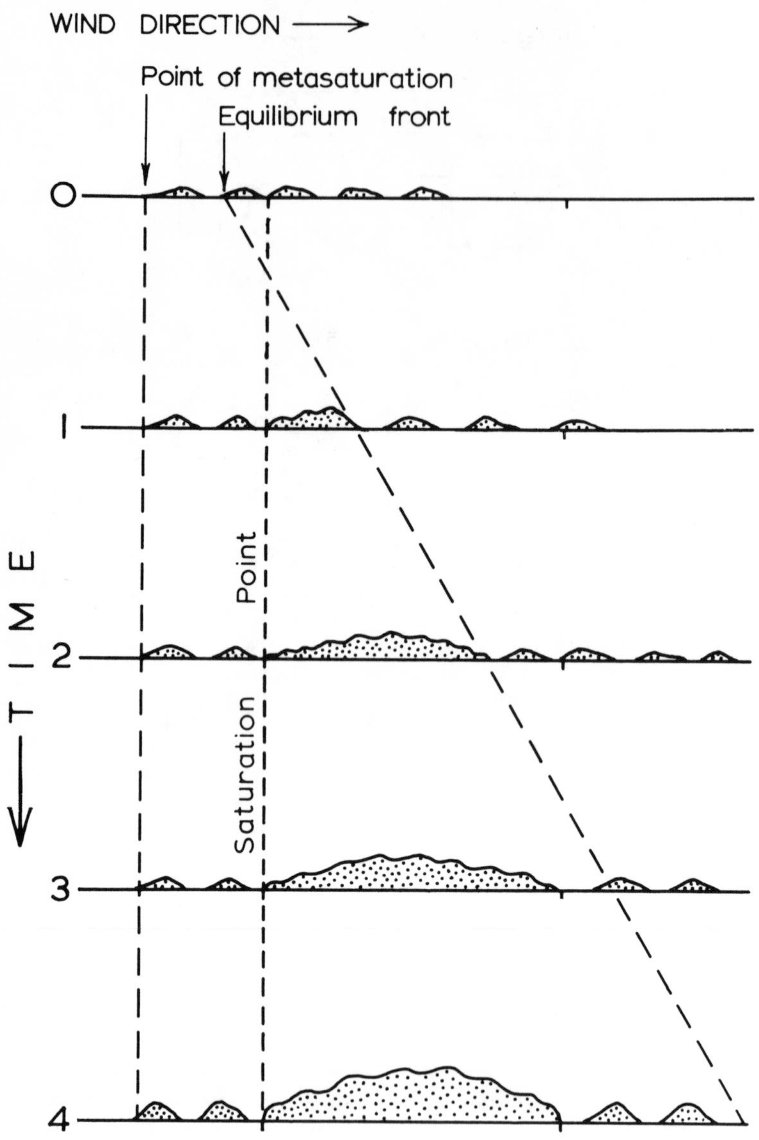

4.44 The development sequence of an erg with bedforms. Deposition of sand in the bedforms starts when the metasaturation point is reached. When the bedforms have grown to equilibrium they stop trapping sand and sand passes on to the next bedform, so that an 'equilibrium front' passes downwind. After stage one the sandflow becomes saturated downwind of the metasaturation point and the erg begins to build upward over a lens of sand (after Wilson, 1971*b*).

In most patterns the first element of complexity arises from the hierarchy. All ergs have at least two orders of aeolian bedforms, namely ripples and dunes, and most large ergs also include draa. Each of these orders will have patterns of their own, although the essential elements of the patterns may be very similar.

Each order will have a pattern composed of four major trends—transverse, longitudinal and left- and right-oblique. The angles of the left- and right-oblique to the main trend will vary from order to order. One or other of these trends may be emphasized by a cross wind in the wind regime. In draa the emphasis will be retained over decades; dunes may well appear rather different from season to season, although the principal alingments will be retained from year to year; ripples (except for megaripples) will change almost daily.

Superimposed on the simple hierarchical and four-trend pattern will be variations assignable to grain-size, and it appears to be the nature of aeolian sorting that different sand types are segregated into different parts of the dune landscape.One will therefore find different dune types in different zones, each type adjusted to a particular sand type. Where sands with a prominent coarse mode are found, dunes will be rounded and generally subdued and will have long wavelengths; ripples may reach megaripple proportions or may not occur at all. Where sands are well sorted and of intermediate size, one would find more accentuated dunes with shorter wavelengths.

We may apply these principles to an area where there are few other independent controls. Plate 4.1 shows a pattern in the southern part of the Ténéré Desert in Niger. Two orders in the hierarchy can be seen: draa-spaced features running from ENE. to WSW. and sief-like dunes cut across there from nearly E. to W.; there are low undulating dunes in the interdraa corridors. The draa appear to be transverse and siefs seem to be oblique (left-oblique) to the draa alignment, while the virtual absence of the right-oblique element suggests the presence of an important seasonal cross wind. The siefs are accentuated and have relatively low wavelength sinuosity along their crests whereas the inter-draa dunes are low, rounded and have a darker albedo. The siefs are probably of medium-grained and well-sorted sand whereas the interdraa dunes are probably of bimodal or coarse sand (they are zibar).

Where other factors intrude we may find dune pattern variations due to topography, vegetation, variable sand supply, position in a growing erg, or palaeoform survival.

References

Alimen, M-H., 1953, Variations granulométriques et morphoscopiques du sable le long de profils dunaires au Sahara occidental, in *Actions Eoliennes, Cent. Nat. de Rech. Sci., Paris, Coll. Int., 35*, 219–35.

Alimen, M-H., Buron, M., and Chavaillon, J., 1958, Caractères granulométriques de quelques dunes d'ergs du Sahara nord-occidental, *Ac. des Sci., Paris, C.R., 247*, 1758–61.

Alimen, M-H., and Fenet, D., 1954, Granulométrie de sables d'erg aux environs de la Saoura (Sahara occidental), *Soc. Géol. de France, C.R., 9–10*, 183–85.

Allen, J. R. L., 1968a, *Current Ripples, their relation to patterns of water and sediment motion* (North Holland, Amsterdam), 433p.

Allen, J. R. L., 1968b, The nature and origin of bedform hierarchies, *Sedimentology, 10*, 161–82.

Allen, J. R. L., 1969a, Erosional current marks of weakly cohesive mud beds, *J. Sed. Pet., 39*, 607–23.

Allen, J. R. L., 1969b, On the geometry of current ripples in relation to stability of flow, *Geog. Ann., 51*, 61–96.

Allen, J. R. L., 1970, *Physical Processes of Sedimentation* (Allen and Unwin, London), 248p.

Allison, L. E., 1964, Salinity in relation to irrigation, *Advances in Agronomy, 16*, 139–80.

Amstutz, G. C. and Chico, R., 1958, Sand size fractions of southern Peruvian barchans and a brief review of the genetic grainsize function, *Bull. Vereinigung Schweizerischen Petroleum—Geologen-und-Ingenieure, 24*, 47–52.

Anstey, R. L., 1965, Physical characteristics of alluvial fans, *U.S. Army Materiel Command, U.S. Army Natick Laboratories, Technical Report, ES-20*, 109p.

Antevs, E., 1928, Wind deserts in Iceland, *Geog. Rev., 18*, 675–76.

Antevs, E., Arroyo–cutting and filling, *J. Geol., 60*, 375–85.

Arkley, R. J., 1963, Calculation of carbonate and water movement in soil from climatic data, *Soil Sci., 96*, 239–48.

Arkley, R. J., 1967, Climates of some great world soil groups of the western United States, *Soil Sci., 103*, 389–400.

Ascher, R., and Ascher, M., 1965, Recognizing the emergence of Man, *Science*, *147*, 243–50.

Aufrère, L., 1928, L'orientation des dunes et la direction des vents, *Ac. des Sci., C.R.*, *187*, 833–35.

Aufrère, L., 1931a, Le cycle morphologique des dunes, *Ann. de Géog.*, *40*, 362–85.

Aufrère, L., 1931b, Classification des dunes, *Int. Geog. Cong. Paris, C.R., Actes 2*, 699–711.

Aufrère, L., 1935, Essai sur les dunes du Sahara algérien, *Geog. Ann.*, *17*, 481–500.

Bagnold, R. A., 1931, Journeys in the Libyan Desert, *Geog. J.*, *78*, 13–39; 524–35.

Bagnold, R. A., 1933, A further journey in the Libyan Desert, *Geog. J.*, *82*, 103–29; 403–4.

Bagnold, R. A., 1941, *The Physics of Blown Sand and Desert Dunes* (Methuen, London), 265p.

Bagnold, R. A., 1951, The sand formations of south Arabia, *Geog. J.*, *117*, 78–86.

Bagnold, R. A., 1953a, The surface movement of blown sand in relation to meteorology, in *Desert Research, Proc. Int. Symp. held in Jerusalem, May 7–14, 1952*, sponsored by the Res. Counc. of Israel and UNESCO, *Research Council of Israel Special Publication, 2*, 89–93.

Bagnold, R. A., 1953b, Forme des dunes de sable et régime des vents, in *Actions Eoliennes, Cent. Nat. de Rech. Sci., Paris, Coll. Int.*, *35*, 23–32.

Bagnold, R. A., 1953c, in discussion following Kampé de Ferriet (1953), 101.

Bagnold, R. A., 1954a, The physical aspects of dry deserts, in Cloudsley-Thompson, J. C. (ed.), *Biology of Deserts* (Institute of Biology, London), 7–12.

Bagnold, R. A., 1954b, Experiments on a gravity-free dispersion of large solid spheres in a Newtonian fluid under shear, *Proc. Roy. Soc., London, Ser. A.*, *225*, 49–63.

Bakker, J. P., and LeHeux, J. W. N., 1952, A remarkable new geomorphological law, *Koninkl. Nederl. Akad. van Wetenschappen Proc., Ser. B.*, *55*, 399–571.

Balchin, W. G. V., and Pye, N., 1955, Piedmont profiles in the arid cycle, *Proc. Geol. Ass.*, *66*, 167–81.

Ball, J., 1927, Problems of the Libyan Desert, *Geog. J.*, *70*, 21–38; 105–28; 209–24.

Ball, J., 1939, *Contributions to the geology of Egypt* (Ministry of Finance, Survey & Mines Department, Cairo, Government Press, Bulâq), 308p.

Barshad, I., 1955, Soil development, in Bear, F. E. (ed.), *The Chemistry of the Soil* (Reinhold, New York), 1–52.

Barshad, I., 1957, Factors affecting clay formation, in *Clays and Clay Minerals, 6th Conference* (Pergamon, New York), 110–32.

Barshad, I., 1962, Soil development, in Bear, F. E. (ed.), *The Chemistry of the Soil* (Reinhold, New York), 2nd edition, 1–52.

Bartowski, T., 1969, Relief linéaire—relief typique des regions de loess actuelles, *Biul. Periglac., 20*, 213–16.

Bashin, O. von, 1899, Die Enstehung Wellenähnlicher Oberflochen—formen, *Zeit. Gessell. für Erdkunde zu Berlin, III, 34,* 408–24.

Bather, F. A., 1900, Wind-worn pebbles in the British Isles, *Proc. Geol. Ass., 16*, 396–420.

Beadnell, H. J. L., 1910, The sand dunes of the Libyan Desert, *Geog. J., 35*, 379–95.

Beadnell, H. J. L., 1934, Libyan Desert dunes, *Geog. J., 84*, 337–40.

Beaty, C. B., 1963, Origin of alluvial fans, White Mountains, California and Nevada, *Ann. Ass. Am. Geog., 53*, 516–35.

Beaty, C. B., 1970, Age and estimated rate of accumulation of an alluvial fan, White Mountains, California, U.S.A., *Am. J. Sci., 268*, 50–77.

Beaumont, P., 1968, Salt weathering on the margin of the Great Kavir, Iran, *Geol. Soc. Am. Bull., 79*, 1683–84.

Beckett, P. H. T., 1968, Soil formation and slope development. I: a new look at Walter Penck's *Aufbereitung* concept, *Zeit. für Geom., 12*, 1–24.

Beheiry, S. A., 1967, Sandforms in the Coachella Valley, Southern California, *Ann. Ass. Am. Geog., 57*, 25–48.

Bellair, P., 1953, Sables désertiques et morphologie éolienne, *Int. Geol. Cong., 19th, Algiers, 1952, fasc.* 7, 113–18.

Bellair, P., 1954, Sur l'origine des dépôts de sulphate de calcium actuels et anciens, *Ac. des Sci., Paris, C.R., 239*, 1059–61.

Belly, P-Y., 1964, Sand movement by wind, *U.S. Army Corps of Engineers, Coastal Engineering Research Centre, Tech. Mem., 1*, 38p.

Bernard, F., 1954, Rôle des insects sociaux dans les terrains du Sahara, in Cloudsley-Thompson, J. C. (ed.), *Biology of Deserts* (Institute of Biology, London), 104–11.

Bettenay, E., 1962, The salt lake systems and their associated aeolian features in the semi-arid regions of Western Australia, *J. Soil Sci., 13*, 11–17.

Bettenay, E., Blackmore, A. V., and Hingston, F. J., 1964, Aspects of the hydrologic cycle and related salinity in the Belka Valley, Western Australia, *Austr. J. Soil Res., 2*, 187–210.

Bettenay, E., and Hingston, F. J., 1964, Development and distribution of soils in the Merredin area, Western Australia, *Austr. J. Soil Res., 2*, 173–86.

Birot, P., 1951, Sur les reliefs granitiques en climat sec, *Bull. Ass. Géog. Fr., 220–221*, 138–41.

Birot, P., 1968, *The Cycle of Erosion in Different Climates* (Batsford, London), 144p.

Birot, P., and Dresch, J., 1966, Pédiments et glacis dans l'Ouest des Etats-Unis, *Ann. de Géog., 411*, 513–52.

Black, R. F., and Barksdale, W. L., 1949, Orientated lakes of northern Alaska, *J. Geol., 57*, 105–18.

Blackwelder, E., 1925, Exfoliation as a phase of rock weathering, *J. Geol., 33*, 793–806.

Blackwelder, E., 1927, Fire as an agent in rock weathering, *J. Geol., 35*, 134–40.

Blackwelder, E., 1929, Cavernous rock surfaces of the desert, *Am. J. Sci., 17*, 393–99.

Blackwelder, E., 1931a, Desert plains, *J. Geol., 39*, 133–40.

Blackwelder, E., 1931b, The lowering of playas by deflation, *Am. J. Sci., 21*, 140–44.

Blackwelder, E., 1933, The insolation hypothesis of rock weathering, *Am. J. Sci., 26*, 97–113.

Blackwelder, E., 1934, Yardangs, *Geol. Soc. Am. Bull., 45*, 159–66.

Blackwelder, E., 1940, Crystallization of salt as a factor in rock weathering, *Geol. Soc. Am. Bull, 51*, 1956 (abstract).

Blackwelder, E., 1948, Historical significance of desert lacquer, *Geol. Soc. Am. Bull., 59*, 1367 (abstract).

Blackwelder, E., 1954, Pleistocene lakes and drainage in the Mojave Region, southern California, *Cal. Div. Mines Bull., 170*, part V, 35–40.

Blake, W. P., 1855, On the grooving and polishing of hard rocks and minerals by dry sand, *Am. J. Sci., 2nd series, 20*, 178–81.

Blake, W. P., 1858, *Report of a Geological Reconnaissance in California* (Ballière, New York).

Blake, W. P., 1904, Origin of pebble-covered plains in desert regions, *Trans. Am. Inst. Min. Eng., 34*, 161–62.

Blandford, W. T., 1877, Geological notes on the great desert between Sind and Rajputana, *Geological Surv. of India Records, 10*, 10–21.

Blissenbach, E., 1954, Geology of alluvial fans in semiarid regions, *Bull. Geol. Soc. Am., 65*, 175–90.

Bluck, B. J., 1964, Sedimentation on an alluvial fan in southern Nevada, *J. Sed. Pet., 34*, 395–400.

Boaler, S. B., and Hodge, C. A. H., 1962, Vegetation stripes in Somaliland, *J. Ecology, 50*, 465–74.

Bocquier, G., 1968, Biogéocénoses et morphogenèse actuelle de certaines pédiments du Bassin Tchadien, *Trans. 19th Int. Cong. of Soil Sci., Adelaide, Australia, 4*, 605–14.

Bolyshev, N. N., 1964, Role of algae in soil formation, *Soviet Soil Sci., 1964*, 630–35.

Bond, G., 1948, The direction of origin of the Kalahari sands of southern Rhodesia, *Geol. Mag., 85*, 305–13.

Bond, R. D., 1968, Water-repellent sands, *Trans. 19th Int. Cong of Soil Sci., Adelaide, Australia, 1*, 339–47.

Boocock, C., and van Straten, O. J., 1962, Notes on the geology and hydrogeology of the central Kalahari Region, Bechuanaland Protectorate, *Trans. and Proc. Geol. Soc. S. Africa, 65*, 125–71.

Borovskii, V. M., 1961, Salt exchange between the sea and the land, and long term dynamics of soil processes, *Soviet Soil Sci., 1961*, 237–44.

Bosworth, T. O., 1922, *Geology of the Tertiary and Quaternary Periods in the Northwestern Part of Peru* (MacMillan, London), 434p.

Boulaine, J., 1954, Le Sebkha Ben Ziane et sa 'lunette' ou bourrelet; exemple de complexe morphologique formé par la dégradation éolienne des sols salés, *Rev. Géom. Dyn., 5*, 102–23.

Boulaine, J., 1961, Sur la rôle de la vegetation dans la formation des carapaces calcaires mediterranéenes, *Academie des Sci., Paris, C.R., 253*, 2568–70.

Bourcart, J., 1928, L'action du vent a la surface de la terre, *Rev. Géog. Phys. et Géol. Dyn., 1*, 26–54; 194–265.

Bourlière, F., and Hadley, M., 1970, The ecology of tropical savannas, *Ann. Rev. of Ecology and Systematics, 1*, 125–52.

Bowman, I., 1924, *Desert Trails of Atacama*, American Geographical Society, *Special Publication, 5*, 362p.

Bradley, W. C., 1963, Large-scale exfoliation in massive sandstones of the Colorado Plateau, *Geol. Soc. Am. Bull., 75*, 519–28.

Bramao, L., 1969, The first draft soil map of the world, *Bull. Int. Soc. Soil Sci., 34*, 4–10.

Branson, F. A., Miller, R. F., and McQueen, I. S., 1961, Soil moisture storage characteristics and infiltration rates as indicated by annual grasslands near Palo Alto, California, *U.S. Geol. Surv. Prof. Pap., 424-B*, 184–86.

Bremer, H., 1965, Musterböden in tropisch-subtropischen Gebieten und Frostmusterböden, *Zeit. für Geom., N.F.9*, 222–36.

Brewer, R., 1956, A petrographic study of two soils in relation to their origin and classification, *J. Soil Sci.*, 7, 268–79.

Bricard, J., 1953, L'electrisation des poussières, application aux vents de sable, in *Action Eoliennes, Cent. Nat. de Rech. Sci., Paris, Coll. Int., 35*, 33–43.

Brice, J. C., 1966, Erosion and deposition in the loess-mantled Great Plains, Medicine Creek drainage basin, Nebraska, *U.S. Geol. Surv. Prof. Pap., 352-H*, 255–339.

Broggi, J. A., 1952, Migracion de arenas a 10 largo de la costs peruana, *Bol. de la Soc. Geol. del Peru, 24*, 25p.

Brookfield, M., 1970, Dune trend and wind regime in Central Australia, *Zeit. für Geom., Suppl. 10*, 121–58.

Brosset, D., 1939, Essai sur les ergs du Sahara occidental, *Bull. Inst. Français de l'Afrique Noire, Dakar, 1*, 657–90.

Brown, C. B., 1924, On some effects of wind and sun in the desert of Tumbez, north-west Peru, *Geol. Mag., 61*, 337–39.

Brown, C. N., 1956, The origin of caliche on the north-east Llano Estacado, Texas, *J. Geol., 64*, 1–15.

Brown, G. F., 1960, Geomorphology of western central Saudi-Arabia, *Int. Geol. Cong., 21st Copenhagen, Report 21*, 150–59.

Brown, I. C., and Drosoff, M., 1940, Properties of soils and their colloids derived from granitic materials in the Mohave Desert, *J. Agricultural Res., 61*, 335–52.

Brüggen, J., 1950, *Fundamentos de la Geologia de Chile* (Inst. Géog. Militar, Santiago), 374p.

Brüggen, J., 1951, Las costras de proteccion en los desiertos, *Revista Universitaria (Chile), 36*, 101–4.

Bryan, K., 1923a, Erosion and sedimentation in the Papago country, *U.S. Geol. Surv. Bull., 730*, 19–90.

Bryan, K., 1923b, Wind erosion near Lees Ferry, Arizona, *Am. J. Sci., Ser. 5, 6*, 291–307.

Bryan, K., 1925a, The Papago Country, *U.S. Geol. Surv. Water-Supply Paper, 499*, 436p.

Bryan, K., 1925b, Date of channel trenching (arroyo cutting) in the arid Southwest, *Science, 62*, 338–44.

Bryan, K., 1926, The San Pedro Valley, Arizona, and the geographical cycle, *Geol. Soc. Am. Bull., 37*, 169–70.

Bryan, K., 1936, The formation of pediments, *Int. Geol. Cong. Report, 16th Session, 2*, 765–75.

Bryan, K., 1940, The retreat of slopes, *Ann. Ass. Am. Geog., 30*, 254–68.

Bryan, K., and Albritton, C. L., 1943, Soil phenomena as evidence of climatic changes, *Am. J. Sci., 241*, 459–90.

Bucher, W. H., 1919, On ripples and related surface sedimentary forms, *Am. J. Sci., 47*, 149–210; 241–69.

Buckham, A. F., and Cockfield, W. E., 1950, Gullies formed by sinking of the ground, *Am. J. Sci., 248*, 137–41.

Büdel, J., 1957, Die 'Doppelten Einebrungsflachen' in den feuchten Tropen, *Zeit. für Geom., N.F. 1*, 201–28.

Büdel, J., 1959, The 'periglacial'-morphologic effects of the Pleistocene climate over the entire world, *Int. Geol. Rev., 1*, 1–16.

Büdel, J., 1963, Klima-genetische geomorphologie, *Geog. Rundschau, 7*, 269–86.

Büdel, J., 1970, Pedimente, Rumpfflächen und Rückland-Steilhänge, *Zeit. für Geom., 14*, 1–57.

Bull, W. B., 1962, Relation of textural (CM) patterns to depositional environment of alluvial-fan deposits, *J. Sed. Pet., 32*, 211–16.

Bull, W. B., 1963, Alluvial-fan deposits in western Fresno County, California, *J. Geol., 71*, 243–51.

Bull, W. B., 1964a, Alluvial fans and near-surface subsidence in western Fresno County, California, *U.S. Geol., Surv. Prof. Pap., 437-A*, 70p.

Bull, W. B., 1964b, Geomorphology of segmented alluvial fans in western Fresno County, California, *U.S. Geol. Surv. Prof. Pap., 532-F*, 89–128.

Bull, W. B., 1968, Alluvial fans, *J. Geol. Ed., 16*, 101–6.

Buntley, G. J., and Westin, F. C., 1965, A comparative study of development color in a chestnut-chernozem-Brunizom soil climosequence, *Proc. Sci. Soc. Am., 29*, 579–82.

Buol, S. W., 1965, Present soil-forming factors and processes in arid and semiarid regions, *Soil Sci., 99*, 45–49.

Buol, S. W. and Hole, F. D., 1961, Clay skin genesis in Wisconsin soils, *Proc. Soil Sci. Soc. Am., 25*, 377–79.

Buol, S. W. and Yessilsoy, M. S., 1964, A genesis study of a Mohave sandy loam profile, *Proc. Soil Sci. Soc. Am., 28*, 254–56.

Butler, B. E., 1950, A theory of prior streams as a causal factor of soil occurrence in the Riverina plain of south-eastern Australia, *Austr. J. Agricultural Res., 1*, 231–52.

Butler, B. E., 1956, Parna, an aeolian clay, *Austr. J. Sci., 18*, 145–51.

Butler, B. E., 1960, Riverine deposition during arid phases, *Austr. J. Sci., 22*, 451–52.

Butler, B. E., and Hutton, J. T., 1956, Parna in the riverine plain of south-eastern Australia and the soils thereon, *Austr. J. Agricultural Res., 7*, 536–553.

Butzer, K. W., 1961, Climatic change in arid regions since the Pliocene, in *A History of Land Use of the Arid Zone* (UNESCO, Paris), *Arid Zone Research, 17*, 31–56.

Butzer, K. W., 1965, Desert Landforms at the Kurkur oasis, Egypt, *Ann. Ass. Am. Geog., 55*, 578–91.

Buursink, J., 1971, *Soils of Central Sudan* (Grafisch Bedrijf Schotanus & Jens, Utrecht), 248p.

Cailleux, A., 1942, Les actions éoliennes périglaciare en Europe, *Soc. Géol. de France, Ser. A., No. 1992*, 176p.

Calkin, P., and Cailleux, A., 1962, A quantitative study of cavernous weathering (taffonis) and its implications to glacial chronology in the Victoria Valley, Antarctica, *Zeit. für Geom., 6*, 317–24.

Campbell, E. M., 1968, Lunettes in southern South Australia, *Trans. Roy. Soc. S. Austr., 92*, 85–109.

Capot-Rey, R., 1939, Pays du Mzab et region des dayas, *Ann. de Géog., 48*, 41–63.

Capot-Rey, R., 1943, La morphologie de l'Erg Occidental, *Inst. de Rech. Sahariennes, Travaux, 2*, 69–104.

Capot-Rey, R., 1945, The dry and humid morphology of the Western Erg, *Geog. Rev., 35*, 391–407.

Capot-Rey, R., 1947, L'Edeyen de Mourzouk, *Inst. de Rech. Sahariennes, Travaux, 4*, 67–109.

Capot-Rey, R., 1953a, Récherches géographiques sur les confins algero-libyens, *Inst. de Rech. Sahariennes, Travaux, 10*, 33–73.

Capot-Rey, R., 1953b, *Le Sahara Français* (P.U.F., Paris), 564p.

Capot-Rey, R., 1957, Le vent et le modelé éolien au Borkou, *Inst. de Rech. Sahariennes, Travaux, 15*, 149–57.

Capot-Rey, R., 1957b, Sur une forme d'érosion éolienne dans le Sahara français, *Tijd. Kon. Nederland Aardnjsk Gen., 74*, 242–47.

Capot-Rey, R., 1965, Remarques sur la désagrégation mécanique dans les grés du Tibesti méridional, *Bull. Assn. Géog. Français, 330–39*, 39–43.

Capot-Rey, R., and Grémion, M., 1965, Remarques sur quelques sables saharien, *Inst. de Rech. Sahariennes, Travaux, 23*, 153–63.

Carson, M. A., 1971. An application of the concept of threshold slopes to the Laramie Mountains, Wyoming, *Inst. Brit. Geog. Spec. Pub., 3*, 31–48.

Carson, M. A., and Kirkby, M. J., 1972, *Hillslope Form and Process*, (Cambridge U.P., Cambridge), 475p.

Chao, S., 1962, Preliminary discussion on the types of gobi in the north-western part of Ho-Si Tsou-lang and its reclamation and utilization, in *Sand Control Group of the Academia Sinica* (ed), translated by Joint Publication Research Service, U.S.A., 189–224.

Chatterji, S. and Jeffery, J. W., 1963, Crystal growth during the hydration of CaSO4. ½H20, *Nature*, 200, 463–64.

Chepil, W. S., 1945, Dynamics of wind erosion: II. Initiation of soil movement, *Soil Sci., 60*, 397–411.

Chepil, W. S., 1950, Properties of soil which influence wind erosion: the governing principle of surface roughness, *Soil Sci., 69*, 149–62.

Chepil, W. S., 1957, Sedimentary characteristics of dust storms: I. Sorting of wind-eroded soil material, *Am. J. Sci., 255*, 12–22.

Chepil, W. S., and Milne, R. A., 1939, Comparative study of soil drifting in the field and in a wind tunnel, *Scientific Agriculture, 19*, 249–.

Chepil, W. S. and Siddoway, F. H., 1959, Strain-gauge anemometer for analysing various characteristics of wind turbulence, *J. Meteorology, 16*, 411–18.

Chepil, W. S. and Woodruff, N. P., 1957, Sedimentary characteristics of dust storms: II. Visibility and dust concentration, *Am. J. Sci., 255*, 104–14.

Chepil, W. S. and Woodruff, N. P., 1963, The physics of wind erosion and its control, *Advances in Agronomy, 15*, 211–302.

Chew, R. M., and Chew, A. E., 1965, The primary productivity of a desert scrub (Larrea tridentata) community, *Ecological Monographs, 35*, 355–75.

Chico, R. J., 1963, Playa mud cracks; regular and king size, *Geol. Soc. Am. Spec. Paper, 76*, 306.

Chico, R. J., 1968, Mud cracks, in Fairbridge, R. W., *The Encyclopedia of Geomorphology* (Reinhold, New York), 761–63.

Chorley, R. J., 1962, Geomorphology and general systems theory, *U.S. Geol. Surv. Prof. Pap., 500-B*, 10p.

Chorley, R. J., 1965, Models in geomorphology, in Chorley, R. J., and Haggett, P. (eds), *Models in Geography* (Methuen, London), 59–96.

Chorley, R. J., Dunn, A. J., and Beckinsale, R. P., 1964, *The History of the Study of Landforms* (Methuen, London), *1*, 678p.

Chudeau, R., 1920, L'étude sur les dunes sahariennes, *Ann. de Geog., 29*, 334–451.

Churchward, H. M., 1961, Soil Studies at Swan Hill, Victoria, Australia, I. Soil layering, *J. Soil Sci., 12*, 73–86.

Churchward, H. M., 1963, Soil studies at Swan Hill, Victoria, IV. Ground-surface history and its expression in the array of soils, *Austr. J. Soil Res., 1*, 242–55.

Clark, A. H., Cooke, R. U., Mortimer, C., and Sillitoe, R. H., 1967, Relationships between supergene mineral alteration and geomorphology, southern Atacama Desert, Chile—an interim report. *Trans. Inst. Min. and Metall., B, 76*, B89–96.

Clark, A. H., *et al.*, 1967, Implications of the isotopic ages of ignimbrite flows, southern Atacama Desert, Chile, *Nature, 215*, 723–24.

Clarke, R. H., and Priestley, C. H. B., 1970, The asymmetry of Australian desert sand ridges, *Search, 1*, 77p.

Clements, T. F., 1952, Wind-blown rocks and trails on Little Bonnie Claire Playa, Nye County, Nevada, *J. Sed. Pet., 22*, 182–86.

Clements, T., *et al.*, 1957, A study of desert surface conditions, *Headquarters Quartermaster Research and Development Command, Environmental Protection Research Division, Technical Report, EP53*, 110p.

Clements, T., Stone, R. O., Mann, J. F., and Eyman, J. L., 1963, A study of windbourne sand and dust in desert areas, *U.S. Army Natick Laboratories, Natick, Mass., Earth Sciences Division, Technical Report, ES–8*, Project Ref. 7x83–01–008, 61p.

Clos-Arceduc, A., 1956, Etudes des photographies aériennes d'une formation végétale sahelienne: la brousse tigrée, *Bull. Inst. français de l'Afrique Noire, Ser. A*, 677–84.

Clos-Arceduc, A., 1966, L'application des methodes d'interpretation des images a des problemes geographiques, exemples et resultats, methodologie, *Rev. de l'Inst. français du Petrole 21*, 1783–800.

Clos-Arceduc, A., 1966*b*, Le rôle déterminant des ondes aériennes stationnaires dans la structure des ergs sahariens et les formes d'érosion aroisinantes,

Acad. des Sci., Paris, C. R., ser. D, 262, 2673-76.

Clos-Arceduc, A., 1967b, La direction des dunes et ses rapports avec celle du vent, *Acad. des Sci., Paris, C. R., ser. D, 264*, 1393–96.

Cloudsley-Thompson, J. L. and Chadwick, M. J., 1965, *Life in Deserts* (Foulis, London), 218p.

Collis-George, N., and Evans, G. N., 1964, A hydrologic investigation of salt-affected soils in an alluvial plain of the Hawkesbury River, New South Wales, *Austr. J. Soil. Res., 2*, 20–28.

Cooke, H. B. S., 1961, The Pleistocene environment in southern Africa, in David, D. H. S. (ed.) *Ecology in Southern Africa* (Junk, The Hague) *Monographae Biologicae, 4*, 1–34.

Cooke, R. U., 1964a, Landforms in the Huasco Valley, Northern Chile, (unpublished M.Sc. Thesis, University of London), 205p.

Cooke, R. U., 1964b, Planation surfaces or 'matureland' in the southern Atacama Desert? *Int. Geog. Cong., 20th, London, Abstracts of Papers*, 87.

Cooke, R. U., 1964c, Landforms in the southern Atacama Desert, *Brit. Ass. Adv. Sci. Proc., E*, 101–2.

Cooke, R. U., 1965, Introduction to surface forms, Chapter 1 of Davis, E. L. (ed.) *Archaeology and Environment of the Panamint Valley, Inyo County, California*, unpublished ms.

Cooke, R. U., 1970a, Morphometric analysis of pediments and associated landforms in the western Mojave Desert, California, *Am. J. Sci., 269*, 26–38.

Cooke, R. U., 1970b, Stone pavements in deserts, *Ann. Ass. Am. Geog., 60*, 560–77.

Cooke, R. U., and Mason, P., Desert Knolls pediment and associated landforms in the Mojave Desert, California. *California Geographer*, in press.

Cooke, R. U. and Mortimer, C., 1971, Geomorphological evidence of faulting in the southern Atacama Desert, Chile. *Rev. de Géom. Dyn., 20*, 71–78.

Cooke, R. U., and Reeves, R. W., 1972, Relations between debris size and the slope of mountain fronts and pediments in the Mojave Desert, California, *Zeit. für Geom.*, 76–82.

Cooke, R. U., and Smalley, I. J., 1968, Salt weathering in deserts, *Nature, 220*, 1226–27.

Cooper, W. S., 1938, Ancient dunes of the upper Mississippi valley as possible climatic indicators, *Bull. Am. Met. Soc., 19*, 193.

Cooper, W. S., 1958, Coastal sand dunes of Oregon and Washington, *Geol. Soc. Am. Mem., 72*, 169p.

Cooper, W. S., 1967, Coastal Dunes of California, *Geol. Soc. Am. Mem., 104*, 131p.

Coque, R., 1962, *La Tunisie Pré-Saharienne, Étude Géomorphologique*, (Colin, Paris), 476p.

Corbel, J., 1963, Pédiments d'Arizona, Centre de Doc. Cart, et Geog., *Memoires et Documents, 9*, 33–95.

Corbel, J., 1963, Pédiments d'Arizona, Centre de Doc. Cart. et Géog., *Memoires et Documents, 9*, 33–95.

Corbel, J., 1964, L'érosion terrestre, étude quantitative, (méthodes, tech-491–521.

Cornet, A., and Pinaud, C., 1957, Sur l'apparition d'un "betoir" dans la daya M'rara a l'Ouest del'Oued Righ, *Inst. de Rech. Sahariennes, Travaux, 15*, 159–61.

Cornish, V., 1914, *Waves of Sand and Snow*, (Fisher-Unwin, London), 383p.

Corte, A. E., 1963, Particle sorting by repeated freezing and thawing, *Science, 142*, 499–501.

Corte, A., and Higashi, A., 1964, Experimental research on desiccation cracks in soil, U.S. Army Materiel Command, Cold Regions Research and Engineering Laboratory, *Research Report, 66*, 72p.

Costin, A. B., 1955, A note on gilgaes and frost soils, *J. Soil Sci., 6*, 32–34.

Cotton, C. A., 1942, *Climatic Accidents in Landscape Making* (Whitcombe and Tombs, Christchurch, N.Z.), 354p.

Cotton, C. A., 1961, The theory of savanna plantation, *Geography, 46*, 89–101.

Cotton, C. A., 1962, Plains and inselbergs of the humid tropics, *Trans. Roy. Soc. New Zealand, 1*, 269–77.

Coursin, A., 1964, Observations et expèriences faites en avril et mai 1956 sur les barkhanes du Souehel el Abiodh (région est de Port-Etiènne), *Bull. Inst. Français de l'Afrique Noire, Ser. A*, 26, 989–1022.

Crittenden, M. D., Jr., 1963, Effective viscosity of the earth derived from isostatic loading of Pleistocene Lake Bonneville, *J. Geoph. Res., 68*, 5517–30.

Crocker, R. L., 1946, Soil and vegetation relationships in the lower south-east of South Australia, a study in ecology, *Trans. Roy. Soc. S. Aust., 68*, 144–72.

Crocker, R. L., 1946, The soil and vegetation of the Simpson Desert and its borders, *Trans. Roy. Soc. of S. Austr., 70*, 235–58.

Culling, W. E. H., 1963, Soil creep and the development of hillside slopes, *J. Geol., 71*, 127–61.

Cumpston, J. H: L., 1964, *The Inland Sea and the Great River* (Angus and Robertson, London), 205p.

Curry, L., 1962, Climatic change as a random series, *Ann. Am. Ass. Geog., 52*, 21–31.

Cvijanovich, B-G., 1953, Sur le rôle des dunes en relation avec le système hydrologique de la nappe souterraine du Grande Erg, *Inst. de Rech. Sahariennes, Travaux, 10*, 131–36.

Dan, J., and Yaalon, D. H., 1968, Pedomorphic forms and pedomorphic surfaces, *Trans. 9th Int. Cong. of Soil Sci., Adelaide, Australia, 3*, 577–84.

Darwin, C., 1891, *Geological Observations on the Volcanic Islands and Parts of*

South America Visited During the Voyage of H.M.S. Beagle, 3rd edition (Smith, Elder, London), 648p.

Daveau, S., 1964, Façonnement des versants de l'Adrar mauritanien, *Zeit. für Géom., Supp., 5*, 120–30.

Daveau, S., 1966 (1967), Le relief du Baten d'Atar (Adrar mauritanien), *Memoires et Documents, Centre de Recherches et Documentation Cartographique et Géographique, 2*, 96p.

Daveau, S., Mousinho, R., and Toupet, C., 1967, Les grandes dépressions fermées de l'Adrar mauritanien, sebkha de Chemchane et Richât, *Bull. Inst. français de l'Afrique Noire, Ser. A, 29*, 414–46.

Davis, E. L., and Winslow, S., 1965, Giant ground figures of prehistoric deserts, *Proc. Am. Phil. Soc., 109*, 8–21.

Davis, W. M., 1930, Rock floors in arid and in humid climates, *J. Geol., 38*, 1–27; 136–38.

Davis, W. M., 1933, Granite domes of the Mohave Desert, California, *San Diego Soc. Nat. Hist. Trans.*, 7, 211–58.

Davis, W. M., 1938, Sheetfloods and streamfloods, *Geol. Soc. Am. Bull, 49*, 1337–416.

Davis, W. M., 1954, The geographical cycle in an arid climate, in Johnson, D. W. (ed.), *Geographical Essays* (Constable, London), 296–322.

Delaney, F. M., 1954, Recent contributions to the geology of the Anglo-Egyptian Sudan, *Trans. 9th Int. Geol. Cong., Algiers, 1952, 20*, 11–18.

de Félice, P., 1956, Processus de soulèvement des grains de sable par le vent, *Acad. Sci., Paris, C. R., 242*, 920–23.

de Martonne, E., and Aufrère, L., 1928, L'extension des régions privées d'écoulement vers l'océan, *Ann. de Géog., 38*, 1–24.

Denny, C. S., 1965, Alluvial fans in the Death Valley region, California and Nevada, *U.S. Geol. Surv. Prof. Pap., 466*, 62p.

Denny, C. S., 1967, Fans and pediments, *Am. J. Sci., 265*, 81–105.

Desio, A., 1968, Short history of the geological, mining and oil exploration in Libya, *Inst. di Geol. dell'Univ. Degli Studi di Milano, Ser. G, Pub. 250*, 123p.

Desio, A., 1969, Pubblicazioni di Ardito Desio, *Universita di Milano, Inst. di Geol.*, 19p.

Devillers, C., 1948, Les dépôts quaternaires de l'Erg Tihodaine (Sahara Central), *Soc. Géol. de France, Bull., Ser 5, 18, C. R. S., 10,* 189–91.

D'Hoore, J. L., 1965, *Soil Map of Africa, Explanatory Monograph*, (Commission for Technical Co-operation in Africa, Lagos), Joint Project No. II, 205p.

Dickson, B. A., and Crocker, R. L., 1954, A chronosequence of soils and vegetation near Mt. Shasta, California, II, *J. Soil Sci., 5*, 173–91.

Dobrolowski, A. B., 1924, Mouvement de l'air et de l'eau sur les accidents du sol, *Geog. Ann., 6*, 300–67.

Doehring, D. O., 1970, Discrimination of pediments and alluvial fans from topographic maps, *Geol. Soc. Am. Bull., 81*, 3109–16.

Doubiansky, V. A., 1928, The sand deserts of the south-east Karakum, *Bull. of Applied Botany, Genetics and Plant Breeding, 19*, 225.

Douglas, I., 1967, Man, vegetation and sediment yields of rivers, *Nature, 215*, 925–28.

Downes, R. G., 1946, Tunnelling erosion in north-eastern Victoria, *J. Counc. Sci. Ind. Res. Austr., 19*, 283–91.

Dragovich, D., 1967, Flaking, a weathering process operating on cavernous rock surfaces, *Bull. Geol. Soc. Am., 78*, 801–4.

Dragovich, D., 1969, The origin of cavernous surfaces (tafoni) in granitic rocks of South Australia. *Zeit. für Geom., 13*, 163–81.

Dregne, H., 1969, Appraisal of research on surface materials of desert environments, in McGinnies, W. G., Goldman, B. J., and Paylore, P. (eds.), *Deserts of the World* (U. Arizona P., Tucson), 287–377.

Dresch, J., 1953, Morphologie de la chaîne d'Ougarta, *Inst. de Rech. Sahariennes, Travaux, 9*, 25–38.

Dresch, J., 1957, Pédiments et glacis d'érosion, pediplains et inselbergs, *L'information Géographique, 22*, 183–96.

Dresch, J., 1961, Observations sur le désert cotier du Perou, *Ann. de Géog, 70*, 179–84.

Dresch, J., 1964, Observations sur les formes et dépôts quarternaire en Syrie, *Bull. Soc. Géol. de France, 5*, 603–7.

Dresch, J., 1968, Reconnaissance dans le Lut (Iran), *Bull. Ass. Géog. Français, 362–3*, 143–53.

Dresch, J., 1970, A propos du désert de Chihuahua, *Ann. de Géog., 431*, 51–57.

Dresch, J., and Rougerie, G., 1960, Morphological observations in the Sahel of the Niger, *Rev. Géom. Dyn., 11*, 49–58.

Drewes, H., 1963, Geology of the Funeral Peak Quadrangle, California, on the east flank of Death Valley, *U.S. Geol. Surv. Prof. Pap., 413*, 78p.

Dubief, J., 1951, Alizés, harmattan, et vents étésiens, *Inst. de Rech. Sahariennes, Travaux*, 7, 187–90.

Dubief, J., 1952, Le vent et le deplacement du sable au Sahara, *Inst. Rech. Sahariennes, Travaux*, 8, 123–62.

Dubief, J., 1953, Les vents de sable dans le Sahara français, in *Actions Eoliennes, Cent. Nat. de Rech. Sci. Paris, Coll. Int., 35*, 45–70.

Duce, J. T., 1918, The effect of cattle on the erosion of canon bottoms, *Science, 48*, 450–52.

Duchaufour, P., 1965, *Précis de Pédologie*, (Masson, Paris), 481p.

Duchemin, G. J., 1958, Essai sur la protection des constructions contre l'ensablement a Port-Etienne, (Mauritanie), *Bull. Inst. Français de Afrique Noire., Ser. A, 20*, 675–86.

Dufour, A., 1936, Observations sur les dunes du Sahara meridional, *Ann. de Géog., 45*, 276–85.

Durand de Corbiac, H. D., 1958, Autant en emporte le vent, ou l'érosion et l'accumulation autour du Tibesti, *Bull. Ass. Ing. Géog., 4*, 147–56.

Durand, J. H., 1953, La vent et sa consequence, l'érosion eolienne, facteur de formation des sols du Sahara, in *Desert Research* (Res. Council of Israel, *Spec. Pub., 2*), 434–37.

Durand, J. H., 1959, Les sols rouges et les croûtes en Algérie, *Direction de l'Hydraulique et de l'Équipement Rural, Clairbois-Birmandreis, Etude Général,* 7, 188p.

Dury, G. H., 1964, General theory of meandering valleys, *U.S. Geol. Surv. Prof. Pap., 452*, parts A, B, and C.

Dury, G. H., 1966a, Pediment slope and particle size at Middle Pinnacle, near Broken Hill, New South Wales, *Austr. Geog. Studies, 4*, 1–17.

Dury, G. H., 1966b, Duricrusted residuals on the Barrier and Cobar pediplains of New South Wales, *J. Geol. Soc. Austr., 13*, 299–307.

Dury, G. H., 1968, Gibber, in Fairbridge, R. W. (ed.), *Encyclopedia of Geomorphology* (Reinhold, New York, 424.

Dury, G. H., and Langford-Smith, T., 1964, The use of the term peneplain in descriptions of Australian landscapes, *Austr. J. Sci., 27*, 171–75.

Dury, G. H., Ongley, E. D., and Ongley, V. A., Attributes of pediment form on the Barrier and Cobar pediplains of New South Wales, *Austr. J. Sci., 30,* 33.

Dylikowa, A., 1969, Le problème des dunes intérieures en Pologne à la lumière des études de structure, *Biul. Periglac., 20*, 48–86.

Dżulyński, S., and Walton, E. K., 1965, Sedimentary features of flysch and greywackés, *Developments in Sedimentology,* 7, 274p.

Edmonds, J. M., 1942, The distribution of the Kordofan sand, *Geol. Mag., 79*, 18–30.

Eggler, D. H., Larson, E. E., and Bradley, W. C., 1969, Granites, grusses, and the Sherman erosion surface, southern Laramie Range, Colorado–Wyoming, *Am. J. Sci., 267*, 510–22.

Elgabaly, M. M., and Khadr, M., 1962, Clay mineral studies of some Egyptian desert and Nile alluvial soils, *J. Soil Sci., 13,* 333–42.

Ellwood, J., Evans, P., and Wilson, I. G., 1972, Small-scale aeolian bedforms, in press.

Emmett, W. W., 1970, The hydraulics of overland flow on hillslopes, *U.S. Geol. Surv. Prof. Pap., 662-A*, 68p.

Engel, C. G., and Sharp, R. P., 1958, Chemical data on desert varnish, *Geol. Soc. Am. Bull., 69*, 487–518.

Enquist, F., 1932, The relation between dune form and wind direction, *Geol. Föreningens i Stockholm Föhandlingar, 54*, 19–59.

Eriksson, E., 1958, The chemical climate and saline soils in the arid zone, in *Climatology, Reviews of Research* (UNESCO, Paris), *Arid Zone Research, 10*, 147–88.

Escande, L., 1953, Similitude des ondulations de sable des modelés reduits et des dunes du desert, in *Actions Éoliennes, Cent, Nat. de Rech. Sci., Paris, Coll. Int., 35*, 71–79.

Estorges, P., 1959, Morphologie du plateau arbaa, *Inst. de Rech. Sahariennes, Travaux, 18*, 21–57.

Evans, I. S., 1970, Salt crystallization and rock weathering: a review, *Rev. de Géom. Dyn., 19*, 153–77.

Evans, J. R., 1962, Falling and climbing sand-dunes in the Cronese ("Cat") Mountains, San Bernardino County, California, *J. Geol., 70*, 107–13.

Everard, C. E., 1963, Contrasts in the form and evolution of hill-side slopes in central Cyprus, *Trans. Inst. Brit. Geog., 32*, 31–47.

Everard, C. E., 1964, Playas and dunes in the Estancia Basin, New Mexico, *Int. Geog. Cong. 20th, London, Abstracts and Papers*, 89–90.

Exner, F. M., 1920, Zur Physik der Dunen, *Akademic der Wissenschaften, Wien, Mathematisch-Naturwissenschafthiche Klasse, Abteilung, 1*, 129, 929–52.

Exner, F. M., 1925, Über die Wechselwirkungzwischen Wasser und Geschiebe in Flüssen, *Sitzungberichte der Heidelberger Akademie der Wissenschaften, Wien, pt. IIa, Bd. 134.*

Fair, T. J. D., 1947, Slope form and development in the interior of Natal, South Africa, *Trans. Geol. Soc. S. Africa, 50*, 105–19.

Faure, H., Manguin, E., and Nydal, R., 1963, Formations lacustres du Quaternaire supérieure du Niger oriental, *Bull. Bur. Rech. Géol. Min., 3*, 41–63.

Feth, J. H., 1961a, A new map of the western coterminous United States showing the maximum known or inferred extent of Pleistocene Lakes, *U.S. Geol. Surv. Prof. Pap., 424-B*, 110–12.

Feth, J. H., 1961b, Effects of rainfall and geology on the chemical composition of water in coastal streams of California, *U.S. Geol. Surv. Prof. Pap., 424-B*, 202–4.

Finkel, H. J., 1959, The barchans of southern Peru, *J. Geol., 67*, 614–47.

Fitzpatrick, E. A., 1967, Soil nomenclature and classification, *Geoderma, 1*, 91–105.

Flach, K. W., Nettleton, W. D., Gile, L. H., and Cady, J. G., 1969, Pedocementation: induration by silica, carbonates, and sesquioxides in the Quaternary, *Soil Sci., 107*, 442–53.

Flach, K. W., and Smith, G. D., 1969, The new system of soil classification as applied to arid-land soils, in McGinnies, W. G., and Goldman, B. J. (eds.), *Arid Lands in Perspective* (U. Arizona P., Tucson), 59–74.

Flammand, G. B. M., 1899, La traversée de l'erg occidental (grand dunes du Sahara oranais), *Ann. de Géog., 9*, 231–41.

Flint, R. F., 1963, Pleistocene climates in low latitudes, *Geog. Rev., 53*, 123–29.

Flint, R. F., and Bond, G., 1968, Pleistocene sand ridges and pans in Western Rhodesia, *Geol. Soc. Am. Bull., 79*, 299–314.

Folk, R. L., 1968, Bimodal supermature sandstones: product of the desert floor. *Int. Geol. Cong., 23rd, Prague Proc., Section 8*, 9–32.

Folk, R. L., 1971a, Longitudinal dunes of the northwestern edge of the Simpson Desert, Northern Territory, Australia, 1, Geomorphology and grain size relationships, *Sedimentology, 16,* 5–54.

Folk, R. L., 1971b, Genesis of longitudinal and oghurd dunes elucidated by rolling upon grease, *Bull. Geol. Soc. Am., 82,* 3461–68.

Fournier, F., 1960, *Climat et Érosion: la Relation Entre l'Érosion du Sol par l'Eau et les Precipitations Atmospheriques* (P.U.F., Paris), 201p.

Fränzle, O., 1965, Klimatische Schwellenwerte der Bodenbildung in Europe und der U.S.A., *Die Erde, 96*, 86–104.

Free, E. E., 1911, Desert pavements and analogous phenomena, *Science, 33*, 355.

Frere, Sir H. Bartle E., 1870, On the Rann of Cutch and neighbouring regions, *J. Roy. Geog. Soc., 40*, 181–207.

Fry, E. J., 1927, The mechanical action of crustaceous lichens on substrata of shale, schist, gneiss, limestone and obsidian, *Ann. of Botany, 41*, 437–60.

Fry, E. J., 1924, A suggested explanation of the mechanical action of lithophytic lichens on rocks (shale), *Ann. of Botany, 38*, 175–96.

Frye, J. C., 1950, Origin of the Kansas Great Plains depressions, *Kansas Univ. State Geol. Surv., Bull., 86, part 1*, 20p.

Fuller, W. H., Cameron, R. E., and Raica, N. Jr., 1960, Fixation of nitrogen in desert soils by algae, *Trans. 7th Int. Cong. of Soil Sci., Madison, 2*, 617–24.

Gabriel, A., 1958, Zur Oberflächengestaltung der Pfannen in Frockenräumen, Zentral Persiens, *Westchr. 60, Lebensjahres Spreitzer, Wien*, 42–57.

Gale, H. S., 1914, Salines in the Owens, Searles and Panamint basins, southeastern California, *U.S. Geol. Surv., Bull., 580-L*, 251–323.

Galli-Olivier, C., 1967, Pediplain in northern Chile and the Andean uplift, *Science, 158*, 653–55.

Galloway, R. W., 1970, The full glacial climate in Southwestern United States, *Ann. Ass. Am. Geog., 60*, 245–56.

Gardner, W. R., and Brooks, R. H., 1957, A descriptive theory of leaching, *Soil Sci., 83*, 295–304.

Gautier, E-F., and Chudeau, R., 1908–1909, *Missions au Sahara* (Colin, Paris), Vol. I: Sahara Algérien, 371p; Vol. II: Sahara Soudanais, 326p.

Gautier, E-F., 1935, *Sahara—the Great Desert* (Columbia U.P., New York), 264p.

Gavrilovič, D., 1970, Die Überschwemmungen im Wadi Bardagué im Jahr

1968 (Tibesti, Republique du Tchad), *Zeit. für Géom., NF14*, 202–18.

Geikie, J., 1898, *Earth Sculpture or the Origin of Land-forms* (Murray, London), 397p.

Gibbs, H. S., 1945, Tunnel-gully erosion on the Wither Hills, Marlborough, *N.Z. J. Sci. Technology, 27*, 135–46.

Gigout, M., 1960, Sur la genèse des croûtes calcaries pleistocènes en Afrique du Nord, *Soc. Geol. de France, C.R.S.*, 8–10.

Gilbert, G. K., 1875, Report on the geology of portions of Nevada, Utah, California and Arizona, part 1 of *Geographical and Geological Explorations and Surveys West of the 100th meridian (Engineers Dept., U.S. Army), 3*, 21–187.

Gilbert, G. K., 1877, *Report on the Geology of the Henry Mountains* (Government Printing Office, Washington, D.C.), 160p.

Gilbert, G. K., 1895, Lake basins created by wind erosion, *J. Geol., 3*, 47–49.

Gilbert, G. K., 1914, The transportation of debris by running water, *U.S. Geol. Surv. Prof. Pap., 86*, 263p.

Gile, L. H., 1961, A classification of *Ca* horizons in soils of a desert region, Dona Ana County, New Mexico, *Proc. Soil Sci. Soc. Am., 25*, 52–62.

Gile, L. H., 1966a, Coppice dunes and the Rotura Soil, *Proc. Soil Sci. Soc. Am., 30*, 657–60.

Gile, L. H., 1966b, Cambic and certain non-cambic horizons in desert soils of southern New Mexico, *Proc. Soil Sci. Soc. Am., 30*, 773–81.

Gile, L. H., 1967, Soils of an ancient basin floor near Las Cruces, New Mexico, *Soil Sci., 103*, 265–76.

Gile, L. H., 1968, Morphology of an argillic horizon in desert soils of southern New Mexico, *Soil Sci., 106*, 6–15.

Gile, L. H., 1967, Soils of an ancient basin floor near Las Cruces, New Mexico, Mexico, *Proc. Soil Sci. Soc. Am., 34*, 465–72.

Gile, L. H., and Grossman, R. B., 1968, Morphology of the argillic horizon in desert soils of southern New Mexico, *Soil Sci., 106*, 6–15.

Gile, L. H., Grossman, R. B., and Hawley, J. W., 1969, Effects of landscape dissection on soils near University Park, New Mexico, *Soil Sci., 108*, 273–82.

Gile, L. H., and Hawley, J. W., 1966, Periodic sedimentation and soil formation on an alluvial fan piedmont in southern New Mexico, *Proc. Soil Sci. Soc. Am., 30*, 261–68.

Gile, L. H., and Hawley, J. W., 1969, Age and comparative development of desert soils at the Gardner Spring radio-carbon site, New Mexico, *Proc. Soil Sci. Soc. Am., 32*, 709–16.

Gile, L. H., Peterson, F. F., and Grossman, R. B., 1965, The *K*-horizon—a master horizon of $CaCO_3$ accumulation, *Soil Sci., 99*, 74–82.

Gile, L. H., Peterson, F. F., and Grossman, R. B., 1966, Morphological and genetic sequences of carbonate accumulation in desert soils, *Soil Sci., 101*, 347–60.

Gilluly, J., 1937, Physiography of the Ajo Region, Arizona, *Geol. Soc. Am. Bull., 48*, 323–48.

Gimmingham, C. H., 1955, A note on water table, sand-movement and plant distribution in a North African oasis, *J. Ecology, 43*, 22–25.

Glasovskaya, M. A., 1968, Geochemical landscapes and types of geochemical soil sequences, *Trans. 19th Int. Cong. of Soil Sci., Adelaide, Australia, 4*, 303–12.

Glennie, K. W., 1970, *Desert Sedimentary Environments* (Elsevier, Amsterdam), 222p, (*Developments in Sedimentology, 14*).

Glennie, K. W., and Evamy, B. D., 1968, Dikaka: plant and plant-root structures associated with aeolian sand, *Palaeogeography, Palaeoclimatology and Palaeoecology, 4*, 77–87.

Goss, D. W., and Allen, B. L., 1968, A genetic study of two soils developed on granite in Llano County, Texas, *Proc. Soil Sci. Soc. of Am., 32*, 409–13.

Goudie, A., 1969, Statistical laws and dune ridges in southern Africa, *Geog. J., 135*, 404–6.

Goudie, A., 1971, *Calcrete as a component of semi-arid landscapes, Unpublished Ph.D. Thesis, University of Cambridge*, 434p.

Goudie, A., Cooke, R., and Evans, I., 1970, Experimental investigation of rock weathering by salts, *Area*, 1970, 42–48.

Graf, W. H., 1970, *Hydraulics of Sediment Transport* (McGraw-Hill, New York), 513p.

Grandet, Cl., 1957, Sur la morphologie dunaire de la rive sud du lac Faguibine, *Inst. de Rech. Sahariennes, Travaux, 16*, 171–80.

Gregory, K. J., and Brown, E. H., 1966, Data processing and the study of landform, *Zeit. für Geom., N.F., 10*, 237–63.

Gregory, J. W., 1914, The lake system of Westralia, *Geog. J., 43*, 656–64.

Greig-Smith, P., and Chadwick, M. J., 1965, Data on pattern within plant communities; III *Acacia-Capparis* semi-desert scrub in the Sudan, *J. Ecology, 53*, 465–74.

Grenier, M. P., 1968, Observations sur les taffonis du désert Chilien, *Bull. Ass. Géog. Français, 364–65*, 193–211.

Griggs, D. T., 1936, The factor of fatigue in rock exfoliation, *J. Geol., 44*, 781–96.

Grove, A. T., 1958, The ancient erg of Hausaland and similar formations on the south side of the Sahara, *Geog. J., 124*, 528–33.

Grove, A. T., 1959, A note on the former extent of Lake Chad, *Geog. J., 125*, 465–67.

Grove, A. T., 1960a, Note following Prescott and White (1960), *Geog. J., 126*, 202–3.

Grove, A. T., 1960b, The geomorphology of the Tibesti Region, *Geog. J., 126*, 18–31.

Grove, A. T., 1969, Landforms and climatic change in the Kalahari and Ngamiland, *Geog., J., 135*, 191–212.

Grove, A. T., and Sparks, B. W., 1952, Le déplacement des galets par la vent sur la glace, *Rev. de Géom. Dyn., 1*, 37–39.

Grove, A. T., and Warren, A., 1968, Quaternary landforms and climate on the south side of the Sahara, *Geog. J., 134*, 194–208.

Gruet, M., 1955, Fentes á gypse des confins sahariens, *Rev. de Géom. Dyn., 6*, 60–68.

Hack, J. T., 1941, Dunes of the western Navajo Country, *Geog. Rev., 31*, 240–63.

Hadley, R. F., 1967, Pediments and pediment-forming processes, *J. Geol. Ed., 15*, 83–89.

Hadley, R. F., and Schumm, S. A., 1961, Hydrology of the Upper Cheyenne River Basin: B. Sediment sources and drainage-basin characteristics in Upper Cheyenne River Basin, *U.S. Geol. Surv. Water-Supply Pap., 1531*, 137–98.

Hagedorn, H., 1968, Uberaolische Abtrangung Formung in der südost-Sahara, *Erdkunde, 22*, 257–69.

Hall, D. N. (ed.), 1971, An expedition to the Aïr Mountains and the Ténéré Desert, *Geog. J., 137*, 445–67.

Hallsworth, E. G., and Beckmann, G. G., 1969, Gilgai in the Quaternary, *Soil Sci., 107*, 409–20.

Hallsworth, E. G., Lemerle, T. H., and Rayner, H. V., 1952, Studies in pedogenisis in New South Wales, I, The influence of rice cultivation on the grey and brown 'gilgai' soils of the Murrumbidgee irrigation areas, *J. Soil Sci., 3*, 89–102.

Hallsworth, E. G., Robertson, G. K., and Gibbons, F. R., 1955, Studies in pedogenesis in New South Wales, VII, The 'gilgai' soils, *J. Soil Sci., 6*, 1–31.

Hallsworth, E. G., and Waring, H. D., 1964, Studies of pedogenesis in New South Wales, VIII, An alternative hypothesis for the formation of solodized-solonetz of the Pilliga district, *J. Soil Sci., 15*, 158–77.

Hammond, E., 1954, A geomorphic study of the Cape Region of Baja California, *University of California Publications in Geography, 10*, 45–112.

Hanna, S. R., 1969, The formation of longitudinal sand dunes by large helical eddies in the atmosphere, *J. Applied Meteorology, 8*, 874–83.

Harradine, F., and Jenny, H., 1958, Influence of parent material and climate on texture and nitrogen and carbon contents of virgin California soils, I, *Soil Sci., 85*, 235–43.

Harbeck, G. E., 1955, The effect of salinity on evaporation, *U.S. Geol. Surv. Prof. Pap., 272-A,* 6p.

Hare, F. K., 1961, The causation of the arid zone, in *A History of Land Use of the Arid Zone*, (UNESCO, Paris), *Arid Zone Research, 17*, 25–30.

Harlé, E., and Harlé, J., 1919, *Mémoire sur les dunes de Gascogne, avec observations sur la formation des dunes,* Extrait du *Bull. Sec. de Géog. du Comité des Travaux Historiques et Scientifiques, 34*, 145p.

Harris, S. A., 1957, The mechanical constitution of certain present day Egyptian dune sands, *J. Sed. Pet., 27*, 421–34.

Harris, S. A., 1958, Differential analysis of aeolian sand, *J. Sed. Pet., 28*, 164–74.

Harris, S. A., 1959, The classification of gilgaied soils: some evidence from northern Iraq, *J. Soil Sci., 10*, 27–33.

Harris, S. A., 1968, Gilgai, in Fairbridge, R. W. (ed.) *The Encyclopedia of Geomorphology* (Reinhold, New York), 425–26.

Hassanein Bey, 1925, *The Lost Oasis* (Butterworth, London) 316p.

Hastenrath, S. L., 1967, The barchans of the Arequipa region, southern Peru, *Zeit. für Geom., 11*, 300–31.

Hastings, J. R., and Turner, R. M., 1965, *The Changing Mile* (U. Arizona P., Tucson), 317p.

Hawker, H. W., 1927, A study of the soils of Hidalgo County, Texas, and the stages of their lime accumulation, *Soil Sci., 23*, 475–83.

Haynes, C. V., Jr., 1968, Geochronology of late-Quaternary alluvium, in Morrison, R. B., and Wright, H. E. Jr. (eds.), *Means of Correlation of Quaternary Successions* (U. Utah P., Salt Lake City), 591–615.

Hedin, S., 1904–5, *The Scientific Results of a Journey in Central Asia, 1899–1902* (Lithographic Institute of the general staff of the Swedish Army, Stockholm). (Vol. I, The Tarim Basin, 521p).

Hefley, H. M. and Sidwell, R., 1945, Geological and ecological observations of some High Plains dunes, *Am. J. Sci., 243*, 361–76.

Heim, A., 1887, Ueber Kantergeschiebe aus dem norddeutschen Diluvium, *Vierteljahrssch. Natur. Gesell., 32*, 383–85.

Heller, S. J., 1932, Sur la morphologie de quelques formations sabloneuses des Karakoum transcaspiens, *Gosundarstvennoe Geograficheskoe Obshchestvo Isvestia, 64*, 387–90.

Hemming, C. F., 1965, Vegetation arcs in Somaliland, *J. Ecology, 53*, 57–67.

Hemming, G. E., and Trapnell, C. G., 1957, A reconnaissance survey of the soils of the south Turkana desert, *J. Soil Sci., 8*, 167–83.

Hewes, L. I., 1948, A theory of surface cracks in mud and lava and resulting geometrical relations, *Am. J. Sci., 246*, 138–49.

Higgins, C. G., 1953, Miniature 'pediments' near Calistoga, California, *J. Geol., 61*, 461–65.

Higgins, C. G., 1956, Formation of small ventifacts, *J. Geol., 64*, 506–16.

Hillel, D., and Tadmor, N., 1962, Water regime and vegetation in the Central Negev Highlands of Israel, *Ecology, 43*, 33–41.

Hills, E. S., 1939, The lunette, a new landform of aeolian origin, *Austr. Geographer, 3,* 15–21.

Högbom, I., 1923, Ancient inland dunes of north and middle Europe, *Geog. Ann., 5*, 113–243.

Hollingworth, S. E., 1964, Dating the uplift of the Andes of Northern Chile, *Nature, 201*, 17–20.

Holm, D. A., 1953, Dome-shaped dunes of the Central Nejd, Saudi Arabia, *Int. Geol. Cong., 19th, Algiers, 1952, C. R., fasc.* 7, 107–12.

Holm, D. A., 1957, Sigmoidal dunes, a transitional form, *Geol. Soc. Am. Bull., 68*, 1746 (abstract).

Holm, D. A., 1960, Desert geomorphology in the Arabian Peninsula, *Science, 132*, 1369–79.

Holmes, C. D., 1955, Geomorphic development in humid and arid regions: a synthesis, *Am. J. Sci., 253*, 377–90.

Hooke, R. LeB., 1967, Processes on arid-region alluvial fans, *J. Geol., 75*, 438–60.

Hooke, R. LeB., 1968, Steady-state relationships on arid-region alluvial fans in closed basins, *Am. J. Sci., 266*, 609–29.

Hooke, R. LeB., Yang, H-Y, and Weiblen, P. W., 1969, Desert varnish: an election probe study, *J. Geol., 77*, 275–88.

Horikawa, K., and Shen, H. W., 1960, Sand movement by wind, *U.S. Army Corps of Engineers, Beach Erosion Board, Technical Mem., 119*, 51p.

Horn, M. E., Rutledge, E. M., Dean, H. C., and Lawson, M., 1964, Classification and genesis of some solonetz (sodic) soils of eastern Arkansas, *Proc. Soil Sci. Soc. Am., 28*, 688–92.

Hörner, N. G., 1933, Geomorphic processes in continental basins of central Asia, *Int. Geol. Cong. Report, 16th Session, 2*, 721–35 (Washington, 1936).

Horton, R. E., 1945, Erosional development of streams and their drainage basins; hydrophysical approach to quantitative morphology, *Geol. Soc. Am. Bull., 56*, 275–370.

Houbolt, J. J. H. C., 1968, Recent sediments in the Southern Bight of the North Sea, *Geol. en Mijnebouw, 47*, 245–73.

Howard, A. D., 1942, Pediment passes and the pediment problem, *J. Geom., 5*, 3-31; 95–136.

Howe, G. M., *et al.*, 1968, Classification of world desert areas, *U.S. Army Natick Laboratories, Technical Report, 69-38-ES*, 106p.

Hoyt, J. H., 1966, Air and sand movements in the lee of dunes, *Sedimentology*, 7, 137–44.

Huffington, R. M., and Albritton, C. C., 1941, Quaternary sands of the High Plains, *Am. J. Sci., 239*, 325–38.

Hume, W. F., 1925, *Geology of Egypt* (Government P., Cairo), I, 408p.

Hunt, C. B., 1961, Stratigraphy of desert varnish, *U.S. Geol. Surv. Prof. Pap., 424-B*, 194–95.

Hunt, C. B., and Mabey, D. R., 1966, Stratigraphy and structure, Death Valley, California, *U.S. Geol. Surv. Prof. Pap., 494-A*, 162p.

Hunt, C. B., Robinson, T. W., Bowles, W. A., and Washburn, A. L., 1966, Hydrologic basin Death Valley, California, *U.S. Geol. Surv. Prof. Pap., 494-B*, 138p.

Hunt, C. B., and Washburn, A. L., 1960, Salt features that simulate ground patterns formed in cold climates, *U.S. Geol. Surv. Prof. Pap., 400-B*, B403.

Hunt, C. B., and Washburn, A. L., 1966, Patterned ground, in *U.S. Geol. Surv. Prof. Pap., 494-B*, B104–B133.

Hunting Technical Services, 1961, *Ghulam Mohammed Barrage Command, Vol. I*, Report No. 4, West Pakistan Water and Power Development Authority, Sukkur-Gudu-Ghalam Mohammed Drainage and Salinity Control Project, 154p.

Hunting Technical Services, 1965, *Lower Indus Valley Report* (West Pakistan Water & Power Development Authority, Karachi).

Huntington, E., 1907, *The Pulse of Asia* (Houghton, Mufflin and Co., Boston), 415p.

Huntington, E., 1914, The climatic factor as illustrated in arid America, *Carnegie Inst. Wash. Pub., 192*, 341p.

Hurault, J., 1966, Etude photo-aérienne de la tendance à la remobilisation des sables éoliens sur la rive nord du Lac Tchad, *Rev. Inst. Français de Pétrole, 21*, 1837–46.

Hutton, J. T., 1968, The redistribution of the more soluble chemical elements associated with soils as indicated by analysis of rainwater, soils and plants, *Trans. 9th Int. Cong. of Soil Sci., Adelaide, Australia, 4*, 313–12.

Hutton, J. T., and Leslie, T. I., 1958, Accession of non-nitrogenous ions dissolved in rainwater to soils in Victoria, *Austr. J. Agricultural Res., 9*, 492–507.

Inglis, D. R., 1965, Particle sorting and stone migration by freezing and thawing, *Science, 148*, 1616–17.

Inman, D. L., Ewing, G. C., and Corliss, J. B., 1966, Coastal sand dunes of Guero Negro, Baja, California, Mexico, *Geol. Soc. Am. Bull.*, 77, 787–802.

Ismail, F. T., 1969, Role of ferrous iron oxidation in the alteration of biotite and its effect on the type of clay minerals formed in arid and humid regions, *Am. Mineralogist, 54*, 1460–66.

Ismail, F. T., 1970, Biotite weathering and clay formation in arid and humid regions, California, *Soil Sci., 109*, 257–61.

Ives, R. L., 1946, Desert ripples, *Am. J. Sci., 244*, 492–501.

Jackson, E. A., 1957, Soil features in arid regions with particular reference to Australia, *Austr. Inst. of Agricultural Sci. J., 25*, 196–208.

Jackson, M. L., 1959, Frequency distribution of clay minerals in major Great Soil Groups as related to the factors of soil formation, in Ingerson, E. (ed.) *Clays and Clay Minerals, Monog. 2.* Earth Science Series (Pergamon, London), 133–43.

Jackson, M. L., 1968, Weathering of primary and secondary minerals in soils, *Trans 9th Int. Cong. of Soil Sci., Adelaide, Australia, 4*, 281–92.

Jaeger, F., 1921, *Deutsch Sudwestafrika* (Wissenschaftliche Geselschaft, Breslau), 251p.

Jauzein, A., 1958, Les formes d'accumulation éolienne liées à des zones inondables en Tunisie septentrionalle, *Acad. des Sci., Paris, C. R., 247*, 2396–99.

Jennings, J. N., 1955, Le complexe des sebkhas: un commentaire provenant des antipodes, *Rev. Géom. Dyn., 6*, 69–72.

Jennings, J. N., 1967, Some karst areas of Australia, in Jennings, J. N., and Mabbutt, J. A. (eds.), *Landform Studies from Australia and New Guinea* (Cambridge U.P., Cambridge), 256–94.

Jennings, J. N., and Mabutt, J. A. (eds.) 1967, *Landform Studies from Australia and New Guinea* (Cambridge U.P., Cambridge), 434p.

Jenny, H., 1929, Relation of temperature to the amount of nitrogen in soils, *Soil Sci., 27*, 169–88.

Jenny, H., 1941, *Factors of Soil Formation* (McGraw-Hill, New York), 281p.

Jenny, H., 1958, Role of the plant factor in the pedogenic functions, *Ecology, 39*, 5–16.

Jenny, H., 1961, Derivation of state factor equations of soils and ecosystems, *Proc. Soil Sci. Soc. Am., 25*, 385–88.

Jessup, R. W., 1960a, The Stony Tableland soils of the southeastern portion of the Australian Arid Zone and their evolutionary history, *J. Soil Sci., 11*, 188–96.

Jessup, R. W., 1960b, Identification and significance of buried soils of Quaternary age in the southeast portion of the Australian arid zone, *J. Soil Sci., 11*, 197–205.

Jessup, R. W., 1961, A Tertiary-Quaternary pedological chronology for the southeastern portion of the Australian arid zone, *J. Soil Sci., 12*, 199–213.

Jewitt, J. N., 1952, The distribution of organic matter in depth in some seasonally flooded soils, *J. Soil Sci., 3*, 63–67.

Johnson, D. H., 1970, The role of the tropics in the global circulation, in Corby, G. A. (ed.), *The Global Circulation of the Atmosphere* (Royal Meteorological Society, London), 113–36.

Johnson, D. W., 1932, Rock planes in arid regions, *Geog. Rev., 22*, 656–65.

Jones, D. G., 1957, The rising water-table in parts of Daura and Katsina emirates, Katsina Province, *Records of the Geological Survey of Nigeria, 1957*, 24–28.

Jordan, W. M., 1964, Prevalence of sand-dune types in the Sahara Desert, *Geol. Soc. Am. Spec. Pap., 82*, 104–5 (abstract).

Jorré, G., 1935, Les formes de relief dans les steppes désertiques de l'Asie Centrale, *Ann. de Géog., 44*, 317–21.

Judson, S., 1950, Depressions of the Nu portion of the Southern High Plains of Eastern New Mexico, *Geol. Soc. Am. Bull, 61*, 253–74.

Jutson, J. T., 1918, The influence of salts in rock weathering in sub-arid Western Australia, *Proc. Roy. Soc. Victoria, 30* (n.s.), 165–72.

Jutson, J. T., 1934, The physiography (geomorphology) of Western Australia, *Geol. Surv. W. Austr. Bull., 95*, 2nd Edition, 366p.

Kádár, L., 1934, A study of the sand sea in the Libyan Desert, *Geog. J., 83*, 470–78.

Kampé de Feriet, J., 1953, La turbulence atmosphérique et les phénomènes d'érosion, in *Actions Eoliennes, Cent. Nat. de Rech. Sci., Coll. Int., 35*, 82–101.

Karcz, I., 1969, Mud pebbles in a flash floods environment, *J. Sed. Pet., 39*, 333–37.

Kàrmàn, T. von, 1947, Sand ripples in the desert, in *Collected Works, 1956, Vol. 4*, 352–56.

Kàrmàn, T. von, 1953, Considerations aérodynamiques sur la formation des ondulations du sable, in *Actions Eoliennes, Cent. Nat. de Rech. Sci., Coll. Int., 35*, 103–8.

Karmeli, D., and Ravina, I., 1968, A study of the Hamra soil association of Israel, *Soil Sci., 105*, 209–15.

Kassas, M., and Imam, M., 1954, Habitat and plant communities in the Egyptian desert, III, the Wadi bed ecosystem, *J. Ecology, 43*, 424–41.

Kawada, S., 1953, Quelques expériences sur l'entrainement du sable par le vent, in *Actions Eoliennes, Cent. Nat. de Rech. Sci., Paris, Coll. Int., 35*, 109–15.

Kawamura, R., 1951, Study on sand movement by wind, *Inst. of Sci. and Tech., Tokyo, Rep. 5*, (3–4), 95–112.

Kawamura, R., 1953, Mouvement du sable sous l'effect du vent, in *Actions Eoliennes, Cent. Nat. de Rech. Sci., Paris, Coll. Int., 35*, 117–51.

Keller, C., 1946, *El Departamento de Arica* (Min. Econ. y Com., Santiago de Chile).

Keller, W. D., 1962, *The Principles of Chemical Weathering* (Lucas, Columbia, Missouri), 111p.

Kelley, W. P., 1951, *Alkali Soils* (Reinhold, New York), 176p.

Kelley, W. P., 1964, Review of investigations on cation exchange and semiarid soils, *Soil Sci., 97*, 80–88.

Kennedy-Shaw, W. B., 1936, An expedition to the south Libyan Desert, *Geog. J., 87*, 193–221.

Kerr, R. C., and Nigra, J. O., 1952, Eolian sand control, *Bull. Am. Ass. Pet. Geol., 36*, 1541–73.

Keyes, C. R., 1908, Rock floor of intermont plains of the arid region, *Geol. Soc. Am. Bull., 19*, 63–92.

Keyes, C. R., 1910a, Relations of present profiles and geologic structures in desert ranges, *Geol. Soc. Am. Bull., 21*, 543–64.

Keyes, C. R., 1910b, Deflation and the relative efficiencies of erosional processes under conditions of aridity, *Geol. Soc. Am. Bull., 21*, 565–98.

Keyes, C. R., 1911, Mid-continental eolation, *Geol. Soc. Am. Bull., 22*, 687–714.

Keyes, C. R., 1912, Deflative scheme of the geographic cycle in an arid climate, *Geol. Soc. Am. Bull., 23*, 537–62.

Keyes, C. R., 1932, Quantitative measure of desert denudation, *Pan-American Geologist, 58*, 285–300.

Kindle, E. M., 1917, Some factors affecting the development of mud-cracks, *J. Geol., 25*, 135–44.

King, D., 1956, The Quaternary stratigraphic record at Lake Eyre North and the evolution of existing topographic forms, *Trans. Roy. Soc. of S. Austr., 79*, 93–103.

King, D., 1960, The sand-ridge deserts of South Australia and related aeolian landforms of Quaternary arid cycles, *Trans. Roy. Soc. of S. Austr., 83*, 99–108.

King, L. C., 1936, Wind-faceted stones from Marlborough, New Zealand, *J. Geol., 44*, 201–13.

King, L. C., 1953, Canons of landscape evolution, *Geol. Soc. Am. Bull., 64*, 721–52.

King, L. C., 1962, *The Morphology of the Earth* (Oliver and Boyd, Edinburgh), 699p.

King, W. H. J., 1916, The nature and formation of sand ripples and dunes, *Geog. J., 47*, 189–209.

King, W. H. J., 1918, Study of a dune belt, *Geog. J., 51*, 16–33.

Kirk, L. G., 1952, Trails and rocks observed on a playa in Death Valley National Monument, California, *J. Sed. Pet., 22*, 173–81.

Kirkby, M. J., 1969, Erosion by water on hillslopes, in Chorley, R.J. (ed.), *Water, Earth and Man* (Methuen, London), 229–238.

Kirkby, M. J., and Chorley, R. J., 1967, Throughflow, overland flow and erosion, *Bull. Int. Ass. Sci. Hyd., 12*, 5–21.

Krinsley, D. B., 1968, Geomorphology of three kavirs in northern Iran, in Neal, J. T. (ed.), Playa surface morphology: miscellaneous investigations, *Air Force Cambridge Research Laboratories, Environmental Research Papers, 283*, 105–30.

Krinsley, D. B., 1970, A geomorphological and paleoclimatological study of the playas of Iran, *U.S. Geol. Surv., Final Scientific Report, Contract PRO CP 70-800*, 2 vols., 486p.

Krinsley, D. B., Woo, C. C., and Stoertz., G. E., 1968, Geologic characteristics of seven Australian playas, in Neal, J. T. (ed.), Playa surface morphology: miscellaneous investigations, *Air Force Cambridge Research*

Laboratories, Environmental Research Papers, 283, 59–103.

Kubiena, W. L., 1955, Uber die Braunlehmrelikte der Atakor (Zentral-Sahara), *Erdkunde, 9*, 115–32.

Kuenen, Ph. H., 1928, Experiments on the formation of wind-worn pebbles, *Leidsche Geol. Meded., 3*, 17–38.

Kuenen, Ph. H., 1960, Experiment Abrasion, 4, Eolian action, *J. Geol., 68*, 427–49.

Kuettner, J., 1969, The band structure of the atmosphere, *Tellus, 11*, 267–94.

Kustnetzova, T. V., 1958, Some properties of takyrs of the Kizyl-Arvat plain in the southwestern part of the Turkmenian S.S.R., *Soviet Soil Sci., 1958*, 497–504.

Kwaad, F. J. P. M., 1970, Experiments on the granular disintegration of granite by salt action, *Fys. Geogr. en Bodemkundig Lab., From Field to Laboratory, Pub., 16*, 67–80.

Lachenbruch, A. H., 1962, Mechanics of thermal contraction cracks and ice-wedge polygons in permafrost, *Geol. Soc. Am. Spec. Paper, 70*, 65p.

LaMarche, V. C., 1967, Spheroidal weathering of thermally metamorphosed limestone and dolomite, White Mountains, California, *U.S. Geol. Surv. Prof. Pap., 575-C*, C32–C37.

Lamb, H. H., 1966, Climate in the 1960's: World's wind circulation reflected in prevailing temperatures, rainfall patterns and the levels of African lakes, *Geog. J., 132*, 183–212.

Lancey-Forth, N. B. de, 1930, More journeys in search of Zerzura, *Geog. J., 75*, 49–64.

Landsberg, S. Y., 1956, The orientation of dunes in Britain and Denmark, with respect to the wind, *Geog. J., 122*, 176–89.

Langbein, W. B., 1961, Salinity and hydrology of closed lakes, *U.S. Geol. Surv. Prof. Pap., 412*, 20p.

Langbein, W. B., and Schumm, S. A., 1958, Yield of sediment in relation to mean annual precipitation, *Trans. Am. Geoph. Union, 39*, 1076–84.

Langer, A. M., and Kerr, P. F., 1966, Mojave playa crusts: physical properties and mineral content, *J. Sed. Pet., 36*, 377–96.

Langford-Smith, T., 1962, Riverine plains chronology, *Austr. J. Sci., 25*, 96–97.

Langford-Smith, T., and Dury, G. H., 1964, A pediment at Middle Pinnacle, near Broken Hill, New South Wales, *J. Geol. Soc. Austr., 11*, 79–88.

Langford-Smith, T., and Dury, G. H., 1965, Distribution, character and attitude of the duricrust in the northwest of New South Wales and the adjacent areas of Queensland, *Am. J. Sci., 263*, 170–90.

Lattimore, O., 1951, *Inner Asian Frontiers of China* (American Geographical Society, *Research Series, 21*), 585p.

Lawson, A. C., 1915, The epigene profiles of the desert, *University of California Bull. Dept. Geol., 9*, 23–48.

Lebon, J. H. G., and Robertson, V. C., 1961, The Jebel Marra, Darfur and its region, *Geog. J., 127*, 30–49.

Leggett, R. F., Brown, R. J. E., and Johnston, G. H., 1966, Alluvial fan formation near Aklavik, Northwest Territories, Canada, *Geol. Soc. Am. Bull., 77*, 15–30.

Lefèvre, M-A., 1952, Note sur les pédiments du désert Mojave, Californie, *Bull. Soc. Belge d'Etudes Géog., 21*, 259–68.

Leliavsky, S., 1955, *An Introduction to Fluvial Hydraulics* (Constable, London), 257p.

Le lubre, M., 1950, Une reconnaissance aerienne sur l'Edeyen de Mourzouk (Fezzán), *Inst. de Rech. Sahariennes, Travaux, 5*, 219–21.

Leopold, A., 1921, A plea for recognition of artificial works in forest erosion control policy, *J. Forestry, 19*, 267–73.

Leopold, L. B., 1951, Rainfall frequency: an aspect of climatic variation, *Trans. Am. Geoph. Union, 32*, 347–57.

Leopold, L. B., Emmett, W. W., and Myrick, R. M., 1966, Channel and hillslope processes in a semiarid area, New Mexico, *U.S. Geol. Surv. Prof. Pap., 352-G*, 193–253.

Leopold, L. B., and Langbien, W. B., 1962, The concept of entropy in landscape evolution, *U.S. Geol. Surv. Prof. Pap., 500-A*, 20p.

Leopold, L. B., and Maddock, T., Jr., 1953, The hydraulic geometry of stream channels and some physiographic implications, *U.S. Geol. Surv., Prof. Pap., 352*, 57p.

Leopold, L. B., and Miller, J. P., 1956, Ephemeral streams—hydraulic factors and their relation to the drainage net, *U.S. Geol. Surv. Prof. Pap., 282-A*, 36p.

Leopold, L. B., Wolman, M. G., and Miller, J. P., 1964, *Fluvial Processes in Geomorphology* (Freeman, San Francisco), 522p.

Lettau, K., and Lettau, H., 1969, Bulk transport of sand by the barchans of the Pampa La Joya in Southern Peru, *Zeit. für Geom., N.F.13*, 182–95.

Lettau, K., and Lettau, H. (in press), Experimental and micrometeorological field studies of dune migration, in Lettau, K. and Lettau, H. (eds.), *Exploring the World's Driest Climate* (U. of Wisconsin Press, Madison), Chapter 9.

Lewis, G. C., and White, J. L., 1964, Chemical and minerological studies on slick spots soils in Idaho, *Proc. Soil Sci. Soc. Am., 28*, 805–8.

Leverett, F., 1942, Wind work accompanying or following the Iowan glaciation, *J. Geol., 50*, 548–55.

Lewis, A. D., 1936, Sand dunes of the Kalahari within the borders of the Union, *S. African Geog. J., 19*, 25–57.

Lewis, P. F., 1960, Linear topography in south western Palouse, Washington-

Oregon, *Ann. Ass. Am. Geog., 50*, 98–111.

Litchfield, W. H., 1962, Soils of the Alice Springs area, in *General Report on Lands of the Alice Springs Area, Northern Territory, 1956–57*, compiled by R. A. Perry (C.S.I.R.O., Melbourne), 185–207.

Litchfield, W. H., 1963, Soils of the Wiluna-Meekatharra area, in *General Report on lands of the Wiluna-Meekatharra area, Western Australia, 1958* (C.S.I.R.O., Melbourne), 123–42.

Livingstone, D. A., 1954, On the orientation of lake basins, *Am. J. Sci., 252*, 547–54.

Lobova, E. V., 1967, *Soils of the Desert Zone of the U.S.S.R.*, (Akademiya Nauk SSSR. Pochvennyi Institut im. v.v. Dokuchaeva, Moskva, 1960, Israel Program for Scientific Translations, Jerusalem), 405p.

Loew, O., 1876, Appendix H.6, in Wheeler, G. M., *Annual Report upon Geographical Surveys West of the 100th meridian, in California, Nevada, Utah, Colorado, Wyoming, New Mexico, Arizona and Montana, being Appendix J. J. of the Annual Report of the Chief Engineers* for 1876.

Lofgren, B. E., and Klausing, R. L., 1969, Land subsidence due to groundwater withdrawal, Tulare-Wasco area, California, *U.S. Geol. Surv. Prof. Pap., 437-B*, 103p.

Logan, R., 1964, The origin of the rock pediments of desert regions, *Acta Geog. Lovaniensia, 3*, 51–57.

Long, J. T., and Sharp, R. P., 1964, Barchan-dune movement in the Imperial Valley, California, *Geol. Soc. Am. Bull., 75*, 149–56.

Longwell, C. R., 1928, Three common types of desert mud-cracks, *Am. J. Sci., 15*, 136–45.

Louis, H., 1959, Beobachtungen über die Inselberge bei Hua-Hin Am Golf von Siam, *Erdkunde, 13*, 314–19.

Lowdermilk, W. C., and Sundling, H. L., 1950, Erosion pavement, its formation and significance, *Trans. Am. Geoph. Union, 31*, 96–100.

Lustig, L. K., 1965, Clastic sedimentation in Deep Springs Valley, California, *U.S. Geol. Surv. Prof. Pap., 352-F*, 131–92.

Lustig, L. K., 1966, The geomorphic and paleoclimatic significance of alluvial deposits in southern Arizona, *J. Geol., 74*, 95–102.

Lustig, L. K., 1968, Inventory of research on geomorphology and surface hydrology of desert environments, in McGinnies, W. G., *et al.* (eds.), *Deserts of the World* (U. Arizona P., Tucson), 95–283.

Lustig, L. K., 1969a, Trend surface analysis of the Basin and Range province and some geomorphic implications, *U.S. Geol. Surv. Prof. Pap., 500-D*.

Lustig, L. K., 1969b, Quantitative analysis of desert topography, in McGinnies, W. G., and Goldman, B. J. (eds.), *Arid Lands in Perspective* (U. Arizona P., Tucson), 47–58.

Lumley, J. L., and Panofsky, H. A., 1964, *The Structure of Atmospheric*

Turbulence (Wiley, New York), 237p.

Lydolph, P. E., 1957, A comparative study of the dry western littorals, *Ann. Ass. Am. Geog.*, *47*, 213–30.

Mabbutt, J. A., 1955a, Pediment land forms in Little Namaqualand, South Africa, *Geog. J.*, *121*, 77–83.

Mabbutt, J. A., 1955b, Erosion surfaces in Namaqualand and the ages of surface deposits in the western Kalahari, *Trans. Geol. Soc. S. Africa*, *58*, 13–30.

Mabbutt, J. A., 1961, A stripped land surface in Western Australia, *Trans. Inst. Brit. Geog.*, *29*, 101–14.

Mabbutt, J. A., 1963, Wanderrie banks: micro-relief patterns in semiarid Western Australia, *Geol. Soc. Am. Bull.*, *74*, 529–40.

Mabbutt, J. A., 1965, Stone distribution in a Stony Tableland soil, *Austr. J. Soil Res.*, *3*, 131–42.

Mabbutt, J. A., 1966, Mantle-controlled planation of pediments, *Am. J. Sci.*, *264*, 78–91.

Mabbutt, J. A., 1967, Denudation chronology in Central Australia. Structure, climate and landform inheritance in the Alice Springs area, in Jennings, J. N., and Mabbutt, J. A. (eds.), *Landform Studies from Australia and New Guinea* (Cambridge U.P., Cambridge), 144–81.

Mabbutt, J. A., 1968, Aeolian landforms of central Australia, *Austr. Geog. Studies*, *6*, 139–50.

Mabbutt, J. A., 1969, Landforms of arid Australia, in Slatyer, R. O., and Perry, R. A. (eds.), *Arid Lands of Australia* (A.N.U.P., Canberra), 11–32.

Mabbutt, J. A., 1971, The Australian arid zone as a prehistoric environment, in Mulvaney, D. J., and Golson, J. (eds.), *Aboriginal Man and Environment in Australia* (A.N.U.P., Canberra), 66–79.

Mabbutt, J. A., and Sullivan, M. E., 1968, The formation of longitudinal dunes: evidence from the Simpson Desert, *Austr. Geographer*, *10*, 483–87.

Mabbutt, J. A., Wooding, R. A., and Jennings, J. N., 1969, The asymmetry of Australian desert sand ridges, *Aust. J. Sci.*, *32*, 159–60.

MacFadyen, W. A., 1950, Vegetation patterns in the semi-desert plains of British Somaliland, *Geog. J.*, *116*, 199–211.

Madigan, C. T., 1936a, *Central Australia* (Oxford, London), 267p.

Madigan, C. T., 1936b, The Australian sand-ridge deserts, *Geog. Rev.*, *26*, 205–27.

Madigan, C. T., 1938, The Simpson Desert and its borders, *Proc. Roy. Soc. New S. Wales*, *71*, 503–35.

Madigan, C. T., 1945, Simpson Desert expedition, scientific reports: introduction, narrative, physiography and meteorology, *Trans. Royal Soc. S. Australia*, *69*, 118–39.

Madigan, C. T., 1946, The sand formations, *Simpson Desert Expedition 1939, Scientific Report 6 : Geology, Trans. Roy. Soc. S. Australia, 70*, 45–63.

Mainquet, M., 1968, Le Borkou, aspects d'un modelé éolien, *Ann. de Géog.*, 77, 296–322.

Makin, J., Schilstra, J., and Theisen, A. A., 1969, The nature and genesis of certain aridisols in Kenya, *J. Soil Sci., 20*, 111–25.

Mammerickx, J., 1964a, Quantitative observations on pediments in the Mojave and Sonoran deserts (Southwestern United States), *Am. J. Sci., 262*, 417–35.

Mammerickx, J., 1964b, Pédiments desertiques et pédiments tropicaux, *Acta Geog. Lovaniensia, 3*, 359–70.

Marbut, C. F., 1935, Soils of the United States, in *Atlas of American Agriculture*, part III (U.S. Government Printing Office, Washington, D.C.).

Martin, P. S., 1963, Geochronology of pluvial lake Cochise, southern Arizona, II, pollen analysis of a 42-meter core, *Ecology, 44*, 436–44.

Martin, W. P., and Fletcher, J. E., 1943, Vertical zonation of great soil groups on Mount Graham, Arizona, as correlated with climate, vegetation and profile characteristics, *University of Arizona Agricultural Experimental Station Technical Bull., 99*, 89–153.

Matschinski, M., 1952, Sur les formations sableuses des environs de Bebi-Abbès, *Soc. Géol. de France, C.R., 9–10*, 171–74.

Matschinski, M., 1962, Sur la distribution des petits mares de l'Ile de France, *Acad. Sci. Français, C.R., 254*, 331–34.

Maxson, J. H., 1940, Fluting and faceting of rock fragments, *J. Geol., 48*, 717–51.

Mayland, H. F., and McIntosh, T. H., 1963, Nitrogen fixation by desert algal crust organisms, *Agronomy Abstracts, 1963*, 33.

McCleary, J. A., 1968, The biology of desert plants, in Brown, G. W. Jr. (ed.), *Desert Biology* (Academic P., New York), 141–94.

McCoy, F. W., Jr., Nokleberg, W. J., and Norris, R. M., 1967, Speculations on the origin of the Algodones dunes southern California, *Geol. Soc. Am. Bull., 78*, 1039–44.

McGee, W. J., 1896, Expedition to Papagueria and Seriland, *Am. Anthr., 9*, 93–98.

McGee, W. J., 1897, Sheetflood erosion, *Geol. Soc. Am. Bull., 8*, 87–112.

McGinnies, W. G., *et al.*, 1968, *Deserts of the World* (U. Arizona P., Tucson), 788p.

McIntyre, D. S., 1958a, Permeability measurements of soil crusts by raindrop impact, *Soil Sci., 85*, 185–89.

McIntyre, D. S., 1958b, Soil splash and the formation of surface crusts by raindrop impact, *Soil Sci., 85*, 261–66.

McKee, E. D., 1962, Origin of the Nubian and similar sandstones, *Sonderbruck aus der Geol. Rundschau, Bd. 52*, 551–87.

McKee, E. D., 1966, Structures of dunes at White Sands National Monument, New Mexico (and a comparison with structures of dunes from other selected areas), *Sedimentology, 7*, 1–69.

McKee, E. D., Douglass, J. R., and Rittenhouse, S., 1971, Deformation of lee-side laminae in eolian dunes, *Geol. Soc. Am. Bull., 82*, 359–78.

McKee, E. D., and Tibbitts, G. C. Jr., 1964, Primary structures of a seif dune and associated deposits in Libya, *J. Sed. Pet., 34*, 5–17.

Meckelein, W., 1957, Une mission scientifique allemande en Libye, *Inst. de Rech. Sahariennes, Travaux, 16*, 213–16.

Meckelein, W., 1959, *Forschungen in der zentralen Sahara* (Westermann, Braunschweig), 181p.

Meckelein, W., 1965, Beobachtungen und Gedanken zu Geomorphologischen Konvergenzen in Polar—und Wärmewüsten, *Erdkunde, 19*, 31–39.

Medinger, M., 1961, La crue de décembre 1960 de l'Oued Mya, *Inst. de Rech. Sahariennes, Travaux, 20*, 203–6.

Meigs, P., 1953, World distribution of arid and semi-arid homoclimates, in *Reviews of Research on Arid Zone Hydrology* (UNESCO, Paris), 203–9.

Meinig, D. W., 1965, The Mormon culture region: strategies and patterns in the geography of the American West, *Ann. Ass. Am. Geog., 55*, 191–220.

Melton, F. A., 1940, A tentative classification of sand dunes, *J. Geol., 48*, 113–73.

Melton, M. A., 1957, An analysis of the relations among elements of climate, surface properties, and geomorphology, *Office of Naval Research Technical Report II* (Project NR. 389–042), 102p.

Melton, M. A., 1958, Geometric properties of mature drainage systems and their representation in an E4 phase space, *J. Geol., 66*, 35–54.

Melton, M. A., 1965a, The geomorphic and palaeoclimatic significance of alluvial deposits in southern Arizona, *J. Geol., 73*, 1–38.

Melton, M. A., 1965b, Debris-covered hillslopes of the southern Arizona desert—consideration of their stability and sediment contribution, *J. Geol., 73*, 715–29.

Mensching, H., *et al.*, 1970, *Sudan—Sahel—Sahara* (Jahrbuch für 1969 der Geographischen Gesellschaft zu Hannover), 219p.

Merriam, R., 1969, Source of sand dunes of southeastern California and northwestern Sonora, Mexico, *Geol. Soc. Am. Bull., 80*, 531–34.

Michel, P., 1969, Morphogenesis and pedogenesis, *Sols Africains, 14*, 109–141.

Miller, R. F., Branson, F. A., McQueen, I. S., and Culler, R. C., 1961, Soil moisture under juniper and pinyon compared with moisture under grassland in Arizona, *U.S. Geol. Surv. Prof. Pap., 424-B*, 233–35.

Miller, R. F., and Ratzlaff, K. W., 1961, Water movement and ion distribution in soils, *U.S. Geol. Surv. Prof. Pap., 422-B*, 45–46.

Miller, V. C., 1971, Discrimination of pediments and alluvial fans from topographic maps: discussion, *Geol. Soc. Am. Bull., 82*, 2375.

Mitchell, C. W., 1959, Investigations into the soils and agriculture, lower Diyala area, eastern Iraq, *Geog. J., 125*, 390–97.

Mitchell, C. W., and Naylor, P. E., 1960, Investigations into the soils and agriculture of the Middle Diyala region, eastern Iraq, *Geog. J., 126*, 469–75.

Mitchell, C. W., and Perrin, R. M. S., 1967, The subdivision of hot deserts of the world into physiographic units, *Rév. de l'Inst. Français du Pétrole, 21*, 1855–72.

Monod, Th., 1954, Modes 'contracte' et 'diffus' de la végétation saharienne, in Cloudsley-Thompson, J. L. (ed.), *Biology of Deserts* (Inst. of Biology, London), 35–44.

Monod, Th., 1958, *Majâbat Al-Koubrâ* (Memoires de l'Institut Francais D'Afrique Noire, *52*), 406p.

Monod, Th., 1962, Notes sur le Quaternaire de la region Tazzmont- el Bayyel (Adrar de Mauritanie), *Congrès Panafricain du Préhistoire et de l'Etude du Quaternaire 4e, Leopoldville, 1959, Actes*, 172–88.

Monod, Th., and Cailleux, A., 1945, Etude de quelques sables et grès du Sahara occidental, *Bull. Inst. Français del'Afrique Noire, 7*, 174–90.

Morrison, R. B., 1964, Lake Lahontan; geology of southern Carson Desert, Nevada, *U.S. Geol. Surv. Prof. Pap., 401*, 156p.

Morrison, R. B., 1965, Quaternary geology of the Great Basin, in Wright, H. E. Jr., and Frey, D. G. (eds.), *The Quaternary of the United States* (Princeton U.P., Princeton), 265–85.

Morrison, A., and Chown, M. C., 1965, Photographs of the Western Sahara from the Mercury MA-4 satellite, *Photogrammetric Engineering, 31*, 350–62.

Mortensen, H., 1927, *Der Formenbschatz de Nordchilenischen Wüste* (Weidmannsche, Berlin), 191p.

Mortensen, H., 1950, Das Gesetz der Wüstenbildung, *Universitas, 5*, 801–814.

Motts, W. S., 1958, Caliche genesis and rainfall in the Pecos Valley area of southeastern New Mexico, *Geol. Soc. Am. Bull., 69*, 1737 (abstract).

Motts, W. S., 1965, Hydrologic types of playas and closed valleys and some relations of hydrology to playa geology, in Neal, J. T., 1965a, Geology, Mineralogy and Hydrology of U.S. playas, *Air Force Cambridge Research Laboratories, Environmental Research Paper, 96*, 73–104.

Moulden, J. C., 1905, Origin of pebble-covered plains in desert regions, *Trans. Am. Inst. Min. Eng., 35*, 963–64.

Mueller, G., 1960, The theory of formation of north Chilean nitrate deposits through 'capillary concentration', *Rep. Int. Geol. Cong. 19th, Nordern, 1960, Part I*, 76–86.

Mueller, G., 1968, Genetic histories of nitrate deposits from Antarctica and Chile, *Nature, 219*, 1131–34.

Mulcahy, M. J., 1961, Soil distribution in relation to landscape development, *Ziet. für Geom.*, *5*, 211–25.

Mulcahy, M. J., 1967, Landscapes, laterites and soils in southwestern Australia, in Jennings, J. N., and Mabbutt, J. A. (eds.), *Landform Studies from Australia and New Guinea* (Cambridge U.P., Cambridge), 211–30.

Murata, T., 1966, A theoretical study of the forms of alluvial fans, *Tokyo Metropolitan University Geographical Report, 1*, 33–43.

Neal, J. T., 1965a, Geology, mineralogy and hydrology of U.S. playas, *Air Force Cambridge Research Laboratories Environmental Research Papers, 96*, 176p.

Neal, J. T., 1965b, Giant desiccation polygons of Great Basin playas, *Air Force Cambridge Research Laboratories Environmental Research Papers, 123*, 30p.

Neal, J. T., 1968, Playa surface changes at Harper Lake, California: 1962–1967, in Neal, J. T. (ed.), Playa surface morphology: miscellaneous investigations, *Air Force Cambridge Research Laboratories Environmental Research Papers, 283*, 5–30.

Neal, J. T. (ed.), 1968, Playa surface morphology: miscellaneous investigations, *Air Force Cambridge Research Laboratories Environmental Research Papers, 283*, 150p.

Neal, J. T., 1969, Playa variation, in McGinnies, W. G., and Goldman, B. J. (eds.), *Arid Lands in Perspective* (U. Arizona P., Tucson; A.A.A.S., Washington), 13–44.

Neal, J. T., Langer, A. M., and Kerr, P. F., 1968, Giant desiccation polygons of Great Basin Playas, *Geol. Soc. Am. Bull., 79*, 69–90.

Neal, J. T., and Motts, W. S., 1967, Recent geomorphic changes in playas of western United States, *J. Geol., 75*, 511–25.

Nettleton, W. D., Flach, K. W., and Brasher, B. R., 1969, Argillic horizons without clay skins, *Proc. Soil Sci. Soc. Am., 33*, 121–25.

Newbold, D., 1924, A desert odyssey of a thousand miles, *Sudan Notes and Records,* 7, 43–92; 104–7.

Newell, N. D., and Boyd, D. W., 1955, Extraordinarily coarse eolian sand of the Ica Desert, Peru, *J. Sed. Pet., 25*, 226–28.

Nikiforoff, C. C., 1937, General trends of the desert types of soil formation, *Soil Sci., 43*, 105–25.

Nir, D., 1964, Les marges méridionales du phénomène karstique en Israël, *Rev. Géog. Alpine, 52*, 533–41.

Norris, R. M., 1966, Barchan dunes of Imperial Valley, California, *J. Geol., 74*, 292–306.

Norris, R. M., 1969, Dune reddening and time, *J. Sed. Pet., 39*, 7–11.

Norris, R. M. and Norris, K. S., 1961, Algodones dunes of southeastern California, *Geol. Soc. Am. Bull., 72*, 605–20.

O'Brien, M. P., and Rindlaub, B. D., 1936, The transportation of sand by wind, *Civil Engineering, 6*, 325–27.

Oertel, A. C., 1968, Some observations incompatible with clay illuviation, *Trans. 9th Int. Cong. of Soil Sci., Adelaide, Australia, 4*, 481–88.

Ollier, C. D., 1960, The inselbergs of Uganda, *Zeit. für Geom., N.F.4,* 43–52.

Ollier, C. D., 1963, Insolation weathering: examples from central Australia, *Am. J. Sci., 261*, 376-18.

Ollier, C. D., 1965, Dirt-cracking—a type of insolation weathering, *Austr. J. Sci., 27*, 236–37.

Ollier, C. D., 1966, Desert gilgai, *Nature, 212*, 581–83.

Ollier, C. D., 1969, *Weathering* (Oliver and Boyd, Edinburgh), 304p.

Ollier, C. D., and Tuddenham, W. G., 1962, Inselbergs in central Australia, *Zeit. für Geom., 5*, 257–76.

Olson, J. S., 1958, Lake Michigan dune development, *J. Geol., 66*, 254–63; 345–51; 437–83.

Opdyke, N. D., 1961, The palaeoclimatalogical significance of desert sandstone, in Nairn, A. E. M. (ed.), *Descriptive Palaeoclimatology* (Interscience, New York), 45–60.

Osborn, H. B., and Lane, L., 1969, Precipitation-runoff relations for very small semiarid rangeland watersheds, *Wat. Res. Res., 5*, 419–25.

Paige, S., 1912, Rock-cut surfaces in the desert ranges, *J. Geol., 20*, 442–50.

Palansi, G., 1955, Au sujet du Niger fossile dans la région de Tombouctou, *Rev. Géom. Dyn., 6*, 217–18.

Parke, J. G., 1857, Report of explorations for railroad routes, U.S. Congress, *House Executive Document 91*, 33rd Congress, 2nd Session.

Parker, G. G., 1963, Piping, a geomorphic agent in landform development of the drylands, *Int. Ass. Sci. Hyd. Pub., 65*, 103–13.

Passarge, S., 1904, *Die Kalahari* (Reimer, Berlin), 2 vols.

Passarge, S., 1930, Ergebrisse einer studienreise nach Sudtunisien in jahre, 1928, *Mitteilungen der Geog. Gesellschaft, Hamburg, 1930, 61*, 96–122.

Peel, R. F., 1966, The Landscape in Aridity, *Trans. Inst. Brit. Geog., 38*, 1–23.

Penck, A., 1905, Climatic features in the land surface, *Am. J. Sci., 19*, 165–174.

Penck, W., 1953, *Morphological Analysis of Land Forms* (MacMillan, London), 429p. Translated from the German by H. Czech and K. Boswell.

Perry, R. A., *et al.*, 1962, General Report on lands of the Alice Springs area, Northern Territory, 1956–67, C.S.I.R.O., *Land Research Series, 6.*

Pesce, A., 1968, *Gemini space photographs of Libya and Tibesti; a geological and geographical analysis* (Petroleum Exploration Society of Libya, Tripoli), 81p.

Petrov, M. P., 1962, Types de déserts de l'Asie centrale, *Ann. de Géog., 71*, 131–55.

Poldervaart, A., 1957, Kalahari sands, in *Pan African Congress on Prehistory, 3rd, Livingstone, Southern Rhodesia, 1955, Proceedings*, 106–14.

Potter, P. E., and Pettijohn, F. J., 1967, *Palaeocurrents and basin analysis* (Springer, Berlin), 296p.

Powell, J., and Pedgley, D. E., 1969, A year's weather at Termit, Republic of Niger, *Weather, 24*, 247–54.

Powell, J. W., 1875, *Exploration of the Colorado River of the West (1869–72)* (Washington), 291p. Reprinted in 1957 by the University of Chicago Press.

Powell, J. W., 1962, *Report on the Lands of the Arid Region of the United States* (Edited by Stegner, W.; Harvard U.P., Cambridge, Mass.), 202p. Based on the original report, dated 1878.

Powers, W. E., 1936, The evidences of wind abrasion, *J. Geol., 44*, 214–19.

Prenant, A., 1951, Morphologie de la Plaine de la Zonsfana, *Inst. de Rech. Sahariennes, Travaux, 1*, 23–68.

Prescott, J. A., 1949, A climatic index for the leaching factor in soil formation, *J. Soil Sci., 1*, 9–19.

Prescott, J. R. V., and White, H. P., 1960, Sand formations in the Niger Valley between Niamey and Bourem, *Geog. J., 126*, 200–3.

Price, W. A., 1963, Physico-chemical and environmental factors in clay dune genesis, *J. Sed. Pet., 33*, 766–78.

Price, W. A., and Kornicker, L. S., 1961, Marine and lagoonal deposits in clay dunes, Gulf Coast, Texas, *J. Sed. Pet., 31*, 245–55.

Prill, R. C., 1968, Movement of moisture in the unsaturated zone in a dune area, southwestern Kansas, *U.S. Geol. Surv. Prof. Pap., 600-D*, D1–D9.

Prince, H. C., 1961. Some reflections on the origin of hollows in Norfolk compared with those in the Paris region, *Rev. Géom. Dyn., 12*, 109–17.

Queney, P., 1945, Observations sur les rides formées par le vent à la surface du sable dans les ergs sahariens, *Ann. de Géoph., 1*, 279–82.

Queney, P., 1953, Classification des rides de sable et théorie ondulatoire de leur formation, in *Actions Eoliennes, Cen. Nat. de Rech. Sci., Paris, Coll. Int., 35*, 179–95.

Queney, P., and Dubief, J., 1943, Action d'un obstacle ou d'un fossé sur un vent charge de sable, *Inst. de Rech. Sahariennes, Travaux, 2*, 169–76.

Rahn, P. H., 1966, Inselbergs and nickpoints in southwestern Arizona, *Zeit. für Geom., 10*, 217–25.

Rahn, P., 1967, Sheetfloods, streamfloods, and the formation of pediments, *Ann. Ass. Am. Geog., 57*, 593–604.

Raverty, H. G., 1878, The Mihran of Sind, *J. Asiatic Soc. of Bengal, 61.*

Ravkovitch, S., Pines, F., and Ben-Yair-M, 1958, Composition of colloids in the soils of Israel, *Ktavim, 9*, 69–82.

Raychaudri, S. P., and Sen, N., 1952, Certain geomorphological aspects of the Rajputana Desert, *Proc. Symposium on the Rajputana Desert, National Institute of Sciences of India, Bull. No. 1*, 249–

Reeves, C. C., 1966, Pluvial lake basins of West Texas, *J. Geol., 74*, 269–91.

Reeves, C. C., and Suggs, J. D., 1964, Caliche of central south Llano Estacado, Texas, *J. Sed. Pet., 34*, 699–72.

Reiche, P., 1937, The Toreva block—a distinctive landslide type, *J. Geol., 45*, 538–48.

Reiche, P., 1945, A survey of weathering processes and products, *New Mexico University Publication in Geology, 1.*

Rempel, P., 1936, The crescentic dunes of the Salton Sea and their relation to vegetation, *Ecology, 17*, 347–58.

Renard, K. G., and Keppel, R. V., 1966, Hydrographs of ephemeral streams in the Southwest, *Proc. Am. Soc. Civ. Eng., J. Hydraulics Division, 92*, 33–52.

Rich, J. L., 1935, Origin and evolution of rock fans and pediments, *Geol. Soc. Am. Bull., 46*, 999–1024.

Richards, L. A. (ed.), 1954, *Saline and Alkali Soils*, United States Department of Agriculture, *Agricultural Handbook, 60.*

Richardson, P. D., 1968, The generation of scour marks near obstacles, *J. Sed. Pet., 38*, 965–70.

Rim, M., 1951, The influence of geophysical processes on the stratification of sandy soils, *J. Soil Sci., 2*, 188–95.

Rim, M., 1958, Simulation by dynamical model of sand tract morphologies occurring in Israel, *Bull. Res. Council of Israel, 7-G(2/3)*, 123–33.

Robinson, G. M., and Peterson, D. E., 1962, Notes on earth fissures in southern Arizona, *U.S. Geol. Surv. Circular, 466*, 7p.

Rodin, L. E., and Bazilevich, N. I., 1965, in Fogg, G. E. (ed.), *Production and Mineral Cycling in Terrestrial Vegetation* (Oliver and Boyd, Edinburgh and London), 184–207.

Rognon, P., 1961, Modification de l'alteration argileuse depuis la fin du Tertiaire dans le massif volcanique de l'Atakor (Hoggar), *C.R.S., Soc. Géol. de France,* 186–88.

Rognon, P., 1967a, *Le Massif de l'Atakor et Ses Bordures (Sahara Central), Étude Géomorphologique* (C.N.R.S., Paris), 551p.

Rognon, P., 1967b, Climatic influences on the African Hoggar during the Quaternary, based on geomorphic observations, *Ann. Ass. Am. Geog., 57*, 115–27.

Roth, E. S., 1965, Temperature and water content as factors in desert weathering, *J. Geol., 73*, 454–68.

Rose, C. W., 1960, Soil detachment caused by rainfall, *Soil Sci., 89*, 28–35.

Rose, C. W., 1961, Rainfall and soil structure, *Soil Sci., 91*, 49–54.

Rozycki, S. Z., 1968, The directions of winds carrying loess dust as shown by analysis of accumulative loess forms in Bulgaria, in Schultz, C. B., and Frye, J. C. (eds.), *Loess and Related Aeolian Deposits of the World* (U. of Nebraska, Lincoln, Nebraska), 235–46.

Rougerie, G., no date, *Systèmes Morphogéniques et Familles de Modelés dans les Zones Arides* (C.D.U., Paris), 167p.

Ruellan, A., 1968, Les horizons d'individualization et d'accumulation du calcaire dans les sols du Meroc, *Trans. 9th Int. Cong. of Soil Sci., Adelaide, Australia, 4*, 501–10.

Ruhe, R. V., 1967, Geomorphic surfaces and surficial deposits in southern New Mexico, *State Bureau of Mines and Mineral Resources, New Mexico Institute of Mining and Technology, Memoir, 18*, 66p.

Russell, R. J., 1932, Landforms of the San Gorgonio Pass, Southern California, *University of California Publications in Geog., 6*, 106–14.

Russell, R. J., 1936, The desert rainfall factor in denudation, *Int. Geol. Cong., 16th Session, 2*, 753–63.

Russell, W. L., 1929, Drainage alignment in the western Great Plains, *J. Geol., 37*, 249–55.

Ruxton, B. P., 1958, Weathering and subsurface erosion on granite at the piedmont angle, Baros, Sudan, *Geol. Mag., 95*, 353–77.

Ruxton, B. P., and Berry, L., 1961, Weathering profiles and geomorphic position on granite in two tropical regions, *Rev. de Géom. Dyn., 12*, 16–31.

Sandford, K. S., 1933, Past climate and early man in the southern Libyan Desert, *Geog. J., 82*, 219–22.

Sandford, K. S., 1935, Geological observations on the N.W. frontier of the Anglo-Egyptian Sudan, *Quart. J. Geol. Soc., 91*, 323–81.

Sandford, K. S., 1937, The geology of north-central Africa, *Quart. J. Geol. Soc., London, 93*, 534–80.

Sandford, K. S., 1953, Notes on sand dunes in Egypt and the Sudan, *Geog. J., 119*, 363–66.

Schattner, I., 1961, Weathering phenomena in the crystalline of the Sinai in the light of current notions, *Bull. Res. Counc. Israel, 10G*, 247–66.

Schick, A. P., 1970, Desert floods: interim results of observations in the Nahal Yael Research Watershed, southern Israel, 1965–1970, IASH-UNESCO Symposium on the results of Research on Representative and Experimental Basins (Wellington, New Zealand), 478–93.

Schmalz, R. F., 1968, Formation of red-beds in modern and ancient deserts: discussion, *Geol. Soc. Am. Bull., 79*, 277–80.

Schoeller, H., 1945, Le Quaternaire de la Saoura et du Grand Erg occidental, *Inst. de Rech. Sahariennes, Travaux, 3*, 57–71.

Schoewe, W. M., 1932, Experiments on the formation of wind-faceted pebbles, *Am. J. Sci., 24*, 111–34.

Schreiber, H. A., and Kincaid, D. R., 1967, Regression models for predicting on-site runoff from short-duration convective storms, *Wat. Res. Res., 3*, 389–95.

Schumm, S. A., 1956a, The role of creep and rainwash on the retreat of badland slopes, *Am. J. Sci., 254*, 693–706.

humm, S., 1956b, The movement of rocks by wind, *J. Sed. Pet., 26*, 4–86.

chumm, S. A., 1960, The shape of alluvial channels in relation to sediment ype, *U.S. Geol. Surv. Prof. Pap., 352-B*, 17–30.

Schumm, S. A., 1961, Effect of sediment characteristics on erosion and deposition in ephemeral-stream channels, *U.S. Geol. Surv. Prof. Pap., 352-C*, 31–70.

Schumm, S. A., 1962, Erosion on miniature pediments in Badlands National Monument, South Dakota, *Geol. Soc. Am. Bull., 73*, 719–24.

Schumm, S. A., 1963a, The disparity between present rates of denudation and orogeny, *U.S. Geol. Surv. Prof. Pap., 454-H*, 13p.

Schumm, S. A., 1963b, A tentative classification of alluvial river channels, *U.S. Geol. Surv. Circular, 477*, 10p.

Schumm, S. A., 1964, Seasonal variations of erosion rates and processes on hillslopes in western Colorado, *Zeit für Geom. Supp., 5*, 215–38.

Schumm, S. A., 1965, *Quaternary paleohydrology,* in Wright, H. E., Jr., and Frey, D. G. (eds.), *The Quaternary of the United States* (Princeton U.P., Princeton), 783–94.

Schumm, S. A., 1968, River adjustment to altered hydrologic regimen—Murrumbidgee River and paleochannels, Australia, *U.S. Geol. Surv. Prof Pap., 598*, 65p.

Schumm, S. A., 1969, River metamorphosis, *Proc. Am. Soc. Civ. Eng., J Hydraulics Division, 95*, 255–73.

Schumm, S. A., and Chorley, R. J., 1964, The fall of threatening rock, *Am J. Sci., 262*, 1041–54.

Schumm, S. A. and Chorley, R. J., 1966, Talus weathering and sca recession in the Colorado plateaus, *Zeit. für Geom., 10*, 11–36.

Schumm, S. A. and Hadley, R. F., 1957, Arroyos and the semiarid cycle erosion, *Am. J. Sci., 255*, 161–74.

Schumm, S. A., and Hadley, R. F., 1961, Progress in the application of lar form analysis in studies of semiarid erosion, *U.S. Geol. Surv. Circular, 4* 14p.

Schumm, S. A., and Lichty, R. W., 1965, Time, space and causality geomorphology, *Am. J. Sci., 263*, 110–19.

Scott, R. M., 1962, Exchangeable bases of mature well-drained soils relation to rainfall in East Africa, *J. Soil Sci., 13*, 1–9.

Segerstrom, K., 1962, Deflated marine terrace as a source of dune cha Atacama Province, Chile, *U.S. Geol. Surv. Prof. Pap., 450-C*, 91–93.

Segerstrom, K., 1963, Matureland of northern Chile and its relationshi ore deposits, *Geol. Soc. Am. Bull., 74*, 513–18.

Segerstrom, K., and Henriquez, H., 1964, Cavities or 'tafoni', in rock f of the Atacama Desert, Chile, *U.S. Geol. Surv. Prof. Pap., 501-C*, C C125.

Seligman, N. G. N., Tadmor, N. H., and Raz, Z., 1962, Range surve[illegible] central Negev, *National and University Institute of Agriculture, Re*[illegible] *Department of Field Crops, Bull., 67*.

Sevenet, Lieut., 1943, Étude sur le Djouf (Sahara occidental), *Bull.* [illegible] *Français del'Afrique Noire, 5*, 1–26.

Shantz, H. L., 1956, History and problems of arid lands development, [illegible] White, G. F. (ed.), *The Future of Arid Lands* (A.A.A.S., Washington), 3–2[illegible]

Sharon, D., 1962, On the nature of hamadas in Israel, *Zeit für Geom., 6*, 129–47.

Sharp, R. P., 1940, Geomorphology of the Ruby-East Humboldt Range, Nevada, *Geol. Soc. Am. Bull., 51*, 337–72.

Sharp, R. P., 1949, Pleistocene ventifacts east of the Big Horn Mountains, Wyoming, *J. Geol., 57*, 175–95.

Sharp, R. P., 1963, Wind ripples, *J. Geol., 71*, 617–36.

Sharp, R. P., 1964, Wind-driven sand in the Coachella Valley, California, *Geol. Soc. Am. Bull., 75*, 785–804.

Sharp, R. P., 1966, Kelso Dunes, Mohave Desert, California, *Geol. Soc. Am. Bull., 77*, 1045–74.

Sharp, R. P., and Nobles, L. H., 1953, Mudflow of 1941 at Wrightwood, southern California, *Geol. Soc. Am. Bull., 64*, 547–60.

Sharp, W. E., 1960, The movement of playa scrapers by the wind, *J. Geol., 68*, 567–72.

Shaw, Napier, 1942, *Manual of Meteorology, vol. III, The Physical Processes of Weather* (Cambridge U.P., Cambridge), 445p.

Shawe, D. R., 1963, Possible wind-erosion origin of linear scarps on the Saga Plain, south-western Colorado, *U.S. Geol. Surv. Prof. Pap., 475-C*, C138–C42.

Shields, L. M., Mitchell, C., and Drouet, F., 1957, Alga- and lichen-stabilized surface crusts as soil nitrogen sources, *Am. J. Botany, 44*, 489–98.

Shotton, F. W., 1954, The availability of underground water in hot deserts, in Cloudsley-Thompson, J. C. (ed.), *Biology in Deserts* (Institute of Biology, London), 13–17.

Sidwell, R., and Tanner, W. P., 1938, Quaternary dune building in central Kansas, *Geol. Soc. Am. Bull., 49*, 139 (abstract).

Simonett, D. S., 1951, On the grading of dune sands near Castlereagh, New South Wales, *Roy. Soc. New S. Wales, J. and Proc., 84*, 71–79.

Simonett, D. S., 1960, Development and grading of dunes in western Kansas, *Ann. Ass. Am. Geog., 50*, 216–41.

Simons, F. S., 1956, A note on Pur-Pur Dune, Viru Valley, Peru, *J. Geol., 64*, 517–21.

Simons, F. S., and Eriksen, G. E., 1953, Some desert features of northwest central Peru, *Soc. Geol. Peru Bol., 26*, 229–46.

Simonson, R. W., and Hutton, C. E., 1954, Distribution curves for loess, *Am. J. Sci., 252*, 99–105.

Sinclair, J. G., 1922, Temperatures of the soil and air in a desert, *Monthly Weather Review, 50*, 142–44.

Simonson, R. W., 1959, Outline of a generalized theory of soil genesis, *Proc. Soil Sci. Soc. Am., 23*, 152–56.

Smalley, I. J., 1964, Flow-stick transitions in powders, *Nature, 201*, 173–34.

Smalley, I. J., 1966, Contraction crack networks in basalt flows, *Geol. Mag., 103*, 110–14.

Smalley, I. J., 1970, Cohesion of soil particles and the intrinsic resistance of simple soil systems to wind erosion, *J. Soil Sci., 21*, 154–61.

Smalley, I. J., and Vita-Finzi, C., 1968, The formation of fine particles in sandy deserts and the nature of 'desert' loess, *J. Sed. Pet., 38*, 766–74.

Smith, B. R., and Buol, S. W., 1968, Genesis and relative weathering intensity studies in three semiarid soils, *Proc. Soil Sci. Soc. Am., 32*, 261–65.

Smith, G. I., 1966, Geology of Searles Lake—a guide to prospecting for buried continental salines, in Rau, J. L. (ed.), *Second Symposium on Salt* (Northern Ohio Geol. Soc., Cleveland), 167–80.

Smith, G. I., 1968, Late-Quaternary geologic and climatic history of Searles Lake, southeastern California, in Morrison, R. B., and Wright, H. E. Jr. (eds.), *Means of Correlation of Quaternary Successions* (U. of Utah P., Salt Lake City), 293–310.

Smith, H. T. U., 1945, Giant grooves in northwestern Canada, *Geol. Soc. Am. Bull., 56*, 1198 (abstract).

Smith, H. T. U., 1954, Eolian sand on desert mountains, *Geol. Soc. Am. Bull., 65*, 1036–37 (abstract).

Smith, H. T. U., 1956, Giant composite barchans of the northern Peruvian desert, *Geol. Soc. Am. Bull., 67*, 1735 (abstract).

Smith, H. T. U., 1963, Eolian geomorphology, wind direction and climatic change in North Africa, *U.S. Air Force, Cambridge Research Laboratories*, Contract No. AF. 19(628)–298. (Also cited as AD. 405.144.)

Smith, H. T. U., 1965, Dune morphology and chronology in central western Nebraska, *J. Geol., 73*, 557–78.

Smith, H. T. U., 1966, Windgeformte Gerollwellen in der Antarktis, *Umshau, 66*, 334.

Smith, H. T. U., 1969, Photo-interpretation studies of desert basins in northern Africa, *U.S. Air Force Cambridge Research Laboratories*, Contract No. AF. 19(628)2486, *Final Report*, 77p.

Smith, J., 1949, Distribution of tree species in the Sudan in relation to rainfall and soil texture, *Sudan Government, Ministry of Agriculture, Bull., 4*.

Smith, K. G., 1958, Erosional processes and landforms in Badlands National Monument, South Dakota, *Geol. Soc. Am. Bull., 69*, 975–1008.

Smith, R., and Robertson, V. C., 1962, Soil and irrigation classification of shallow soils overlying gypsum beds, northern Iraq, *J. Soil Sci., 13*, 106–15.

Smith, R. M., Twiss, P. C., Krauss, R. K., and Brown, M. J., 1970, Dust deposition in relation to site, season and climatic variables, *Proc. Soil Sci. Soc. Am., 34*, 112–17.

Snyder, C. T., Hardman, G., and Zdenek, F. F., 1964, Pleistocene lakes in the Great Basin, *U.S. Geol. Surv., Miscellaneous Geologic Investigations, Map I-416*.

Sokolow, N. A., 1894, *Die Dünen, Bildung, Entwicklung, und innerer Bau* (translated from the Russian by A. Arzruni, Springer, Berlin), 298p.

Spencer, B. (ed.), 1896, *Report on the work of the Horn Scientific Expedition to Central Australia* (Dulan, London and Melbourne), 4 vols.

Springer, M. E., 1958, Desert pavement and vesicular layer of some desert soils in the desert of the Lahontan Basin, Nevada, *Proc. Soil Sci. Soc. Am., 22*, 63–66.

Stahnke, C. R., Rogers, J. R., and Allen, B. L., 1969, A genetic and mineralogical study of a soil developed from granitic gniess in the Texas central basin, *Soil Sci., 108*, 313–20.

Stanley, G. M., 1955, Origin of playa stone tracks, Racetrack Playa, Inyo County, California, *Geol. Soc. Am. Bull., 66*, 1329–50.

Steensrup, K. J. V., 1894, Om Klitterns Vandrung, *Meddelelser, Dansk, Geol. Forening, Copenhagen, 1*, 1–14.

Stephens, C. G., 1946, Comparative morphology and genetic relationships of certain Australian, North American and European Soils, *J. Soil Sci., 1*, 123–49.

Stephen, I., Bellis, E., and Muir, A., 1956, Gilgai phenomena in tropical black clays of Kenya, *J. Soil Sci., 7*, 1–9.

Stevens, J., 1969, *The Sahara is Yours—a handbook for desert travellers* (Constable, London), 200p.

Stoddart, D. R., 1969, Climatic geomorphology: review and re-assessment, *Progress in Geography, 1*, 161–222.

Stokes, W. L., 1964, Incised, wind-aligned stream patterns of the Colorado Plateau, *Am. J. Sci., 262*, 808–16.

Stone, R. O., 1967, A desert glossary, *Earth-Science Reviews, 3*, 211–68.

Stone, R., 1968, Deserts and desert landforms, in Fairbridge, C. W. (ed.), *Encyclopedia of Geomorphology* (Reinhold, New York), 271–79.

Strahler, A. N., 1950, Equilibrium theory of erosional slopes approached by frequency distribution analysis, *Am. J. Sci., 248*, 673–96; 800–14.

Striem, H. L., 1954, The seifs on the Israel-Sinai border and the correlation of their alignment, *Research Council of Israel, Bull., 4*, 195–98.

Stuart, D. M., Fosberg, M. A., and Lewis, C. G., 1961, Caliche in southwestern Idaho, *Proc. Soil Sci. Soc. Am., 25*, 132–35.

Suslov, S. P., 1961, *The Physical Geography of Asiatic Russia* (Freeman, San Francisco and London), 594p.

Sutton, O. G., 1953, *Micrometeorology* (McGraw-Hill, New York), 333p.

Syers, J. K., Jackson, M. L., and Berkheiser, V. E., Clayton, R. N., and Rex, R. W., 1969, Eolian sediment influence on pedogenesis during the Quaternary, *Soil Sci., 107*, 421–27.

Symmons, P. M., and Hemming, C. F., 1968, A note on wind-stable stone-mantles in the southern Sahara, *Geog. J., 134*, 60–64.

Tackett, J. L., and Pearson, R. W., 1965, Some characteristics of soil crusts formed by simulated rainfall, *Soil Sci., 99*, 407–13.

Tator, B. A., 1952, 1953, Pediment characteristics and terminology, *Ann. Ass. Am. Geog., 42*, 295–317; *43*, 47–53.

Tator, B. A., 1953, The climatic factor and pedimentation, *Int. Geol. Cong., Alger, C.R., 7*, 121–30.

Teakle, L. J. H., and Samuel, L. W., 1930, The reaction of Western Australian soils, *Royal Soc. W. Austr. Proc., 16*, 69–79.

Tedrow, J. C. F., 1966, Polar desert soils, *Proc. Soil Sci. Soc. Am., 30*, 381–87.

Thesiger, W., 1949, A further journey across the Empty Quarter, *Geog. J., 113*, 21–46.

Thomas, M. F., 1966, Some geomorphological implications of deep weathering patterns in crystalline rocks in Nigeria, *Trans. Inst. Brit. Geog., 40*, 173–193.

Tilho, J., 1911, *Documents scientifique de la mission Tilho* (Ministère des Colonies, Imprimeure Nationale), 3 vols: I, 269p; II, 632p; III, 484p.

Tremblay, L. P., 1961, Wind stations in northern Alberta and Saskatchewan, Canada, *Geol. Soc. Am. Bull., 72*, 1561–64.

Tricart, J., 1954, Une forme de relief climatique: les sebkhas, *Rev. Géom. Dyn., 5*, 97–101. Translated by *Am. Met. Soc.*, 1967.

Tricart, J., 1956, Étude éxpedimentale du problème de la gelivation, *Bull. Periglac., 4*, 285–318.

Tricart, J., 1958, Observations sur le rôle ameublisseur des termites, *Rev. Géom. Dyn., 8*, 170–72.

Tricart, J., 1959, Géomorphologie dynamique de la moyenne vallée du Niger (Soudan), *Ann. Géog., 68*, 333–43.

Tricart, J., 1967, Étude des faciès d'une dune gypseuse (sud d'Oran, Algérie), *Bull. Soc. Géol. de France, 7e ser., 9*, 865–75.

Tricart, J., and Brochu, M., 1955, Le grand erg ancien du Trarza et du Cayor, *Rev. Géom. Dyn., 4*, 145–76.

Tricart, J., and Cailleux, A., 1961 and 1964, *Le Modelé des Régions Sèches* (C.D.U., Paris), 2 vols. Subsequently published with revisions in a single volume under the same title by S.E.D.E.S., Paris (1969), 472p.

Tricart, J., and Cailleux, A., 1965, *Introduction a la Géomorphologie Climatique* (S.E.D.E.S., Paris), 306p.

Tricart, J., and Mainguet, M., 1965, Caractéristiques granulometriques de quelques sables éoliens du desert péruvien; aspects de la dynamique des barchans, *Rev. de Géom. Dyn., 15*, 110–21.

Trichet, J., 1963, Description d'une forme d'accumulation de gypse par voie éolienne dans le sud Tunisien, *Bull. Soc. Géol. de France, 7e ser., 5*, 617–21.

Troeh, F. R., 1965, Landform equations fitted to contour maps, *Am. J. Sci., 263*, 616–27.

Tsyganenko, A. F., 1968, Aeolian migration of water soluble matter and its probable geochemical and soil formation significance, *Trans. 9th Int. Cong. of Soil Sci., Adelaide, Australia, 4*, 333–42.

Tuan, Yi-Fu, 1959, Pediments in southeastern Arizona, *University of California Publications in Geography, 13*, 140p.

Tuan, Yi-Fu, 1962, Structure, climate and basin land forms in Arizona and New Mexico, *Ann. Ass. Am. Geog., 52*, 51–68.

Twidale, C. R., 1962, Steepened margins of inselbergs from north-western Eyre Peninsula, South Australia, *Zeit. für Geom., N.F. 6*, 51–69.

Twidale, C. R., 1964, A contribution to the general theory of domed inselbergs: conclusions derived from observations in South Australia, *Trans. Inst. Brit. Geog., 34*, 91–113.

Twidale, C. R., 1967, Origin of the piedmont angle as evidenced in South Australia, *J. Geol., 75*, 393–411.

Twidale, C. R., 1968, *Geomorphology with Special Reference to Australia* (Nelson, Melbourne and Sydney), 406p.

Twidale, C. R., 1971, *Structural Landforms* (M.I.T. Press, Cambridge, Mass.), 247p.

Twidale, C. R., and Corbin, E. M., 1963, Gnammas, *Rev. de Géom. Dyn., 14*, 1–20.

Tyurin, I. V., Antipov-Karataer, I. N., and Chizhevskii, M. G., 1960, *Reclamation of Solonetz Soils in the U.S.S.R.* (International Program for Scientific Translations, Jerusalem, 1967).

Udden, J. A., 1894, Erosion, transportation and sedimentation by the atmosphere, *J. Geol. 2*, 318–31.

UNESCO, 1953, *Directory of Institutions Engaged in Arid Zone Research* (UNESCO, Paris), 110p.

United States Department of Agriculture, Soil Survey Staff, Soil Conservation Service, 1960, *Soil Classification, a comprehensive system, 7th approximation* (U.S. Government Printing Office, Washington), 265p.

Urvoy, Y, 1933, Les formes dunaires à l'Ouest du Tchad, *Ann. de Géog., 42*, 506–15.

Urvoy, Y., 1936, Structure et modelé du Soudan Français (Colonie du Niger), *Ann. de Géog., 45*, 19–49.

Urvoy, Y., 1942, Les bassins du Niger, étude de géographie physique et paleogéographie, *Mem. Inst. Français del'Afrique Noire, 41*, 139p.

Vaché-Grandet, G., 1959, L'erg du Trarza, notes de géomorphologie dunaire, *Inst. de Rech. Sahariennes, Travaux, 18*, 161–72.

Van der Hoven, I., 1957, Power spectrum of horizontal windspeed in the frequency range from 0·0007 to 900 cycles per hour, *J. Meteorology, 14*, 160–64.

van der Merwe, C. R., 1954, The soils of the desert and arid regions of South Africa, *Committee for Technical Co-Operation in Africa, Interafrican Conference of Soils, Leopoldville, Congo, 2*, 827–34.

van der Merwe, C. R., and Heystek, H., 1955, Clay minerals of South African soil groups III. Soils of the desert and adjoining semarid areas, *Soil Sci., 80*, 479–94.

van Houten, F. B., 1961, Climatic significance of red beds, in Nairn, E. M. (ed.), *Descriptive Palaeoclimatology* (Interscience, New York), 89–139.

Verger, F., 1964, Mottureaux et gilgais, *Ann. de Géog., 73*, 413–30.

Verlaque, C., 1958, Les dunes d'In Salah, *Inst. de Rech. Sahariennes, Travaux, 17*, 12–58.

Verstappen, H. Th., 1968, On the origin of longitudinal (seif) dunes, *Zeit. für Geom. N.F. 12*, 200–20.

Verstappen, H. Th., and Zuidam, R. A. van, 1970, Orbital photography and the geosciences, a geomorphological example from the central Sahara, *Geoforum, 2*, 33–47.

Vita-Finzi, C., 1969, *The Mediterranean Valleys* (Cambridge U.P., Cambridge), 140p.

Volkov, I. A., 1955, Airflow patterns behind barchans (in Russian), *Vsesoiuznoe Geograficheskoe Obschchestvo, S.S.S.R., Leningrad, Isvestia 87*, 284–87, Quoted in *Meteorological Abstracts 8*, 1957, 334.

Wade, A., 1910, On the formation of Dreikante in desert regions, *Geol. Mag.*, 7, 394–98.

Walker, P. H., 1964, Sedimentary properties and processes on a sandstone hillside, *J. Sed. Pet., 34*, 328–34.

Walker, T. R., 1967, Formation of red beds in modern and ancient deserts, *Geol. Soc. Am. Bull., 78*, 353–68.

Walker, T. R., 1968, Formation of red beds in modern and ancient deserts—reply, *Geol. Soc. Am. Bull., 79*, 281–82.

Walker, T. R., and Honea, R. M., 1969, Iron content of modern deposits in the Sonoran desert: a contribution to the study of red beds, *Geol. Soc. Am. Bull., 80*, 535–44.

Walker, T. R., Ribbe, P. H., and Honea, R. M., 1967, Geochemistry of hornblende alteration in Pliocene red beds, Baja California, Mexico, *Geol. Soc. Am. Bull., 78*, 1055–60.

Wallén, C. C., 1967, Aridity definitions and their applicability, *Geog. Ann., 49*, 367–84.

Walther, J., 1924, *Das Gesetz der Wüstenbildung in gegenwart und vorzeit* (Von Quelle und Meyer, Leipzig), 421p.

Walther, W., 1951, L'influence des facteurs physiques sur la morphologie des sables éoliens et des dunes, *Rev. Géom. Dyn., 2*, 242–58.

Warnke, D. A., 1969, Pediment evolution in the Halloran Hills, Central Mojave Desert, California, *Zeit. für Geom., 13*, 357–89.

Warren, A., 1966, *The Qoz Region of Kordofan*, Unpublished Ph.D. Thesis, University of Cambridge, 272p.

Warren, A., 1968, Dune volume and trend measurements in the Nebraska Sandhills, *Proc. Nebraska Acad. Sci., 78*, 23.

Warren, A., 1969, A bibliography of desert dunes and associated phenomena, in McGinnies, W. G., and Goldman, B. J. (eds.), *Arid Lands in Perspective* (U. of Arizona P., Tucson), 75–99.

Warren, A., 1970, Dune trends and their implications in the central Sudan, *Zeit. für Geom., Supp., 10*, 154–80.

Warren, A., 1971, Dunes in the Ténéré Desert, *Geog. J., 137*, 458–61.

Warren, A., 1972, Observations on dunes and bi-modal sand in the Ténéré Desert, *Sedimentology*, in press.

Washburn, A. L., 1956, Classification of patterned ground and review of suggested origins, *Geol. Soc. Am. Bull., 67*, 823–66.

Webb, B. P., and Wopfner, H., 1961, Plio-pleistocene dunes northwest of lake Torrens, South Australia, and their influence on the erosional pattern, *Aust. J. Sci., 23*, 379–81.

Weinert, H. H., 1965, Climatic factors affecting the weathering of igneous rocks, *Agricultural Meteorology, 2*, 27–42.

Weise, O. R., 1970, Zur Morphodynamik der Pediplanation, *Zeit. für Geom., Supp., 10*, 64–87.

Wellington, J. H., 1955, *Southern Africa, A Geographical Study, vol. I, Physical Geography* (Cambridge U.P., Cambridge), 528p.

Wellman, H. W., and Wilson, A. T., 1965; Salt weathering, a neglected erosive agent in coastal and arid environments, *Nature, 205*, 1097–98.

Wells, L. A., 1902, *Journal of the Calvert Scientific Exploring Expedition, 1896–7, Western Australia Parliamentary Paper, 46*.

Whitaker, R. H., Buol, S. W., Niering, W. A., and Havens, Y. H., 1968, A soil and vegetation pattern in the Santa Catalina Mountains, Arizona, *Soil Sci., 105*, 440–51.

White, E. M., 1961, Calcium-solodi or planosol genesis from solodised-solonetz, *Soil Sci., 91*, 175–77.

White, E. M., and Bonestall, R. G., 1960, Some gilgaied soils in South Dakota, *Proc. Soil Sci. Soc. Am., 24*, 305–9.

White, E. M., and Papendick, R. I., 1961, Lithosolic solodised-solonetz soils in southwestern North Dakota, *Proc. Soil Sci. Soc. Am., 25*, 504–6.

White, L. P., 1971, The ancient erg of Hausaland in southwestern Niger, *Geog. J., 137*, 69–73.

White, L. P., and Law, R., 1969, Channelling of alluvial depression soils in Iraq and Sudan, *J. Soil Sci., 20*, 84–90.

Whittaker, R. H., and Niering, W. A., 1965, Vegetation of the Santa Cataline Mountains, Arizona: a gradient analysis of the southern slope, *Ecology, 46*, 429–52.

Wiegland, C. L., Lyle, S. L., and Carter, D. L., 1966, Interspersed salt affected and unaffected dryland soils of the Lower Rio Grande Valley, II occurrence of salinity in relation to infiltration rates and profile characteristics, *Proc. Soil Sci. Soc. Am., 30*, 106–9.

Wild, A., 1958, The phosphate content of Australian soils, *Austr. J. Agricultural Res., 9*, 193–204.

Wilhelmy, H., 1964, Cavernous rock surfaces (tafoni) in semiarid and arid climates, *Pakistan Geog. Rev., 19*, 9–13.

Williams, C. B., 1954, Bioclimatic observations in the Egyptian desert, in Cloudsley-Thompson, J. C. (ed.), *Biology of Deserts* (Institute of Biology, London), 18–27.

Williams, G., 1964, Some aspects of the eolian saltation load, *Sedimentology, 3*, 257–87.

Williams, G. E., 1967, Characteristics and origin of a pre-Cambrian pediment, *J. Geol., 77*, 183–207.

Williams, M. A. J., 1966, Age of alluvial clays in the Western Gezira, Republic of the Sudan, *Nature, 211*, 270–71.

Williams, M. A. J., 1968, A dune catena on the clay plains of the west central Gezira, Republic of the Sudan, *J. Soil Sci., 19*, 367–78.

Williams, M. A. J., 1969, Prediction of rainsplash erosion in the seasonally wet tropics, *Nature, 222*, 763–64.

Wilson, I. G., 1970, *The External Morphology of Wind-laid Sand Deposits*, unpublished Ph.D. Thesis, University of Reading, 181p.

Wilson, I. G., 1971a, Journey across the Grand Erg oriental, *Geog. Mag., 43*, 264–70.

Wilson, I. G., 1971b, Desert sandflow basins and a model for the development of ergs, *Geog. J., 137*, 180–97.

Wilson, I. G., 1972a, Aeolian bedforms—their development and origins, *Sedimentology*, in press.

Wilson, I. G., 1972b, Sand Waves, *New Scientist, 23*, March, 634–37.

Wiman, S., 1963, A preliminary study of experimental frost weathering, *Geog. Ann., 45*, 113–21.

Winkler, E. M., and Wilhelm, E. J., 1970, Salt burst by hydration pressures in architectural stone in urban atmosphere, *Geol. Soc. Am. Bull., 81*, 567–72.

Wirth, E., 1958, Morphologische und Bodenkundliche Beobachtungen in der Syrischirakischen Wüste, *Erdkunde, 12*, 26–42.

Wolman, M. G., and Miller, J. P., 1960, Magnitude and frequency of forces in geomorphic processes, *J. Geol., 68*, 54–74.

Wood, W. H., 1970, Rectification of wind-blown sand, *J. Sed. Pet.*, *40*, 29–37.

Woodruff, N. P., and Siddoway, F. H., 1965, A wind erosion equation, *Proc. Soil Sci. Soc. Am.*, *29*, 602–8.

Wopfner, H., and Twidale, C. R., 1967, Geomorphological history of the Lake Eyre basin, in Jennings, J. N., and Mabbutt, J. A., (eds.) *Landform Studies from Australia and New Guinea* (Cambridge U. P., Cambridge), 119–43.

Worrall, G. A., 1959, The Butana grass patterns, *J. Soil Sci.*, *10*, 34–53.

Worrall, G. A., 1969, The red sands of the southern Sahara, *Bull. de Liason, Association Sénégalaise pour L'Etude du Quaternaire de l'Ouest Africain*, *21*, 36–39.

Wright, H. E., Jr., 1958, An extinct wadi system in the Syrian Desert, *Bull. Res. Council of Israel*, *7G*, 53–59.

Wright, A. C. S., and Urzúa, H., 1963, Meteorización en la region costera del desierto del Norte de Chile, *Com. y Res. y Trabajos*, Cont. Latino-americana para el estudo de las regiones aridas, 26–28.

Yaalon, D. H., 1963, On the origin and accumulation of salts in groundwater and soils in Israel, *Bull. Res. Council of Israel*, *11G*, 105–31.

Yaalon, D. H., 1965, Downward movement and distribution of anions in soil profiles with limited wetting, in Hallsworth E. G., and Crawford, D. V. (eds.), *Experimental Pedology* (Butterworths, London), 157–64.

Yaalon, D. H., 1970, Parallel stone cracking, a weathering process on desert surfaces, Geol. Inst. Bucharest, *Tech. and Ec. Bull.*, *18*, 107–11.

Yaalon, D. H., and Ganor, E., 1966, The climatic factor of wind erodibility and dust blowing in Israel, *Israel J. of Earth Sci.*, *15*, 27–32.

Yaalon, D. H., and Ganor, E., 1968, Chemical composition of dew and dry fallout in Jerusalem, Israel, *Nature*, *217*, 1139–40.

Yaalon, D. H., and Lomas, J., 1970, Factors controlling the supply and the chemical composition of aerosols in a near-shore and coastal environment. *Agricultural Meteorology*, 7, 443–54.

Yabukov, T. F., and Bespalova, R. Y., 1961, Soil-formation processes during the invasion of sands by plants in the northern deserts of the Caspian region, *Soviet Soil Sci.*, *6*, 651–58.

Young, R. G., 1964, Fracturing of sandstone cobbles in caliche cemented terrace gravels, *J. Sed. Pet.*, *34*, 886–89.

Zakirov, R. S., 1969, Some regularities in sand-transport of a wind-sand stream (in Russian), *Problemy Osrvoeniya Puolyn*, *1*, 73–77.

Zingg, A. W., 1953, Some characteristics of eolian sand movement by saltation process, in *Actions Éoliennes, Cent. Nat. de Rech. Sci., Coll. Int.*, *35*, 197–208.

Index